Environmental Psychology

ENVIRONMENTAL PSYCHOLOGY

Charles J. Holahan

University of Texas at Austin

Random House New York

First Edition
987654321
Copyright © 1982 by Random House, Inc.

Library of Congress Cataloging in Publication Data

Holahan, Charles J.
    Environmental psychology.

    Bibliography: p.
    Includes index.
    1. Environmental psychology. I. Title.
BF353.H66   155.9   82–378
ISBN: 0-394-32320-3   AACR2

    Text design: Teresa Harmon
    Cover design: Ronald F. Hall

Manufactured in the United States of America

To my father
Charles J. Holahan, Sr.
(1918–1980)
In Memory

Preface

The growth of environmental psychology, although it is a young field, has been exceptionally rapid. This success reflects both its unique subject matter and its place in the historical development of psychology. The field is inherently interesting and exciting; it confronts real human problems with which each of us shares some personal experience. The subject matter of environmental psychology forms the fabric of our daily lives—the effects of the settings where we live and work, the environmental demands of crowding and noise, the spatial dimension in our relationships with one another. The salience of the new field has been further enhanced by concerns of society with the quality of our physical environment and with the long-term consequences of industrial pollution, careless waste disposal, and poor management of natural resources. Moreover, environmental psychology has been especially responsive to demands that psychologists play a meaningful role in working toward the resolution of major environmental problems.

This book is intended as a primary text in introductory and survey courses in environmental psychology and the related areas of environment and behavior, architectural psychology, ecological psychology, and social ecology. It is also appropriate to courses in environmental design and planning where the emphasis is on the human and social dimensions in design. While the test is primarily envisioned for upper-division undergraduate courses, it may also be used as either a primary or adjunct text in graduate courses in environmental psychology, environmental planning, and urban studies.

The text is oriented toward meeting a number of needs of instructors and students in courses on environmental psychology. For example, the teacher of environmental psychology needs a text having a breadth of coverage consistent with the diversity and vitality of the discipline. In addition, because environmental psychology has evolved in response to pressing social concerns in a range of disciplinary fields, its development has been dramatically varied and diversified in terms of the research questions posed, the analytical strategies employed, and the implications drawn. A viable course in environmental psychology must rely on a text that is able to weave an integrating fabric around the field's disparate content, presenting a coherent definition and identity for the new discipline. Finally, the history of environmental psychology is alive with a sense of urgency about the societal application of scientific knowledge. Students of environmental psychology deserve a text that reflects a concern with the contemporary problems of society, a balance between sound science and social need, a blend of scholarship and excitement. This book attempts to meet these needs, offering an introduction to the field of environmental psychology that is comprehensive, thematically coherent, and informed by social interest.

■ The book provides a comprehensive survey of the breadth and diversity of inquiry that characterize the field of environmental psychology. It includes, for example, discussions of perceptual and cognitive processes as they relate to the environment, human performance in designed settings, environmental stressors, and dyadic and group social processes in the environment. It is responsive to the requirements of typical university courses in environmental psychology that present a broad survey of the field, and frees instructors from the need to assign a variety of books to adequately cover the discipline.

■ A key feature of this text is the thematic focus it provides from which to view the field of environmental psychology. The rapid and highly diversified manner in which environmental psychology has evolved has tended to make a thematic focus in the discipline elusive. The somewhat discursive manner in which the field has often been presented leaves students without a guiding focus for organizing knowledge in the area. This book focuses on the *psychological processes* in which people engage in interacting with and coping with the physical environment as a framework for organizing and appreciating the breadth of knowledge in the field.

- This thematic focus is evident, for example, in the text's chapter organization. The first half of the book explores at a predominantly individual level the psychological processes that mediate the influences of physical settings on people's activities (e.g., environmental perception, environmental cognition, environmental attitudes, environmental performance, and coping with ambient environmental stressors). The second half of the text examines at a predominantly social level the psychological processes that mediate between environment and behavior (e.g., coping with crowding, privacy, territoriality, personal space, and affiliation and support in the urban environment).

- The thematic organization is further developed by its focus on the *adaptational value* of each of the psychological processes examined. It emphasizes the adaptive ways in which people cope with environmental challenges; the individual is viewed as an active and dynamic agent in dealing with the environment. The final chapter of the text explicitly develops a unifying framework for studying environment and behavior based on a model of the total person engaged in a transactional relationship with a holistic environmental context.

- The book also strives to achieve an integration of sound scholarship with the intrinsic excitement and social relevance that characterize the field of environmental psychology. The format of each chapter is designed to encourage the student to perceive environmental psychology as a process of discovery, progressing from personal experience, to scientific inquiry, and finally to practical application. For example, each psychological process in the environment is first presented at a personal, experiential level. The nature of the process is then considered, along with the central methodological concerns involved in examining the process scientifically. Next, the psychological functions served by the process are discussed, and theoretical perspectives of the process are explored. Finally, the consideration of each psychological process concludes by examining the practical application of research findings in the environmental planning enterprise.

Charles J. Holahan

Acknowledgments

This book would not have been possible without the advice, assistance, and encouragement of many people. Special thanks are due to Gary Evans whose thoughtful and detailed suggestions on two earlier versions of the manuscript were instrumental in revising the text. Carl Greenberg also provided constructive and facilitative advice on two earlier drafts of the manuscript, while Arthur Patterson and Sheldon Cohen offered helpful suggestions on an initial draft. The major part of the manuscript was completed while I was on a leave of absence from the University of Texas at the Social Ecology Laboratory at Stanford University. I am grateful to Rudolf Moos, Director of the Social Ecology Laboratory, for making the leave possible and for offering me the resources and intellectual stimulation of the laboratory while I worked on this book.

For their assistance with background library research for the text, I am grateful to Michelle Kean, Sherri Evans, Patty Griffin Heilbrun, Marguerite Ponder, and Diane Spearly. I am thankful, too, to the individuals who typed the manuscript at its various stages: Louise Doherty, Mary-Margaret Byerman, Gay Passel, Patty Ardies, and Jean Roberts. In addition, special thanks are extended to the editorial staff of Random House for their experienced guidance throughout this enterprise. Virginia Hoitsma provided the initial encouragement to undertake this task and gave constant support while the manuscript was being completed; Fred Burns skillfully ferried the text through the arduous publication process.

Finally, and most significantly, is my appreciation to my wife, Carole, for her unwavering encouragement and support while I worked on this book; here, words are insufficient.

Contents

XI

BOXED INSERTS

Environmental Psychology

1 The Nature and History of Environmental Psychology

We rarely stop to consider how the physical environments in which we live, study, work, and play affect our lives. Yet the physical settings that surround and support our daily lives exert a major influence on the way we think, feel, and behave. For example, a college student's day may begin with a feeling of contentment that derives in part from waking in a pleasant apartment in an attractive and friendly neighborhood. Or the student may wake unhappily in a dormitory room that is too small, too noisy, and uncomfortably furnished. For students who drive to school, the travel environment in their area may provide a pleasant and efficient trip, or the hassle of bumper-to-bumper traffic. Even people's feelings about college may be affected by the physical design of the university they attend. Students may enjoy a leisurely walk between classes along pathways that are surrounded by natural greenery. Or they may find themselves rushing frantically between classes across a scattered campus dominated by expanses of concrete and asphalt.

The physical settings that encompass our daily lives may also reflect very personal and meaningful aspects of ourselves. For exam-

1

ple, people often use the physical environment to make a statement about who they are—to demonstrate their unique tastes, interests, and attitudes. The pictures or posters in a person's room, the souvenirs and knickknacks accumulated over time, and even the arrangement of room furniture all express an individual's unique personality (Figure 1-1). The physical environment may even influence our choice of friends. Most students' best friends at school are neighbors, people who live in the same apartment building or dormitory, or even their roommates.

Yet, remarkably, despite the important role of the physical environment in our daily lives, we typically take it for granted and are unaware of its influence on us. In fact, until very recently psychologists, too, ignored the important ways in which the physical environment shapes people's lives. Now, however, a new and exciting field of study, *environmental psychology*, is engaged in studying the complex relationships between people and the physical settings in which they conduct their daily lives. Environmental psychologists believe that an important way to learn about the nature of human behavior is to study the ways in which people respond to the demands of their physical environments and use those environments to meet their personal needs.

Environmental psychologists have, for example, surveyed residents of high-rise university dormitories to find out how students feel about dormitory living and how the

Figure 1-1
The highly personalized environment in this dormitory room reflects its occupant's interests, opinions, and personality.

© *Ellis Herwig 1980/Stock, Boston.*

dormitory environment affects their lives. They have studied how the stresses of crowding and excessive noise in the urban environment can affect people's health, social relationships, and morale. They have investigated the ways people's attitudes toward the environment affect the major environmental problems we face today—the crisis of diminishing energy reserves, the pollution of our air and water, and the wanton destruction of natural landscapes. Researchers have even gone inside mental hospitals to learn how institutional environments can affect patients' recovery.

Although environmental psychology is a new field of study, it has grown with amazing speed. Environmental psychology's interest in studying human behavior in the familiar, everyday physical environments where people live and work, as well as its relevance to environmental design and social planning, has made it especially responsive to the demands of today's world. Psychologists have come to recognize a need to learn more about the ways normal people behave in everyday contexts. Informed citizens and policy makers alike have demanded that science provide practical knowledge that can be used to solve major problems of society. Students have insisted that their college courses be relevant to the real world outside the classroom. Let us turn now to a discussion of the nature and character of the exciting and challenging field of environmental psychology.

A DEFINITION OF ENVIRONMENTAL PSYCHOLOGY

As environmental psychology is a broad area of inquiry that is continuing to evolve, a definition of the field must be general enough to encompass both its breadth and its changing nature. While most research in environmental psychology has concerned the psychological effects of the built or architectural environment, interest has also been directed toward research questions that transcend specific environments, such as crowding and privacy. Because environmental psychology evolved in response to social concerns, it tends to focus on socially relevant problems, to maintain a holistic level of analysis, and to emphasize the practical application of its theoretical knowledge. The label *psychology* is employed here in a problem-definition sense rather than a disciplinary one, for since its inception the field of environmental psychology has involved many disciplines. With these concerns in mind, we may state: *Environmental psychology is an area of psychology whose focus of investigation is the interrelationship between the physical environment and human behavior and experience.* As we shall see, this emphasis on the *interrelationship* of environment and behavior is important; not only do physical settings affect people's behavior, but individuals actively influence the environment.

CHARACTERISTICS OF ENVIRONMENTAL PSYCHOLOGY

Because environmental psychology is a new and complex field of study, it is important for us to consider some of the characteristics that describe the way environmental psychologists *approach* their research. The approach a field of study takes to its subject matter shapes the kinds of investigative questions asked, and how and where answers are sought. Our approach to research will also color the kinds of data we seek, are willing to accept, or are inclined to ignore. Because the manner in which investigators approach their research plays such an important role in shaping the scientific enterprise, I shall attempt in this book, in addition to providing a comprehensive discussion of knowledge in environmental psychology, to reflect the approach to its subject matter that has come to characterize this field of study.

ADAPTATIONAL FOCUS

An important characteristic of the approach that environmental psychology takes toward its subject matter is its focus on *processes of adaptation*. Environmental psychologists are especially interested in the varied psychological processes by which people adapt to the complex demands of the physical environment. For example, researchers have examined the social and behavioral consequences of people's efforts to adapt to the excessive stimulation of overcrowded settings. Investigators have also studied the adaptive functions served by the processes involved in perceiving the physical environment and in developing mental images of physical settings. This underlying interest in processes of adaptation has helped to shape the character of inquiry in environmental psychology, suggesting appropriate research questions and central theoretical concerns. In this book an adaptational focus serves as a guiding theme, helping to integrate the diverse research topics and theoretical issues we shall consider.

Robert White (1974) contends that adaptation should be broadly defined to encompass *all of the processes engaged in by living systems in interacting with their environment*. He explains that such a definition can include a wide array of strategies of adaptation, from the most simple ways of dealing with minor environmental irritations to the most complex efforts to cope with major environmental challenges. The definition's emphasis on living *systems* encourages a view of adaptation that envisions a total organism interacting with a holistic environment. Finally, White's definition suggests that living organisms play an *active* role in the adaptation process. Living systems interacting with the environment exhibit a substantial degree of internal governance and independence of action. In summary, the adaptational focus in environmental psychology emphasizes (1) the *processes* whereby living systems interact with the environment; (2) a *holistic* view of the organism and its environment; and (3) the *active* role of living organisms in relation to their environment. Let us consider each of these aspects of the adaptational focus in environmental psychology in more detail.

Psychological processes   The adaptational focus of environmental psychology emphasizes the psychological processes that *mediate* the effects of physical settings on human activity. For example, in studying the effects of classroom noise on students' grades, the environmental psychologist not only asks *whether* noise affects grades, but also *how* such effects come about. The researcher seeks to learn about the psychological processes that mediate between environmental conditions (excessive noise) and behavioral outcomes (students' grades). For instance, the investigator may study changes in students' ability to concentrate, in their attention to social cues, and in their willingness to persist at difficult tasks when noise is introduced into a previously quiet classroom. Such changes in concentration, attention, and persistence can help the environmental psychologist understand *how* an increase in classroom noise may be linked to a drop in students' grades.

In this book we shall focus on a range of psychological processes in which people engage while interacting with the physical environment. We shall explicitly consider the psychological functions served by these psychological processes. The first half of the book discusses psychological processes that mediate between environment and be- havior at an *individual* level. We shall learn how people perceive the environments in which they live and work. We shall discover that each of us develops a unique men- tal map of the environment, including a highly personalized map of our own neigh- borhood and of the town or city in which we live. We shall learn how environmen- tal attitudes, such as those toward conservation, are formed and changed. Finally, we shall consider some of the ways in which the physical environment influences people's performance in school and on the job, along with some of the strategies people have evolved to cope with environmental stressors, such as noise and air pollution.

The second half of the book considers psychological processes that mediate the influences of physical settings on people's activities at a *social* level. We shall discover that people develop complex coping strategies to deal with the potential negative effects of crowding in their lives. We shall consider some of the ways in which people attempt to maintain a desired level of privacy and a territory of their own. We shall also learn that each of us strives to maintain an appropriate distance between ourselves and other people. Finally, we shall find that, in the challenging environment of the central city, people establish ways to build meaningful friendships and to provide social support for one another.

Holistic view   Historically, psychological researchers have tended to analyze features of the environment in very small or molecular units. When psychologists discussed the environment in the past, they often referred to very restricted stimuli, such as a flashing red light or a beeping buzzer, rather than the complex settings that encompass people's daily lives. Environmental psychologists believe that if human behavior is to be under- stood properly, environment and behavior must be seen as interrelated parts of an indi- visible whole.

Kenneth Craik (1970) suggests that it is this holistic approach that makes environ- mental psychology a distinctive field of research. He explains that environmental psy- chologists are interested in studying the physical environment holistically, as it is ordi- narily experienced in people's daily lives. Environmental psychologists may study a street in a central city environment, a room in a household, or a beautiful feature of the natural environment.

Harold Proshansky (1972, 1976) has argued that environmental psychology's con- cern with the interrelationship between the person and the physical setting makes it essential for environmental psychologists to respect and maintain the naturally occur- ring properties of physical settings, the people who reside in those settings, and the ac- tivities that occur in them. Urie Bronfenbrenner (1977) proposes similarly that environ- mental research methods preserve the natural integrity of the settings studied, avoiding the introduction of artificial elements that would distort the meaning of settings to their participants. The holistic view does not suggest that environmental psychologists should avoid laboratory studies or analytical styles of inquiry. Rather, as Irwin Altman

(1976) has noted, the interpretation of laboratory findings needs to be informed by an awareness of the holistic nature of environment-behavior relationships, and the synthesis of discrete research findings must be pursued with a vigor equal to that traditionally devoted to dissecting a problem into its component parts.

Active role   A third aspect of environmental psychology's adaptational focus is an emphasis on the positive and adaptive ways in which people *cope* with environmental challenges. This view encourages us to see the active, varied, and creative ways in which people have learned to deal with the environment. If we fail to look at the ways people actively cope with the environment, we will tend to see the environment as an overwhelming force and people as passive pawns that are merely acted upon. Such a deterministic model of environment and behavior may be depicted visually as follows:

In this deterministic model, human behavior is seen as a direct and passive reaction to controlling environmental conditions. Such a deterministic view suggests that a negative environmental condition, such as overcrowding, leads inevitably to negative human reactions, such as physical or emotional distress. Environmental psychologists now recognize that such a deterministic model presents a misleading and overly simplified picture of the relationship between environment and behavior.

In fact, we shall discover that even when faced with difficult environmental challenges, such as the demands of high-rise urban housing or a poorly designed institutional environment, people are able to develop positive and often creative ways to meet their personal and social needs. Such an adaptational model of environment and behavior may be represented graphically as follows:

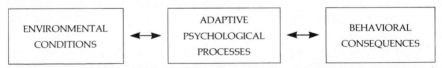

According to this adaptational model, the effects of environment on behavior are seen as mediated by a variety of adaptive psychological processes. The adaptational model shows that the direction of effect in the environment-behavior relationship is reciprocal; that is, people may act on environmental conditions while the environment also acts on human behavior. According to the adaptational model, the potentially negative psychological effects of a stressful situation, such as overcrowding, may be reversed through effective coping processes. Such coping might involve a direct effort either to alter the stressful environmental conditions (e.g., moving to a less crowded setting), or to change the psychological meaning or significance of the stressor (e.g., deciding the crowded setting has more "group spirit"). Of course, the adaptational model does not suggest that people will always be *successful* in reversing the potentially adverse effects of stressful environmental conditions. Some environments are so badly designed at the scale of human needs that even the most vigorous efforts at adaptation cannot wholly overcome their negative social and behavioral effects. It is important to recognize the real environmental obstacles that people confront in their lives, and to avoid blaming the victims of poor environmental conditions for failures and frustrations over which they have no control (Caplan and Nelson, 1973; Ryan, 1971). In summary, many environmental psychologists believe the adaptational model provides the fullest understanding of the environment-behavior relationship. The adaptational model guides our discussion of environment and behavior throughout this book.

INTERDISCIPLINARY INVOLVEMENT

Since its inception, environmental psychology has attracted scholars, researchers, and practitioners from a variety of disciplines, including sociology, geography, anthropology, medicine, architecture, and planning, as well as psychology (Craik, 1970; Proshansky and Altman, 1979). The study of human behavior in physical settings requires the work of researchers in many social sciences as well as that of architects and planners responsible for the design of human settings.

The label "environmental *psychology*" should be understood to describe the problem area of the field rather than a disciplinary restriction. Researchers in environmental psychology investigate a wide range of questions that involve psychological content—spatial behavior patterns, mental images, environmental stress, attitude change. The researchers themselves, however, represent many disciplines in addition to psychology. Joachim Wohlwill (1970) has encouraged environmental psychologists to strengthen and broaden their interdisciplinary ties by improving the avenues for effective communication and collaboration among the varied disciplines engaged in work in this area.

APPLIED ORIENTATION

Research in environmental psychology is oriented toward both the resolution of practical problems and the formulation of new theory. Environmental psychologists have sought to blend these two concerns in such a way that each enhances the other. Balancing these two needs can sometimes present the environmental psychologist with a serious challenge, as when theoretical refinement requires further research while practical concerns demand solutions to present problems (Altman, 1975; Studer, 1973). Some environmental psychologists (Proshansky, 1972; Sommer, 1977) have suggested the relevance of Kurt Lewin's notion of *action research* as a useful model for the blending of practical and theoretical needs in environmental psychology.

Lewin (1947) advanced action research as a model that would both generate new theoretical knowledge and apply that knowledge to the resolution of social problems. He believed that by monitoring the effects of social interventions, psychologists could get practical feedback to use in refining their psychological theories. He also felt that psychological theory was of practical value because it could guide the planning and execution of social interventions in complex problem areas.

The study of university housing environments is one research area where environmental psychologists have successfully combined theoretical and practical objectives (Figure 1-2). Studies of students in crowded living environments have provided theoretical knowledge of how the information overload associated with such a setting can make people less friendly and responsive to one another. At the same time this the-

Figure 1-2
Environmental psychologists are interested in finding ways to make dormitories more satisfactory living environments.

© *Lawrence Frank 1981.*

oretical knowledge has been applied to make practical suggestions to designers about
some ways the architecture and interior design of student living environments might be
improved. For example, Andrew Baum and his associates (Baum, Davis, and Valins,
1979) have suggested that dividing long corridors in dormitories into smaller residential
clusters might help to reduce some of the social stress associated with the long-corridor
design. Similarly, Holahan (1977) has described how the erection of partitions in the
cafeteria of a high-rise dormitory served to foster social contact among the student residents who ate there.

VARIED RESEARCH METHODS

Psychological research has primarily involved experimental studies carried out in the
laboratory. Environmental psychologists, in contrast, have attempted to develop diverse
research methods in a variety of field and laboratory settings, tailoring their research
approach to the particular features of the setting under study and the specific requirements of the research question being posed. Thus, while a great many research studies
in environmental psychology have involved correlational studies in naturalistic settings,
environmental psychologists have also used experimental methods in both laboratory
and field settings when such methods have been appropriate to the research questions
being asked. This varied and flexible research approach is especially important in en-

vironmental psychology because the complex nature of environment-behavior relationships confronts researchers with unique methodological challenges.

Arthur Patterson (1977) has discussed some examples of the varied research
methods used by environmental psychologists. He explains, for example, that in order
to learn how psychiatric patients actually behave in hospital environments, investigators (Ittelson, Proshansky, and Rivlin, 1976) have used naturalistic observation methods
in field settings. In contrast, other researchers (Freedman, Klevansky, and Ehrlich, 1971)
employed an experimental method in a laboratory in order to analyze precisely how
crowding affected people's behavior in an experimentally controlled setting. Finally,
Oscar Newman (1972) chose a quasi-experimental design (Campbell and Stanley,
1966) to learn how urban crime rates were influenced by the contrasting features of the
physical designs of two housing projects.

HISTORY OF
ENVIRONMENTAL PSYCHOLOGY

MIDWEST AND ITS CHILDREN

Midwest Psychological Field Station Although environmental psychology grew
most rapidly during the 1960s, the historical roots of the field can be traced to 1947,
when two psychologists at the University of Kansas established the Midwest Psychological Field Station in Midwest, Kansas, population 800. Their goal was to learn how
real-world environmental settings affect people's behavior, with a special interest in the
behavior and development of children.

Allan Wicker (1979) has provided a vivid portrait of the field station: its high ceilings with oscillating electric fans, its old-fashion oak furniture, its old black bank vault.
He notes that the field station, which was located in Midwest's red-brick bank building,
looked more like a nineteenth-century newspaper office than an important research institute.

Yet the founding of the Midwest Psychological Field Station marked a major innovation in psychology. There Roger Barker and Herbert Wright legitimized psychological research conducted in real-word settings as well as in the psychological laboratory. The subjects of their research were neither college sophomores nor psychiatric
patients, but ordinary people engaged in their daily affairs. The two psychologists
tackled real-world activities in their natural complexity and diversity, recording behavior as it is lived and experienced—walking to school, buying groceries, engaging in
conversation. Over the twenty-five years that the field station was in operation, Barker
and Wright and their colleagues published a series of groundbreaking books (Barker,
1963, 1968; Barker and Associates, 1978; Barker and Gump, 1964; Barker and Schoggen, 1973; Barker and Wright, 1951, 1955) and papers (Barker, 1960, 1965, 1969; Barker
and Wright, 1949; Wright, 1956, 1960) that played a major role in revolutionizing the
way psychological research was thought about and conducted.

Behavior settings When Barker and Wright studied people's behavior as it occurred
in natural contexts, they defined the environmental unit they were interested in as a *be-*

havior setting. A behavior setting included a particular pattern of behavior, along with the environmental and temporal features that surrounded the behavior. For example, the behavior setting of Clifford's Drugstore in Midwest included the patterns of behavior that are typical of drugstores (purchasing pharmacy items, drinking a soda at the soda fountain), the physical environment of the drugstore (pharmacy, soda fountain, variety department), and the temporal limits within which the drugstore functioned (from opening time at 8:00 A.M. to closing time at 6:00 P.M.). Notice that a behavior setting is defined in terms of the *characteristic* patterns of behavior that occur in a setting, and not according to the particular individuals who use the setting.

Barker and Wright explained that in a behavior setting, the behavioral and environmental features are interdependent and fit together in a natural and comfortable way. They added that when the fit between behavior and environment in a behavior setting is "out of step," the situation will appear quite unusual (Figure 1-3). Other behavior settings that Barker and Wright identified in Midwest included the worship service at the Presbyterian church, Burgess's Beauty Shop, a Rotary Club meeting, Chaco's Garage and Service Station, a Halloween dance, and a high school varsity basketball game. Barker and Wright developed a procedure for identifying and precisely describing the variety of behavior settings that make up a particular environmental context, such as a school or even a whole town; they called this procedure a *Behavior Setting Survey.*

Ecological psychology   On the basis of their observations in Midwest, Barker and Wright proposed a new field of psychological research, which they termed *ecological psychology.* A major goal of ecological psychology was to learn how people's behavior and development are influenced by the physical environments that are a part of their everyday lives. Barker and Wright emphasized that psychologists could no longer assume that behavior can be adequately predicted solely by psychological tests of individual differences in background or personality. They contended that to predict how people will behave in a particular situation, we need to know something about the nature of the particular environmental setting in which they will be behaving. To acquire this knowledge about the behavioral properties of environmental settings, psychologists would have to leave the laboratory to engage in field research in the natural environments where people conduct the commerce of daily life—homes, schools, work environments, stores, and recreational settings.

Wicker (1979) has recently applied the principles of ecological psychology to study the behavioral consequences of overpopulation in a variety of human settings. He studied the relationships between the number of persons who wanted to participate in a behavior setting and the limits of a setting's capacity to handle large numbers of people while still maintaining its ongoing programs. For example, following up on earlier work conducted by Roger Barker (Barker and Gump, 1964), Wicker analyzed the behavioral differences between large and small high schools in terms of ecological psychology. He found that students in small high schools entered a wider range of school behavior settings, assumed more positions of responsibility, and felt more needed than did their peers in large high schools.

Figure 1-3

Charlie Brown encounters a situation in which behavior and environment are "out of step."

© 1966 United Feature Syndicate, Inc.

ENVIRONMENTAL DESIGN RESEARCH

Problem focus   Although the work of Barker and Wright played an important role in sensitizing psychologists to some of the ways in which physical settings shape human behavior, the scattered interests of psychologists studying environmental influences on behavior did not coalesce into a distinct and independent field of study until the 1960s. Interestingly, the catalyst that brought environmental psychology into being was not theoretical concerns on the part of academic psychologists, but rather practical questions posed by people actually engaged in the design of physical settings. Proshansky (1972) points out that environmental psychology is problem-oriented in that it attempts to answer a wide range of practical questions asked by architects, interior designers, and city planners.

The complexity of design decisions in modern society has confronted architects and designers with a staggering task (Alexander, 1964; Craik, 1970). During the 1960s a growing number of people in the design professions came to recognize the important role that psychology could play in design practice. They realized that collaboration with psychologists and other social scientists might help to answer many design questions, such as how to design settings to support and facilitate particular types of human activity (Craik, 1970; Well, 1965; Studer, 1966; Studer and Stea, 1966).

Architecture and behavior   Central to the concern of the architects and designers who first sought the involvement of psychologists in the design enterprise was a conviction that architecture and human behavior are interrelated. Kiyoshi Izumi (1965) described some of the ways in which physical features and human activities are interrelated in architectural design. At one extreme are buildings designed primarily to contain machinery, equipment, and other inanimate objects. At the other extreme are buildings designed solely to contain human beings—nursing homes, penitentiaries, psychiatric hospitals, and housing in general. Between these extremes are buildings used to contain both people and objects in varying proportions: libraries, laboratories, stores, offices, and so on. As we move from less to more human-oriented buildings, our evaluation of a building's success will be increasingly based on how well it meets the needs and activity preferences of its occupants (Deasy, 1970; Sommer, 1969). A great many contemporary buildings fail to meet the behavioral needs of their users (Watson, 1970). Consider, for example, the new urban neighborhoods of faceless high-rise structures that inhibit rather than support residents' feelings of social identity and group belonging. Similarly, many of the high-rise megadorms increasingly constructed as "progressive" university housing tend to stifle rather than foster meaningful social participation among students. Serge Boutourline (1970:496) has asserted, "The dominant situation in modern life is individuals living in a setting which was not built for them."

The interrelationship between architecture and behavior is a strong and stable one. Harold Proshansky and his colleagues (Proshansky, Ittelson, and Rivlin, 1976) explain that every architectural setting has characteristic patterns of behavior that are associated with it. These characteristic activity patterns are stable and enduring over time, even when the particular people involved in the setting change. They add, however, that common sense is often a poor guide to an understanding of the relationship between design and behavior, and that a careful empirical assessment of how an architectural setting is functioning can be instructive and even surprising. For example, their empirical evaluation of a psychiatric hospital revealed that dining rooms were used more for social games than for eating, and dayrooms were used more for sleeping than for recreation.

New directions in psychology   The architects and designers who first turned to psychologists for assistance in approaching complex design decisions were initially disappointed. Psychology's research history as a predominantly laboratory science left it poorly prepared to answer questions about everyday behavior in real-world settings. Many psychologists initially shied away from what they perceived as unmanageable research questions (Winkel, 1970; Wohlwill, 1970). Disillusioned designers found that existing psychological knowledge was limited to restricted physiological reactions in unusual environments that most designers would never encounter (Dyckman, 1966; Ventre, 1966). They found, for example, that psychologists knew considerably more about the ways individuals respond to the environments of space capsules and submarines than about how people react to the urban or suburban environment (Blackman, 1966).

Persistent pressure from the design professions, however, along with the increasing social concern with environmental issues that characterized the 1960s, brought about a change in this state of affairs. Environmental psychologists began to set new directions in the questions they asked and in the research methods they employed. The ensuing decade saw a growing number of psychologists and other social scientists joining with practitioners in the design professions to form the new discipline of environmental psychology.

Harold Proshansky and Irwin Altman (1979) have provided a rich description of the emergence and growth of environmental psychology. They note that among the first developments were research projects dealing with architectural influences on behavior in psychiatric hospitals in various countries, including Canada (Osmond, 1957; Sommer and Ross, 1958), France (Sivadon, 1970), and the United States (Good, Siegal, and Bay, 1965; Ittelson, Proshansky, and Rivlin, 1976). An opportunity for researchers and practitioners in the field of environment and behavior to meet and to share common concerns occurred at the first annual conference of the Environmental Design Research Association (EDRA), held in 1969 in Chapel Hill, North Carolina (Sanoff and Cohen, 1970). New professional journals, including *Environment and Behavior* and *Man-Environment Systems*, provided a forum for this new area of investigation. Graduate programs and undergraduate courses in environmental psychology were initiated at universities in the United States and in other parts of the world. By the 1970s the new field of environmental psychology had earned a respected place among the established fields of study in the social and behavioral sciences.

Social concerns   Proshansky and Altman explain that the rapid growth experienced by environmental psychology during the 1960s was also encouraged by the prevailing social concerns that characterized that decade. The quality of the physical environment and the long-term environmental consequences of industrial pollution, careless waste disposal, and poor resource management were topics of heated public debate. Earth Day in 1970 represented a dramatic expression of the social consciousness around environmental issues that typified this period. Proshansky and Altman add that the social concern with people's effects on the environment grew to encompass a similar concern with the long-term effects of the physical environment on human beings; in effect, people's mistreatment of the environment creates adverse environmental conditions, such as serious air and water pollution, that come to represent long-term threats to the quality of human life. As people became more sensitive to surrounding environmental conditions, they came to recognize more fully the many subtle ways in which the environment can affect human functioning. Citizens demanded that scientists accept a meaningful role in resolving the major environmental problems in our society. Government funding agencies added further inducements to the development of environmental studies by offering research support for applied and problem-oriented research. Finally, academic psychologists were confronted with graduate and undergraduate students who demanded an approach to psychology that was relevant to the social concerns that were apparent on all sides in their own lives (Figure 1-4).

Institutional environments   In response to persistent pressures from practicing designers, informed citizens, and concerned students, a body of research in environmental psychology evolved that attempted to evaluate the psychological effects of institutional environments. Research was initiated in a broad range of institutional settings—psychiatric and general hospitals, nursing homes, residential treatment settings for children, correctional facilities, and schools and universities. Research in psychiatric hospitals showed that positive forms of social interaction among patients could be substantially increased by physically remodeling a hospi-

Figure 1-4
Social concern with the environment furthered environmental psychology's growth during the 1960s.

© *Lawrence Frank 1970.*

tal ward to make it more attractive and cheerful (Holahan and Saegert, 1973). Investigators in Canada demonstrated that rearranging ward furniture in more social patterns could encourage significantly more social participation among psychiatric patients (Sommer and Ross, 1958). Other researchers discovered that the elongated tunnels and corridors that are typical of many psychiatric hospitals can cause patients to experience distortions of auditory and visual perception (Spivack, 1967).

Studies concerned with quality of life in student residential environments have found more dissatisfaction and less group cohesion in high-rise megadorms than in low-rise dormitory settings. For example, students in crowded housing have been found to be less oriented toward social contact with their fellow residents than students in uncrowded settings (Valins and Baum, 1973), and less cooperative and helpful to their living mates (Bickman, Teger, Gabriele, McLaughlin, Berger, and Sunaday, 1973).

The urban environment   Environmental psychologists have also assessed the social and psychological impact of the urban environment. For example, researchers have investigated people's "mental maps" of the city to assess how easy or difficult it is for residents to understand particular urban environments (Lynch, 1960). Other investigators focused on the ways in which features of the urban environment affect the lives of city residents. They found, for example, that Boston residents who were forced to leave their homes to make way for urban renewal suffered a severe grief reaction that was similar to mourning for the dead (Fried, 1963). Environmental psychologists have also found that many of the design features of high-rise public housing impede the development of positive forms of group life among residents, while encouraging crime and vandalism (Newman, 1973; Yancy, 1971).

The natural environment   An important body of applied research in environmental psychology has attempted to assess people's attitudes toward features of the natural environment (Craik and Zube, 1976a; Zube, Brush, and Fabos, 1975). Such evaluations can be applied, for example, in decisions concerning the development of scenic and recreational sites and preservation programming for natural landscapes (Craik, 1972a). User-based environmental assessments might also be used in the management of natural water resources and in decisions involving the scenic, recreational, or industrial development of water resources (Coughlin, 1976).

THE FORMULATION OF THEORY

While the 1960s saw the rapid growth of design-related research in environmental psychology, the 1970s saw, in addition to continued problem-oriented investigation, the first major steps in the formulation of theory in environmental psychology. Environmental psychologists initiated their efforts at theory construction by drawing on existing theory in other branches of psychology that might be extended to help us understand the complex relationship between environment and behavior (Wohlwill, 1970; Proshansky, 1973). For example, environmental psychologists developed theories of environmental stress that were based on general theories of psychological stress (Selye, 1956; Lazarus, 1966). Environmental stress models have been applied in efforts to understand how people react to a wide range of environmental stressors, such as excessive heat, high noise level, and crowding (Figure 1-5). Other environmental psychologists, drawing on earlier work in social psychology (Brehm, 1966), have proposed theories involving the notion that certain types of environmental settings, such as those that are overcrowded or low in privacy, restrict people's behavioral freedom (Proshansky, Ittelson, and Rivlin, 1976; Stokols, 1976).

Several investigators have proposed theories to explain the negative effects that crowded settings sometimes have on people. Some theorists have suggested that high levels of crowding produce excesses of social and sensory information that overload an individual's capacity to process them (Milgram, 1970; Saegert, Mackintosh, and West, 1975). An alternative theory holds that negative effects occur when a setting becomes "over staffed," and can no longer accommodate all the people who wish to participate in the setting (Wicker, 1979). Finally, a few theorists have proposed broad environmental theories that provide a conceptual basis for understanding a wide range of environmental behaviors (Altman, 1975; Michelson, 1976).

Figure 1-5

Environmental psychologists are concerned with understanding the human consequences of long-term exposure to noise and overcrowding.

© *Menschenfreund.*

RESEARCH METHODS IN ENVIRONMENTAL PSYCHOLOGY

The research methods employed by environmental psychologists are quite diverse, and it is important to recognize that no research method is universally superior to others; rather, each method is designed to answer a particular set of questions. Thus, while each research method is ideally fitted to the specific types of questions it has been designed to answer, it is less effective than other methods in answering other types of questions. Before examining how each research method is uniquely fitted to a specific set of questions, we must develop a vocabulary to describe the type of research questions with which environmental psychologists are concerned.

INTERNAL VALIDITY

Donald Campbell and his associates (Campbell and Stanley, 1966; Cook and Campbell, 1979) have provided a vocabulary to describe the underlying research questions that determine the appropriateness of alternative research methods. They suggest that research methods need to be evaluated in terms of both internal and external validity. *Internal validity* concerns the question: Did the research treatments investigated make a difference in this specific study? For example, in studying the effects of class size on learning, an environmental psychologist might compare test performance in a large class of 150 students and a small class of 20 students. Imagine that after one semester the researcher discovers that students' grades in the small class are significantly higher than those in the large class. The question of internal validity asks whether the grade differences in this particular study can in fact be attributed to the variable of class size. If the investigators had been able to control the operation of all extraneous variables during the study, they could claim with a high degree of certainty that the observed grade differences were due to class size; internal validity would be high.

If, however, the researchers had been unable to control the operation of all extraneous variables, they could not confidently conclude that the grade difference was due to class size; internal validity would be low. Imagine that the investigators had been unable to assign students randomly to the large and small classes, and instead used intact class groups. Here it is possible that the *selection* of subjects for the two conditions had been biased. More highly motivated students may, for instance, have made a greater effort than less motivated ones to enroll in small classes. Thus the better performance of the small class might be attributed to this initial difference in motivation as well as to class size. Imagine further that the investigators had been unable to control the manner in which teachers in the two types of classes taught their courses, and that during the semester the teacher in the small class adopted an updated textbook, while the teacher in the large class stayed with the standard text. Here the *histories* of the large and small classes (the specific events they experienced in addition to the class-size variable) differ, and the observed grade differences might be due to their different histories as well as to class size.

EXTERNAL VALIDITY

External validity concerns the question: To what groups and settings can this research effect be generalized? Again consider a study of the effects of class size on learning, in which the research data indicate that learning is superior in a small class. The question of external validity asks to which groups and settings the finding that class size is inversely related to performance may be generalized. If the study had been carried out in such a way that the students behaved in the same manner they would behave in an actual classroom, the researchers could confidently generalize the findings to other groups of students in real classroom settings; external validity would be high. If, however, the study had been conducted in a laboratory under highly artificial experimental arrangements, so that students were constantly aware they were taking part in a research study, the investigators could not safely generalize their findings to groups of students outside of the laboratory setting; external validity would be low.

Campbell and his colleagues argue that investigators should strive to employ research methods that have as much internal and external validity as possible. While internal validity is essential to the interpretation of research findings, external validity is important to the broader significance and potential applications of research results. They point out, however, that the requirements for internal and external validity are often at odds, and research methods that strengthen one type of validity tend to weaken the other. For example, while increased experimental control will enhance internal validity by limiting the influence of extraneous variables, it tends at the same time to make the research situation more artificial, thereby reducing external validity. Thus, while the environmental psychologist must attend to the need for both internal and external validity, the particular balance between the two must be determined by the unique research questions investigated by each research study. With these underlying concerns in mind, let us explore the major types of research methods used by environmental psychologists.

LABORATORY EXPERIMENTATION

When environmental psychologists wish to establish a high level of internal validity in their research, they use *experimental* methods in a *laboratory* setting. The experimental method allows the investigator to manipulate the experimental (or *independent*) variable systematically, and to assess its effects on outcome (or *dependent*) measures. The laboratory setting permits the researcher to control the influence of other extraneous variables to ensure that they do not affect the dependent measures while the independent variable is manipulated. In addition, laboratory experimentation generally involves the random assignment of subjects to experimental conditions, assuring the investigator that the subjects in the different experimental conditions are equivalent. These requirements—systematic manipulation of the independent variable, control of extraneous variables, and random assignment of subjects to experimental conditions—provide the investigator with a high degree of confidence that observed outcomes are due to the independent variable. Notice, however, that the high level of control that characterizes the laboratory experiment limits the external validity of laboratory findings. While researchers can be confident of the relationships between the independent and dependent variables in the laboratory context, they cannot be certain of the relationships between these variables in the multivariate complexity of naturalistic settings (see Bem and Lord, 1979).

FIELD CORRELATIONAL STUDIES

When the goal of environmental psychologists is to maintain a high degree of external validity in their research findings, they turn to *field correlational* studies (Figure 1-6). Field correlational studies are designed to yield information about naturally occurring activities in real-world settings that are unaffected by the psychologist-investigator. Unlike experimental methods, which provide a causal picture of the relationship between variables, field correlational studies cannot demonstrate that one variable is causally related to another. Rather, they provide information about the *correlation* or association between variables. For example, while field correlational studies can show that high levels of density and high degrees of social pathology often occur together in the central city, they cannot demonstrate that density causes pathology. It is possible that both density and pathology are caused by some other factor, such as poverty. Thus, although field correlational studies are characterized by a high degree of external validity, their internal validity is often low. Most of the field correlational studies we shall review in this book are of two types—naturalistic observation and survey research. *Naturalistic observation* involves the direct observation of behavior in naturalistic set-

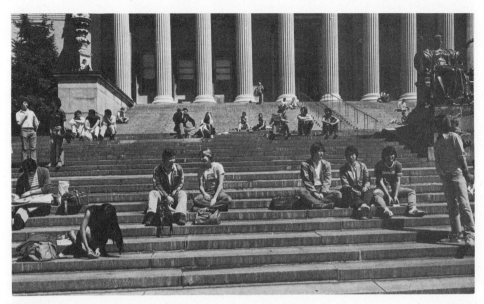

Figure 1-6
In order to learn how people behave in real-world settings, environmental psychologists have conducted field correlational studies in naturalistic settings such as this campus environment.

Lawrence Frank 1981.

tings; *survey research* consists of either paper-and-pencil or personal interviews conducted to assess the personal attitudes and experiences of participants in particular environments.

FIELD EXPERIMENTATION

While most experimental studies are carried out in controlled laboratory settings and most field studies are correlational in nature, it is possible to use the experimental method in field settings. In fact, when environmental psychologists want to achieve a balance between internal and external validity in their research, they employ the *field experiment* (see Seashore and Bowers, 1963). In the field experiment, the investigator systematically manipulates the experimental variable under study while permitting extraneous factors that occur in the field setting to operate naturally. Part of the control possible in the laboratory is traded off to gain some of the contextual richness possible in the field. At the same time, some of the naturalness of the field setting is relinquished to obtain more control of the experimental variable than can occur in field correlational studies. Although there have been many fewer field experiments in environmental psychology than either laboratory experiments or field correlational studies, they indicate the wide range of research strategies available to the environmental psychologist and the close relationship between internal and external validity (see box, "Ethical Concerns in Field Experimentation").

Ethical Concerns in Field Experimentation

Environmental psychologists sometimes use field experimental methods because they enable the investigator to learn how people are affected by an experimental manipulation in a real-world setting when they are unaware that they are subjects in a research study. The very fact that individuals are unaware that they are participating in a psychological study, however, raises some ethical concerns. Typically subjects in field studies are not asked whether or not they wish to participate in a study, and they are rarely debriefed about the nature and purpose of the study. Let us consider an example of a field experiment conducted by Eric Schaps (1972). The experimenter and a colleague (posing as shoppers) visited shoe stores during crowded periods. One "shopper" presented a shoe with a broken heel and asked to see a variety of shoes; she then proceeded to reject each pair of shoes the salesman showed her. During the episode, the second "shopper" surreptitiously took notes on the salesman's behavior.

Irwin Silverman (1975) asked two lawyers for their opinions about the study as well as several other field experiments. While the opinions of the two lawyers differed on many of the studies, both felt there might be grounds for civil action on the part of the shoe store or the salesman, and one lawyer suggested that the experimenters were liable for trespass. In a later survey, David Wilson and Edward Donnerstein (1976) asked almost 200 ordinary citizens for their views of the shoe store study and a variety of other field studies. Seventy-two percent of respondents reported that they would feel harassed or annoyed to be a subject in the shoe store study; 26 percent felt that the study would invade their privacy; 31 percent believed the study was unethical; and 55 percent said they would mind being a subject in the experiment.

The issue of ethical concerns in field experiments is complex; we must weigh the importance of the study and the potential social benefits of its findings in assessing its overall worth. Yet, as Wilson and Donnerstein point out, even when most persons do not find a field study objectionable, the feelings of a substantial minority who are offended by the study should be considered. Wilson and Donnerstein argue that investigators using field experimental methods should take into account the public's attitudes toward the methods employed. This view is in accord with the American Psychological Association's Committee on Ethical Standards in Psychological Research (1973), which proposes that ethical consultants be used in planning and conducting disguised field studies. It is important for environmental psychologists who conduct field experiments to pay serious attention to the ethical and moral aspects of their research, and to respect participants' personal feelings and right to privacy.

In summary, it is essential to realize that no single research strategy is ideal for all of the research questions posed by environmental psychologists. Rather, the environmental psychologist must adopt a flexible approach to the selection of research methods, choosing the strategy that is best fitted to the particular question being asked. For example, in investigating the behavioral effects of aversive noise,

David Glass and Jerome Singer (1972b) desired a high level of internal validity, so that the contrasting effects of various aspects of noise (such as its predictability and controllability) could be precisely understood. They selected a highly controlled laboratory experiment in which various noise parameters could be systematically varied and extraneous influences carefully controlled. Donald Appleyard and Mark Lintell (1972), in contrast, sought a high degree of external validity in order to understand the social correlates of noise in real community settings. They chose a field correlational study that entailed a survey of San Francisco residents who lived on streets that differed in the amount of noise caused by automobile traffic. Finally, Lawrence Ward and Peter Suedfeld (1973) wished to balance the requirements for internal and external validity in systematically studying the effects of noise in a real-world setting. They employed a field experiment in which they played taped traffic noise over loudspeakers on the Rutgers University campus, while carefully recording its effects on the behavior of students and teachers. Notice that while each of these research methods is ideally suited to answering one set of research questions, it is less fitted than the other methods to respond to alternative questions. Whenever possible it is especially useful for environmental psychologists to employ a combination of research methods, so that the inherent weaknesses of one method are balanced by the natural strengths of another.

SUMMARY

Environmental psychology is an area of psychology that focuses on the interrelationship of the physical environment and human behavior and experience. An important characteristic of environmental psychology's approach toward its subject matter is its focus on *processes of adaptation.* Adaptation may be broadly defined to encompass all of the processes engaged in by living systems as they interact with their environment. The adaptational focus in environmental psychology emphasizes (1) the *processes* whereby living systems interact with the environment, (2) a *holistic view* of the organism and the environment, and (3) the *active role* of living organisms in relation to the environment.

Environmental psychology is an *interdisciplinary* field. Since its inception, environmental psychology has attracted scholars, researchers, and practitioners from a variety of disciplines. Its approach to research is characterized by an effort to blend theoretical and practical goals in such a way that each objective enhances the other. A research model for the field is *action research,* which involves a simultaneous effort to generate new theoretical knowledge and to apply that knowledge to the resolution of social problems. Environmental psychology takes a *varied* and *flexible* approach to research methodology in response to the needs of the particular research question and environmental circumstances under study.

The historical roots of environmental psychology can be traced to the research carried out by Barker and Wright at the Midwest Psychological Field Station in the 1950s. Their work marked a major innovation in psychological research in that it involved the study of ordinary people engaged in the daily affairs of life in natural contexts. A *behavior setting* is defined as a particular pattern of behavior along with the environmental and temporal features that surround the behavior. Barker and Wright proposed a field of research they termed *ecological psychology* to learn how human behavior and development are influenced by the physical environments that are a part of people's everyday lives.

Practical questions posed by people engaged in the design of physical settings brought environmental psychology into being as a distinctive and independent field of study during the 1960s. Environmental psychology is problem-oriented in that it attempts to answer a wide range of practical questions asked by architects, interior designers, and city planners. Central to the concern of the architects and designers who first sought the involvement of psychologists in the design process was a conviction that architecture and human behavior are interrelated. Associated with every architectural setting are characteristic patterns of behavior that are stable and enduring over time. The rapid growth of environmental psychology during the 1960s was facilitated by prevailing social concerns with the quality of the physical environment. The 1970s saw, in addition to continued problem-oriented investigation, the first major steps in the formulation of theory in environmental psychology. Among the theories developed by environmental psychology were theories of environmental stress, behavioral freedom, and information overload.

Internal validity concerns the question: Did the research treatments investigated make a difference in this specific study? *External validity* concerns the question: To what groups and settings can this research effort be generalized? The requirements for internal and external validity are often at odds, and research methods that strengthen one type of validity tend simultaneously to weaken the other. When environmental psychologists wish to establish a high level of internal validity in their research, they use *experimental* methods in a *laboratory* setting. The experimental method allows the investigator to manipulate the experimental variable systematically, while the laboratory setting permits the researcher to control the influence of other extraneous variables. When the goal of environmental psychologists is to maintain a high degree of external validity in their research findings, they turn to *field correlational* studies. Field correlational studies are designed to yield information about naturally occurring activities in real-world settings that are unaffected by the psychologist-investigator. When environmental psychologists want to achieve a balance between internal and external validity in their research, they employ the *field experiment*. In the field experiment, the investigator systematically manipulates the experimental variable under study while permitting extraneous factors that occur in the field setting to operate naturally.

2 Environmental Perception

W e learned in Chapter 1 that the activities that comprise our daily lives are intertwined with the physical environments in which we live and work. Many of our daily activities—relaxing, eating, studying, sleeping—are influenced by the architecture and interior design of our home environment. The physical design of the setting where students attend college plays a role in shaping the learning experiences, extracurricular activities, and social friendships that comprise college life. In this chapter we shall discover that all of these diverse daily activities are dependent on our ability to perceive accurately the varied environments that are a part of our lives.

Environmental perception is the bedrock on which environmental behavior is founded. In order to understand, navigate, and effectively use the physical environment, we must first perceive it clearly and accurately. Yet, while environmental perception is essential to our ability to conduct the affairs of daily life, we tend generally to take this process for granted. In fact, environmental psychologists have found that one way to study the important role

23

of environmental perception in people's lives is to put them in novel environments with which they are unfamiliar. In this way, researchers have been able to observe at firsthand the perceptual processes that have become second nature in more familiar settings.

Environmental psychologists have discovered that the process of perceiving the physical environment is complex and dynamic. Environmental perception is an active process, not a passive one. We shall discover that by better understanding the process of environmental perception, we can learn to design settings that are more congruent with people's psychological needs. We shall see also that an understanding of environmental perception may enable us to help people cope with the threats of natural hazards, such as floods or earthquakes. Let us turn now to a consideration of the remarkable, though often neglected, process by which people perceive the physical environments that make up daily life.

THE NATURE OF
ENVIRONMENTAL PERCEPTION

Environmental perception is a marvelous and unique psychological process. Through environmental perception, the diversity of stimulation from the environment that impinges on us from all sides is organized to form a coherent and integrated picture of our world. Before we discuss environmental perception, however, we must distinguish it from environmental cognition and environmental attitudes, which will be examined in Chapters 3 and 4, respectively. Environmental perception involves the process of apprehending through sensory input the physical environment that is immediately present. Environmental cognition concerns the storage, organization, reconstruction, and recall of images of environmental features that are not immediately present. Environmental attitudes are the favorable or unfavorable feelings that people have toward features of the physical environment.

These three processes do not operate in isolation from one another. In fact, the psychological processes by which people cope with the physical environment are interrelated (see Ittelson, 1976; Lowenthal, 1972). Our perception of the environment provides information that is essential to our ideas about the environment and our attitudes toward it. Environmental cognitions and attitudes, in turn, form a set of expectations about the environment that shape our perceptions of it. Consider a tourist who is visiting Boston for the first time. His or her initial perceptions of the city may be somewhat confusing and disorienting. Continued perception of a variety of aspects of the city, however, may eventually offer a basis for a clear and well-organized mental image of it. This clearer image may then enable the tourist to get around Boston more effectively, thus contributing to a more positive attitude toward it. The combination of a clearer image of the city and a more favorable attitude toward it may in turn help the visitor to perceive new areas of Boston more effectively and efficiently. Here and in following chapters we shall discuss psychological processes in the environment separately because this approach facilitates learning about each process. In real life, however, these psychological processes never operate in isolation, but occur in interaction with one another and constantly influence each other.

THE UNIQUENESS OF ENVIRONMENTAL PERCEPTION

Object perception   A valuable discussion of the unique nature of environmental perception has been provided by William Ittelson (Ittelson, 1970, 1973, 1976; Ittelson, Franck, and O'Hanlon, 1976). Ittelson explains that psychologists have historically tended to ignore the process by which people perceive the large-scale, or molar, physical environment. While psychologists have devoted considerable attention to the study of perception, they have typically studied the way people perceive isolated *objects* rather than the way they perceive the environment, which consists of a complex array of many objects. For example, traditional psychological studies of perception have generally examined perceptual processes, such as the ways people perceive size, distance, and movement, as they relate to isolated objects.

Ittelson explains that environmental psychologists are interested in learning how people perceive complex, molar environments, such as a living room, an office setting, or even a neighborhood. This is not to suggest that environmental psychologists cannot learn from earlier research on object perception. Rather, the environmental psychologist must go beyond object perception to consider some of the ways the unique demands of the large-scale physical environment shape the nature of the perceptual process.

Irving Biederman (1972) conducted an intriguing laboratory experiment designed to demonstrate how the perception of objects in the real world is affected by the overall environmental context in which the object is embedded. Subjects briefly viewed slides of various environmental scenes, such as a university campus, a street, or a kitchen. Each scene was presented in two versions, one coherent and one jumbled (Figure 2-1). Subjects were asked to identify particular objects in the scenes, such as a dog. The object to be identified was the same in both the coherent and jumbled versions of the scene, and the section of the scene in which it was located always remained in its original position.

Biederman found that subjects were able to identify the object more accurately in the coherent scene than in the jumbled scene—even when the subjects were told where to look on the slide. He concluded that an object's meaningful context enhances its perceptual recognition. He emphasized that this finding is especially relevant to our understanding of how objects are perceived in real-world settings, because—in sharp contrast to the isolated objects used in traditional laboratory studies—real-world objects are always perceived in a meaningful setting or context.

Environments surround   Ittelson points out that environments are large in relation to people, surrounding those who perceive them. Since people are surrounded by the environment, they have to move about in order to perceive all aspects of it. Unlike an object that can often be adequately perceived from a single vantage

Figure 2-1
People are better able to identify the dog in this campus scene when it is shown in a coherent picture (*top*) than when the picture is jumbled (*bottom*).

From I. Beiderman, "Perceiving Real-World Scenes," Science, *July 7, 1972, 177:77–80. Copyright 1972 by the American Association for the Advancement of Science. Reprinted by permission.*

point, the environment must be experienced from multiple perspectives to be fully perceived. For example, a person who moves to a new apartment may walk through the new setting a number of times, experiencing its unique features, such as the "feel" of different rooms, areas for special decoration or storage, and contrasting views from different windows. Ittelson points out that the surrounding quality of the environment makes environmental perception more like *exploration* than simple observation. Thus an important aspect of environmental perception involves *motoric experience*—an active, physical interchange with the environment. Action in and toward the environment provides the individual with a variety of sensory cues or *feedback* (e.g., visual, auditory, and tactile sensations) about the nature of the environment.

Environments provide an abundance of information   Environments provide people with such an abundance of perceptual information that they cannot possibly process all of it at once. For example, our tourist exploring a neighborhood of Boston for the first time may feel overwhelmed by perceptual information that is often ambiguous and occasionally contradictory. Ittelson points out that the abundance of perceptual information provided by the environment arrives simultaneously through a variety of sensory modalities. Our tourist will be confronted simultaneously with the novel sights, sounds, and smells of the unfamiliar neighborhood. Ittelson notes also that because the perceptual information provided by environments is so abundant, we are presented at any given time with both central and peripheral information. When we direct our attention toward one part of the environment, we simultaneously receive additional perceptual information from areas outside of our central focus.

Environmental perception involves purposive actions   Ittelson emphasizes that environmental perception involves purposive action. The scale and complexity of environments make it impossible for us to perceive them passively. We must actively explore, sort, and categorize the vast array of sensory inputs from the environment. Environments also provide *messages* that help to direct our actions in them. In this sense, Ittelson points out, our actions in regard to the environment are never blind or purposeless. Our Boston visitor must have some plan for exploring; even if a guidebook is not consulted, he or she will at least note street signs and other distinctive features of the setting.

DIMENSIONS OF ENVIRONMENTAL STIMULATION

In order to study environmental perception, environmental psychologists need to identify the dimensions of environmental stimulation that are appropriate to research in this area. Donald Berlyne (1960) proposed several *collative* variables of environmental stimulation that are especially relevant to this task. Collative variables, which include the *novelty, complexity, surprisingness,* and *incongruity* of stimulation, generate a degree of perceptual conflict that leads the perceiver to draw comparisons between the present stimulus and other stimuli. Joachim Wohlwill (1966) further developed the relevance of these collative variables to the study of environmental perception. He found that the manner in which an individual explores a setting will be affected by the novelty of its features. For example, San Francisco's famed cable cars contribute to the city's attractiveness and interest to sightseers. People's differential perceptions of urban and rural settings may be partially influenced by the quite different levels of stimulus complexity in the two environments. Urban settings are composed of a much greater variety of environmental elements than are rural areas. Surprising and unexpected environmental features, Wohlwill points out, can have a pleasing effect on the perceiver. Finally, while excessive incongruity, as when structures that bear no relationship to one another are placed together, can be jarring to an observer, an optimal level of incongruity might constructively heighten an observer's attention.

MEASURING ENVIRONMENTAL PERCEPTION

Psychologists interested in studying how people perceive the large-scale physical environment have faced a formidable methodological challenge. Wohlwill (1966) points out that studies of environmental perception in real-world settings cannot achieve the experimental control over environmental stimulation that is possible in a laboratory setting. Psychologists who study perception in naturalistic settings must use "ready-made" stimuli such as an urban scene or a natural landscape. Because, as Ittelson (1970, 1973, 1976) explains, real-world environments are highly complex, the environmental psychologist is faced with unique challenges in defining and operationalizing environmental stimulation. While it is possible to use photographs or small scale models of real-world environments in controlled laboratory settings, we shall see that such environmental simulations often present a threat to external validity. It is equally difficult to measure the complex activities in which people engage in the process of perceiving the physical environment. Let us look at some of the ways in which environmental psychologists have attempted to cope methodologically with the environmental and response sides of environmental perception.

Environmental stimulation   Because the real-world settings that environmental psychologists study do not allow for controlled manipulation of independent variables, it is difficult to provide an objective index of the stimulus dimensions under study. Wohlwill (1966) explains that one way environmental psychologists have dealt with this issue is by obtaining subjective ratings of particular stimulus dimensions from trained judges. Wohlwill reports a study (Leckart and Bakan, 1965) that used judges' ratings of the stimulus complexity of landscape scenes to demonstrate a positive relationship between the complexity of scenes and the amount of time subjects spend looking at them. A related approach is to collect perceptual judgments from a large number of "naive" or untrained observers. By employing statistical methods, such as multidimensional scaling (Green and Rao, 1972), that are able to describe the interrelationships of complex sources of data, investigators have been able to identify those environmental characteristics (e.g., diversity, warmth, size, complexity, and familiarity) that are common to many people's perception of a setting (see Betak, Brummell, and Swingle, 1974; Hall, Purcell, Thorne, and Metcalfe, 1976; Nasar, 1980).

Another research strategy employed by environmental psychologists to cope with the lack of control available in real-world settings is the use of *simulations* of real-world settings. Gary Winkel and Robert Sasanoff (1976) designed a "simulation booth" to study how people move through and view the various features of an environmental setting. The booth was outfitted with three projectors arranged to show a series of color photographs of the setting being explored. A subject, seated in front of three display screens, could tour the setting—in this case a museum of history and industry—by informing the projector operators of the direction in which he or she wished to proceed. Another richly detailed simulation has been developed by Donald Appleyard and Kenneth Craik (1974, 1978) at the Environmental Simulation Laboratory at the University of California, Berkeley (see box, "An Environmental Simulation Laboratory").

Some investigators (Danford and Willems, 1975; Lowenthal, 1972) have cautioned, however, that while environmental simulation permits experimental control and better measures of statistical reliability than does research in naturalistic contexts (i.e., internal validity is strengthened), it is useful only to the extent that investigators can be certain that the behavioral responses generated are similar to those elicited by real-world settings (i.e., external validity must be adequate). Carl Greenberg and his associates (Firestone, Karuza, Greenberg, and Kingma, 1978; Greenberg and Chambers, 1979) demonstrated, for instance, that the model room simulation used in some crowding studies (see Desor, 1972) may not provide a valid index of reactions to crowding in real-world settings. One problem with simulations relying on small-scale models and photographs is that they do not allow the motoric experience that is essential to perception in real-world settings (Evans, 1980). Clearly, it is important for investigators who use simulation techniques to try to evaluate the relevance of simulation results to human behavior in naturalistic contexts.

Winkel and Sasanoff (1976) did, in fact, compare people's reactions in their simulation booth to the behavior of visitors in the actual museum of history and industry. They found many similarities between people's responses in the simulated and real museum environments. Some differences were also apparent, however, such as a tendency for people to view more of the museum in the comfort of the simulator than when they were actually walking about the museum. Similarly, Kenneth Craik (1978) and George McKechnie (1977a) have described a systematic effort to evaluate the external validity of the Berkeley Environmental Simulation Laboratory. Responses of subjects who viewed films and videotapes of a simulated tour through the scale model environment were compared with those of persons who were driven along the identical tour in the real environment or who viewed a film of the real-world town. Craik and McKechnie's preliminary findings indicated that correlations between individuals' responses to the simulated and real environments were uniformly high on a variety of measures.

Perceptual responses Psychologists studying environmental perception have also been challenged to develop measures of perceptual responses that are able to reflect the richness of the perceptual process. Many studies of environmental perception have used questionnaires or interviews in which subjects can describe verbally the way they perceive various environmental settings. David Lowenthal (1972) has pointed out, however, that such "semantic" measures are able to cap-

An Environmental
Simulation Laboratory

Donald Appleyard and Kenneth Craik (1974, 1978) have designed a unique environmental simulation at the Berkeley Environmental Simulation Laboratory. The simulation consists of a physical scale model of an environmental region and a remotely guided periscope with a tiny lens (1/10-inch radius). The periscope is supported by a gantry system that can move it through the simulated environment at the "eye level" of a perceiver. During the journey through the scale model the periscope can follow various routes, look in any direction, and proceed at variable speeds. It can be made to "walk" through a residential environment or to "drive" through a highway environment. In addition, it can project the environmental scenes it "perceives" on closed-circuit television, videotape, or super-8 or 16-mm colored movie film. The periscope can also take color slides or still photographs from a variety of viewpoints and in multiple sequences.

The films, videotapes, slides, and photographs taken in the simulated environment can be used for a range of scientific or practical purposes. The simulation allows environmental psychologists to study the ways people perceive and comprehend a variety of environmental features in a controlled laboratory setting, where the presentation of environmental stimuli can be systematically manipulated. It may also be used to permit members of a community group or advisory committee to "tour" an envisioned project, so that they may participate in environmental design decisions.

This scale model of a suburban environment with a movable optical probe allows researchers to prepare a richly detailed movie tour through a simulated environment.

Photo courtesy of Kenneth Craik.

ture only those aspects of environmental perception that can be filtered through language. He suggests that there may be benefits to research that relies on both semantic and nonlinguistic responses. For example, as Edward Hall (1966) has explained, the physical distance maintained between two people is closely linked to differences in the way they perceive each other.

One measurement strategy that permits us to glimpse the richness of nonlinguistic perceptual responses is the recording of eye movements. Stephen Carr and Dale Schissler (1969) used an eye-movement recorder to investigate how subjects perceived the urban scene as they approached the center of Boston while driving along an elevated expressway. The apparatus was attached to each subject's head by means of plastic bands and a bite bar. Mounted on the apparatus was a 16-mm movie camera attached to two fiber-optic cables that simultaneously recorded where the subject looked and the exact movements of the eyes as the subject viewed the scene. The investigators found a remarkably high level of agreement on where subjects looked as they drove along the expressway. They concluded that the physical form of the environment visible from the expressway structured the way in which people scanned the environment visually, and determined the physical features they selected for close attention. Environmental features that were looked at by a great many subjects included the city skyline, houses and buildings, overpasses, and billboards.

PSYCHOLOGICAL FUNCTIONS OF ENVIRONMENTAL PERCEPTION

Since we generally take environmental perception for granted, it may be surprising to discover that our perception of the physical environment is one of the most essential psychological processes by which we adapt to it. In fact, environmental perception provides the foundation for all of our knowledge about the world around us and for all of our activities in the environment. One of the chief psychological functions of environmental perception is to direct and manage the many activities that make up our daily lives. Ittelson (1970, 1973, 1976; Ittelson, Franck, and O'Hanlon, 1976) contends that human survival itself would be impossible without our ability to perceive the environment around us. Environmental perception provides the basis for our knowledge about the world in which we live, and this knowledge is essential to our ability to function adaptively in the world. For instance, our perception of the world around us helps us to manage our communication and social interaction with other persons, to identify important features of our everyday environment, and to enjoy a range of aesthetic experiences.

DIRECTING ENVIRONMENTAL ACTIVITY

One important way in which environmental perception helps to direct our daily activity is by providing information necessary to orient ourselves in the environment. Orientation involves establishing a place or series of places in the physical environment from which we can direct our activities (Ittelson, Franck, and O'Hanlon, 1976). Without this ability to orient ourselves in the environment we would be unable to carry out the activities and functions that make up our daily lives. We would quite literally become lost in each new environment we encountered. Imagine what would happen if, when you visited a city for the first time, you were unable to acquire the perceptual information that would allow you to orient yourself (see Ross, 1975). You would be unable to find your way about, and would find yourself making wrong turns at almost every street corner. In fact, of course, we do not find ourselves totally lost each time we visit a new city; environmental perception provides us with the information necessary to orient ourselves in the new setting, allowing us to determine the appropriate directions to turn and to select the most efficient routes between different sites (Golledge and Zannaras, 1970).

On the basis of his studies of urban perception in Boston and Venezuela, Donald Appleyard (1970) discusses the *operational* role of environmental perception in helping people to orient themselves and travel about efficiently in the urban environment. He points out that many features of the urban environment are perceived because of their operational importance. Such urban details as intersections, pedestrian islands, and traffic circles are often strongly perceived because they are so relevant to efficient travel about the city. Similarly, otherwise inconspicuous buildings at important decision points are strongly perceived because they serve to "anchor" the decision point in the traveler's memory.

COPING WITH NOVEL ENVIRONMENTS

The process of environmental perception becomes so automatic in our lives that the perceptual cues we use in adapting to the environment become progressively less obvious to an outside observer. One way to make the process involved in perceiving the environment more open to observation is to observe how people respond to novel environments. Lucille Nahemow (1971) notes that when we travel in a foreign country, we are usually more alert and responsive to environmental cues than when we traverse the familiar path from home to school or work (Figure 2-2).

In order to study perception of a novel environment under conditions that would permit some degree of experimental control, Nahemow and her co-workers reproduced in a laboratory a highly novel environment originally constructed by Robert Whitman for an art exhibition at the Jewish Museum in New York. The environment was composed of eight large mirrors, each associated with a stroboscopic light. A speaker was attached to the back of each mirror to create unusual surface vibrations. The environment was arranged so that the lights and speakers could be operated in various complex sequences.

Nahemow introduced subjects individually to the novel environment for a six-minute period, and then interviewed them concerning their experiences. Although subjects were told they could leave the environment before the six minutes were up, almost 80 percent of subjects remained for the full six minutes. She discovered that people tended to employ two quite different perceptual strategies to cope with the novel environment. One group of subjects tended to view the environment in a *structural* way, as completely separate from themselves. One subject reported that he entertained a number of hypotheses about the way in which the environment operated. In order to time the intervals between the various light sequences, he went through various mental exercises, such as counting to himself.

Figure 2-2
People are particularly aware of environmental cues when they travel to new and un-familiar places.

© *Leonard Speier, 1980.*

A second group of subjects, in contrast, viewed the environment in a more *experiential* manner, seeing themselves as involved in and a part of the environment. One subject reported that he felt especially curious about the environment and about his responses to it. He lay on the floor, stood on one foot, and closed his eyes in an attempt to perceive the environment in a variety of ways. He imagined a sound-and-light show and a ride on the subway. Nahemow found that people who responded to the novel environment in an experiential way were less likely to become bored in the setting than those who responded in a structural manner. Later research (Ittelson and Krawetz, 1975) arranged the novel environment so that a predictable sequence of light and sound followed specific behaviors on the part of subjects. The investigators found that subjects who employed a structural approach were more active in the setting and more aware of the relationships between their own behavior and the sequence of light and sound in the environment than were those who used an experiential approach.

CONTEXTUAL INFLUENCES ON ENVIRONMENTAL PERCEPTION

Because environmental perception is so closely tied to our adaptive functioning in the environment, our *style* of perceiving the environment will, over time, become tailored to the unique characteristics and demands of the settings where we habitually function. Ittelson (1970, 1973, 1976) explains, for example, that two contrasting types of environment will produce two different styles of environmental perception, each fitted to the unique features of its respective environmental context. For instance, as we shall see, people's ability to perceive particular spatial forms (especially under marginal perceptual conditions) is positively related to the availability of those forms in the cultural context with which the subjects are familiar. Environmental psychologists have devised two research strategies to investigate how environmental perception differs in contrasting environmental settings. One strategy has involved studies in environments that differ naturally from one another—environments in different cultural contexts, for example. The second strategy has concerned investigations of environmental perception as a function of experimentally induced environmental differences.

Culture   Interesting studies have been carried out in contrasting cultural contexts (see Altman and Chemers, 1980; Segall, Campbell, and Herskovits, 1966). A classic study in this area was conducted by two Harvard psychologists, Gordon Allport and Thomas Pettigrew (1957). They selected for their study a perceptual illusion that involves a rotating trapezoidal window (Ames, 1951). The window is proportioned in such a way that as it turns, it seems to sway back and forth rather than to rotate. The illusion of sway is explained by the fact that people who are accustomed to rectangular windows assume that this window, too, is rectangular, although it is in fact trapezoidal.

Allport and Pettigrew speculated that if the illusion is due to experience with rectangular windows and with rectangular architectural forms in general, the illusion should be more pronounced in a Western culture where people are familiar with rectangular windows than in a primitive culture where they are not. To investigate this hypothesis, the researchers tested the illusion with urban children (including Europeans and Africans), who were familiar with rectangular windows, and with rural African children from two Zulu reserves, who were unfamiliar with rectangular windows. In fact, rectangular forms in general tend not to exist in the rural Zulu culture. Zulu huts and villages are round, doors are generally round, windows do not exist, and, while the Zulus have a word for "circle," they have no word for "square" or "rectangle."

Allport and Pettigrew discovered, as they had hypothesized, that the rural Zulu children were "tricked" less often by the window illusion than were African or European urban children. Table 2-1 shows the experimental results for the two groups of children from the rural reserves and for the two groups of children from urban areas. The number of "yes" responses indicates the number of times the children were fooled by the window illusion, which was viewed four times under a variety of observational conditions. The table shows that the urban children were tricked by the illusion more often than were the rural children, and that children on the Polela Reserve were fooled more frequently than those on the Nongoma Reserve. The investigators explained that the Polela Reserve was closer to an urban area than was the Nongoma Reserve, and that the Polela children had somewhat more contact with Western architecture.

Allport and Pettigrew speculated that the differences in styles of perceiving the environment they found between urban and rural Zulu children were due to the contrasting cultural experiences of the two groups. They noted that the differences in perception between the groups occurred under those observational conditions where conflicting cues were available concerning "true rotation" and "rec-

Table 2-1.   The number of times children from two rural African cultures (Nongoma and Polela reserves) and from African and European urban cultures were "tricked" ("yes" responses) on the trapezoidal window illusion.

Sample	Number of "yes" responses					Average
	4	3	2	1	0	
Nongoma Reserve	2	4	10	3	1	2.15
Polela Reserve	4	10	3	2	1	2.70
Urban African	12	4	2	2	0	3.30
Urban European	11	5	3	0	1	3.25
Total (N = 80)	29	23	18	7	3	2.85

Source: G. W. Allport and T. F. Pettigrew, "Cultural Influence on the Perception of Movement: The Trapezoidal Illusion Among Zulus," Journal of Abnormal and Social Psychology, 1957, 55:104–13. Copyright 1957 by the American Psychological Association. Reprinted by permission of the publisher and author.

tangular windowness." They speculated that the urban children resolved the perceptual conflict by drawing on the familiar experience of rectangular windows. The rural Zulu children, unable to rely on the experience of rectangular windows, resolved the conflict by relying on the perceptual cues of rotation.

Distorted worlds   An alternative research strategy for investigating contextual influences on environmental perception has involved experimentally induced differences in people's visual environments. This strategy has characteristically involved fitting subjects with distorting glasses that dramatically alter the visual environment (Epstein, 1967; Rock, 1966). After the subject has lived in the artificially distorted environment for a period of time, the investigator assesses how the individual's style of perceiving the environment has changed. Researchers have discovered that people are able to adapt remarkably well to the perceptual world created by the distorting glasses. One approach has involved the use of inversion lenses that literally turn the subject's world upside down. Investigators have discovered that after wearing the distorting lenses for a period of time, subjects learn to carry on their normal activities quite effectively. In fact, subjects may eventually become totally unaware of the artificial distortions in their private world, and may even see the normal world as upside down for a while after the distorting lenses are removed (Kohler, 1962; Snyder and Pronke, 1952).

THEORIES OF ENVIRONMENTAL PERCEPTION

Environmental psychologists interested in understanding how people perceive the physical environment have been able to draw on theories of environmental perception that were established earlier in psychology. For example, the Gestalt theory of perception, which has roots in Germany during the early part of this century, has had an important influence on later researchers' approach to the study of environmental perception. The two most prevalent theories of environmental perception today, which may be traced to the middle years of this century, represent two quite different schools of thought on how we come to perceive the world around us. One theoretical position, which has been termed an *ecological* theory, explains the process of environmental perception in terms of the nature of properties of environmental stimulation. The second theoretical view, which has been labeled a *probabilistic* theory, emphasizes the active role people assume in the perceptual process. Although these theories of environmental perception have often been pitted against each other, both are of value in helping us to appreciate the environmental perception process.

GESTALT THEORY

The Gestalt theory of perception can be traced to experimental research begun early in this century by three German psychologists: Max Wertheimer (1945), Wolfgang Köhler (1929), and Kurt Koffka (1935). Essential to the Gestalt theory of perception was the study of perception as a *holistic* process. In fact, a significant part of the intellectual energy behind the Gestalt movement derived from a reaction against the *reductionistic* approach to perception characteristic of other schools of psychology at the turn of the century; that is, the Gestaltists opposed the view that human perception could be studied by reducing or analyzing the perceptual process into separate, basic elements. They contended that human perception could be understood only as a holistic process, in which "the whole is greater than the sum of its parts." Wertheimer pointed out, for example, that in perceiving a moving picture, we do not see a series of separate still photographs of an actor in various positions, but a "dynamic whole" in which the actor is perceived as engaged in a unified pattern of movement.

A particular contribution of the Gestalt theory of perception was an explication of the underlying "principles of organization" that enable the individual to perceive a collection of discrete stimuli as a holistic pattern. The Gestalt psychologists proposed that the principle of *proximity* allows the perceiver to see elements that are close to one another spatially as related in a pattern. The principle of *similarity* permits the individual to see elements that are alike in shape or color as related in a pattern. Several elements may be grouped together in a row or a smooth curve according to the principle of *continuity*. Finally, the principle of *closure* enables the perceiver to overlook or "close" small gaps in a figure and see it as a whole. These four Gestalt principles of organization are illustrated in Figure 2-3.

At a general level, Gestalt theory has made important contributions to the investigation of environmental perception. The Gestalt emphases on perception as a holistic process and on the dynamic, organizing aspect of perception have influenced much of the later research and theorizing in this area. As a specific theory of perception, however, the Gestalt approach has been criticized in more recent years (see Allport, 1955). Gestalt theory tends to see organization as inherent in the perceptual process itself, for example, and to minimize the organizing role played by past learning and higher-order intellectual processes. As we shall see, later

Figure 2-3

Gestalt theory proposes that four principles of organization—proximity, similarity, continuity, and closure—allow us to perceive discrete stimuli as related in a holistic pattern.

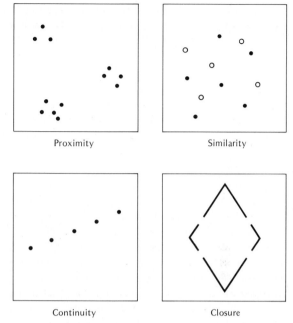

Proximity

Similarity

Continuity

Closure

theories of environmental perception have emphasized the important role of learning in perception, and, especially in recent years, have underscored the influence on perception of higher-order cognitive processes of organization.

AN ECOLOGICAL THEORY

The theory that environmental perception is a product of the ecological characteristics of environmental stimulation has been most fully developed by James J. Gibson (1958, 1960, 1963, 1966, 1979). Gibson argues that environmental perception is a direct product of the stimulation that reaches us from the environment. He believes that all of the information that people need to perceive the environment is already contained in the pattern of stimulation that impinges on us from the environment. Thus it is not necessary for us to construct meaning from the sensations that reach us from the environment. Rather, we directly perceive the meaning that already exists in a patterned environment; that is, meaning is directly perceived in environmental stimulation, and does not require intervening processes of reconstruction and interpretation on the part of the perceiver.

Gibson views environmental perception holistically: people perceive meaningful patterns of stimulation in the environment rather than separate points of stimulation. Thus he views perception not in terms of the response of separate sensory cells, but as the patterned response of groups of cells. Such integrated patterns of response are quite complex, and may involve cells that are not situated together anatomically. Gibson explains that the same perception can emerge from many different response patterns, as when we scan an environmental scene visually, activating a series of patterned responses, and yet perceive a holistic image of the environment.

Because meaning can be directly perceived from the environment, ecological theorists contend that many basic aspects of environmental perception do not have to be learned, and are a part of an organism's response repertoire from infancy. In support of this position, ecological theorists point to research findings by Eleanor Gibson (Gibson and Walk, 1960; Walk and Gibson, 1961) involving the "visual cliff." The visual cliff is a flat surface that is designed to give the illusion of an abrupt drop from what appears to be an edge. Researchers have found that animals that can walk from birth, such as goats, steer clear of the edge from earliest infancy. Human babies avoid the cliff's edge by the age of crawling, a finding that is consistent with the ecological theory, as it is not until the organism is mobile that this kind of perceptual information is needed.

This is not to say that ecological theorists discount the role of learning in environmental perception. In fact, James Gibson believes that through environmental experience the perceiver learns to discriminate more stimulus variables in the environment and to discriminate more important stimulus variables from less important ones. Thus, through learning, the perceiver is able to achieve a progressively more accurate picture of the environment. For example, a child with minimal perceptual experience may be unable to use subtle sensory cues, such as gradients in degree of light or shading, that to an adult readily communicate nearness or distance. Note that the adult in this case does not create the more subtle perceptions, but has simply acquired the necessary perceptual experience to make fuller use of sensory information that is available in the environment.

Gibson's view of environmental perception also recognizes the importance of the individual's active exploration of the environment. He proposes that an environmental object has *invariant functional properties;* that is, physical characteristics that do not change, such as the roundness, hardness, and solidity of a stone. An individual can perceive more of the invariant properties of the objects in his or her world by exploring those objects from a variety of perspectives. According to Gibson, this active exploration serves an important adaptive function, since by experiencing the different functional properties of objects, the individual can determine how best to make use of the varied objects in the environment. Gibson terms the functional properties of objects that are discovered through exploration of the environment *affordances;* that is, such characteristics tell us the range of useful functions that a particular object affords. For example, an object that is sturdy, nonporous, and has an interior space that is somewhat larger than a person is able to afford protection against rain, cold, and wind.

A PROBABILISTIC THEORY

The theory that environmental perception is a function of the active role people play in interpreting sensory input from the environment has been most fully developed by Egon Brunswik (1956, 1969). Brunswik's theory of *probabilistic functionalism* proposes, in sharp contrast to the view of Gibson, that the sensory information that reaches us from the environment is never perfectly correlated with the real environment. In fact, Brunswik contends that we are often presented with complex and even misleading cues about the environment. For example, when a visual stimulus from an object at some distance presents itself to the retina, the observer must determine whether the object is a large one at some distance or a small one close at hand. The perceptual ambiguity in such a situation is resolved when the observer comes to a *probabilistic* estimate of the true situation. Such a probabilistic judgment may be thought of as a "best bet" or an "educated guess" about the true nature of the environment.

In Brunswik's probabilistic model of environmental perception, the individual plays an especially active role in the perceptual process. In order to cope with the ambiguities and inconsistencies of the sensory cues that reach us from the environment, the individual must build up a repertoire of probabilistic statements about the environment. These probabilities are derived from samplings of sensory cues from a great many environmental settings. As we can never sample all possible environments, however, our judgments about any given environment cannot be absolutely certain, but only probabilistic estimates. The individual can test the accuracy of the probabilistic statements by trying out a series of actions in the environment and assessing their functional consequences.

Brunswik proposes a *lens model* to describe the active role of the individual in environmental perception. He explains that the perceptual process operates somewhat like a lens that captures and focuses light rays in a single plane. Brunswik's lens model is illustrated in Figure 2-4: the process of environmental perception captures the scattered stimuli that emanate from the environment (at the left of the figure) and recombines and focuses them in a unified percept (at the right). Brunswik emphasizes that stimuli from the environment vary in their ecological validity; that is, some stimuli provide a more accurate or trustworthy representation of the true environment than do others. In order to improve one's probabilistic estimate of the true environment, one must hierarchically order these environmental stimuli according to their relative trustworthiness, while accumulating and combining a great many environmental cues.

The notion that the individual plays an especially active part in perceiving the

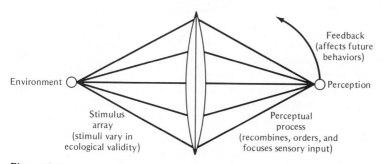

Figure 2-4
Egon Brunswik proposes a "lens model" to explain how the individual "catches" and focuses the stimuli that are dispersed in the environment.

Adapted from E. Brunswik, "The Conceptual Framework of Psychology," in O. Neurath, R. Carnap, and C. Morris (eds.), Foundations of the Unity of Science: Toward an International Encyclopedia of Unified Science, *Vol. 1, 1969, p. 678, by permission of The University of Chicago Press. Copyright 1938, 1939, 1946, 1951, 1952, 1955, 1969 by the University of Chicago. All rights reserved.*

physical environment has been further developed in the *transactional* psychology of Adelbert Ames (see Ittelson, 1960; Ittelson and Kilpatrick, 1952; Kilpatrick, 1952). Ames emphasizes that the role of the individual in the perceptual process is a dynamic and creative one. The probabilistic judgments about the environment that an individual builds up are highly personal, reflecting that person's unique perspective, needs, and environmental goals. "The world each of us knows is a world created in large measure from our experience in dealing with the environment" (Ittelson and Kilpatrick, 1952:175).

Support for the transactional view of environmental perception has come from highly creative experiments involving perceptually distorted environments (see Ames, 1951; Kilpatrick, 1954). Allport and Pettigrew's experiment with the rotating trapezoidal window, for example, is consistent with a transactional view. Transactional theorists have also used the "Ames distorted room"—a room with a sharply slanted floor and rear wall that appears normal when viewed from a certain position. Researchers have found that the illusion of the room is reduced by experience with it—either personal activity in the room or observations of others' activity there (Kilpatrick, 1954). Transactional theorists explain that such perceptual illusions work because the observer interprets the sensory information they present in terms of probabilities or "best bets" developed from past experience with normal environments. These established probabilities are altered when new evidence arises that indicates that the initial judgment is no longer tenable.

The probabilistic theory of perception has tended to play a more prominent role than the ecological theory in contemporary models of environmental perception. For instance, current perceptual theories tend to emphasize the importance of memory and of processing environmental information in explaining how people perceive the world. Kenneth Craik and Donald Appleyard (1980) have recently argued that Brunswik's lens model can serve as a framework for integrating research that has tended to be classified into distinct categories, such as environmental perception, cognition, and attitudes. They provided an example of the integrative role of the lens model by applying it to analyze a panel of judges' estimates of socio-environmental conditions (e.g., family income and residents' concern about crime) on residential streets in San Francisco. Lewis Petrinovich (1979) has suggested, more broadly, that Brunswik's theory of probabilistic functionalism can serve as a comprehensive framework to describe and investigate organism-environment relationships. He applied the lens model to formulate a research paradigm that emphasizes the importance of *situation* sampling, in contrast to the more traditional focus on *subject* sampling. He explained that since the environment is an important factor in shaping people's behavior, researchers interested in learning about a particular behavior need to sample a variety of representative environmental situations to which the behavior under study may be generalized. The probabilistic model is also especially congruent with the adaptational focus emphasized in this book. Both the probabilistic and the adaptational perspectives underscore the active and dynamic psychological processes that people engage in as they cope with the physical environment.

APPLICATIONS TO
ENVIRONMENTAL PLANNING

How might our knowledge of the way people perceive the physical environment be applied to environmental planning? Jon Lang (1974) points out that many design objectives are visual, and that many basic design principles have been influenced by psychological theories of environmental perception. He adds, however, that designers have generally failed to keep abreast of current research in the area of environmental perception, and that many design principles need to be reformulated in terms of current knowledge in this field. He argues that a greater understanding of the processes through which people perceive the physical environment would help to improve the quality and effectiveness of environmental design.

Lang provides a general discussion of some of the applications of the various theories of environmental perception to environmental design. Design principles have been especially strongly influenced by the Gestalt theory of environmental perception. For example, architects have been interested in applying the Gestalt principles of organization to understand the types of visual features that tend to be perceived together in a spatial configuration. Lang adds that the transactional position later encouraged designers to appreciate the important role of personal experience in environmental perception, and to recognize that different people will perceive settings in different ways and that these personal perceptions may be at variance with the designer's own notions.

Lang suggests that although the ecological theory of environmental perception has historically been less influential in formulating design principles, it is of particular relevance to the design field. The ecological theory's emphasis on holistic patterns of stimulation as they are perceived as one scans or moves through an environment is especially applicable to the architect's concern with three-dimensional forms. The ecological theory can encourage designers to see their work as a unified visual layout rather than as isolated elements perceived separately. Lang adds that design education should emphasize the complex visual characteristics of surfaces, contours, and textures as they are perceived by someone moving through the environment, rather than maintain the traditional concern with static two-dimensional forms. In a similar vein, Philip Thiel (1961) argues that perception of the designed environment is a dynamic process, and proposes a notational system that designers could use to describe environmental perception as a sequential process. Stephen Carr and Dale Schissler (1969) add that people's perceptual behavior can be predicted through relatively simple techniques that designers can use without formal training in psychology.

DESIGNING FOR PERCEPTUAL CLARITY

Our knowledge of the way people perceive the physical environment can be put to good use in the design of settings that can be clearly and efficiently perceived by their users. This concern is especially important in the design of environments, such as psychiatric hospitals, where the setting's users may suffer deficits in their perceptual abilities. Humphry Osmond (1957, 1959, 1966) has pointed out that the long corridors prevalent in many psychiatric hospitals can cause visual and auditory perceptual distortions to people with no mental problems, let alone the mentally ill. Such perceptual ambiguity can be especially frightening to schizophrenic patients, who sometimes experience difficulty in determining size-distance relations and may occasionally suffer from hallucinations (see box, "Some Design Problems in Psychiatric Hospitals").

Perceptual clarity in the design of occupational and transportation environments may also have safety implications. Human factors psychologists, for example, have investigated ways to prevent accidents and injuries in occupational settings by improving workers' ability to perceive potentially dangerous environmental features and to avoid perceptual distractions while engaged in potentially hazardous tasks (see Bennett, 1977). Adequate lighting and easily visible warning signs can reduce threats to safety from low overhangs, dangerous steps, and sliding glass doors. Minimizing visual and auditory environmental distractors in the vicinity of industrial machinery can help machine operators and assembly-line workers avoid injuries. Clearly marked and readily perceivable emergency exits, fire alarms, and firefighting equipment can reduce injuries during emergencies. Two laboratory studies (Holahan, Culler, and Wilcox, 1978; Johnston and Cole, 1976) have suggested that distracting visual stimuli in the roadside environment may present a traffic safety problem by interfering with driving performance. These studies note, however, that the performance decrements in the laboratory were small and might not indicate a safety problem under actual driving conditions.

DESIGNING FOR OPTIMAL VISUAL COMPLEXITY

The importance of designing for perceptual clarity does not mean that all designs should strive to be as visually simple as possible. In fact, the question of how visually simple or complex a design should be has been a central concern in the environmental design field. While turn-of-the-century design favored complexity and ornate detail, contemporary theories of design encourage architectural forms with simple lines.

Yet recent psychological knowledge of the way people perceive the physical environment suggests that some degree of visual complexity can afford an interesting and satisfying perceptual experience. An appreciation of the dynamic role of the perceiver in the perceptual process suggests that people are prepared to deal with some degree of visual challenge in the environment. Research has suggested that people favor an *optimal* level of visual complexity in the environment, preferring settings that are visually interesting and compelling without becoming bewildering or disorienting (Wohlwill, 1966). Stephen Kaplan (1975, 1976) has provided an information-based view of this issue, explaining that people prefer environmental features that both offer them the opportunity to acquire additional information and help them to make sense of the environment. The issue of achieving an optimal balance between simplicity and complexity in the designed environment is a central issue, and we shall return to it in discussing both environmental cognition and environmental attitudes.

Some Design Problems in Psychiatric Hospitals

When Mayer Spivack (1967) assessed the perceptual clarity of the design of two psychiatric facilities in the Boston area, he found design elements that grossly distorted the perception of psychologically disturbed patients and even of nonhospitalized, "normal" persons. The highly reflective glossy surfaces of hospital corridors, for example, created a stream of reflections and shadows that marched toward the observer along the walls, floor, and ceiling. When glaring light entered the corridor from a window at the far end, people at a distance seemed to float over the floor in fuzzy outlines, without feet, wrists, or necks. In one hospital Spivack found a tunneled ramp that actually mimicked some of the perceptual effects of the Ames distorted room. While the tunnel's ceiling was horizontal, its floor slanted sharply downward, creating a trapezoidal shape that appeared rectangular to a naive observer. To someone standing at the top of the ramp, a person descending the ramp seemed to shrink, and eventually to disappear through a doorway that looked no bigger than a rabbit hole.

Long hospital corridors with glaring lights and reflective surfaces can cause perceptual distortions.

© *William C. Koechling/ Black Star.*

Amos Rapoport and his associates have addressed considerable attention to the question of how perceptually complex design should be (Rapoport and Hawkes, 1970; Rapoport and Kantor, 1967). He has argued that the overly simple lines of contemporary design are generally uninteresting to observers. Psychologists have found that people prefer some degree of complexity and ambiguity in visual patterns. Rapoport argues for environmental designs that achieve an "optimal range" in perceptual input, avoiding both the boredom of overly simple designs and the potential chaos of excessively complex settings. The ideal level of visual complexity, he believes, should be construed in terms of the maximum rate of information that observers can deal with effectively. One way to achieve an optimal balance between simplicity and complexity in design is to create visual variety within an overall unifying pattern.

Bernard Pyron (1971), who measured visual scanning with films of design models, also concludes that people's perceptual behavior benefits from some degree of visual diversity in both the buildings and the landscaping of designed settings. The work of architect Robert Venturi (1966) exemplifies an approach to design that strives for a degree of visual complexity. Venturi favors architectural designs that reflect the complexity and contradiction that are inherent in modern experience. He adds, however—echoing Rapoport and Kantor's (1967) concern for a unifying pattern—that he is speaking of a type of complexity in design that is responsive to and informed by the architectural work as a whole (see box, "Learning from Las Vegas").

PLANNING FOR NATURAL DISASTERS

Our knowledge of the way people perceive the physical environment can also be applied to planning for natural disasters. A major concern in disaster planning relates to people's tendency to build in and around regions that are subject to natural hazards—floods, tornados, earthquakes—without making adequate preparations to deal with such hazards when they come. Planning in this area might be facilitated by a better understanding of the way residents of such areas perceive the natural hazards to which they can expect to be exposed.

An extensive body of research dealing with people's perceptions of natural hazards has been conducted by Robert Kates and his colleagues, who began to study perceptions of natural hazards in the late 1950s (Burton and Kates, 1964; Burton, Kates, and White, 1978; Kates, 1976). Kates was especially interested in learning how people's perceptions of natural hazards were reflected in the ways in which they prepared to cope with them. He found that residents' personal experiences with past hazards dictated how they coped with potential future hazards. People tend not to see potential consequences of natural hazards with which they have had no personal experience. A recent study (Hanson, Vitek, and Hanson, 1979) has suggested, however, that in the case of tornados, people's willingness to take precautionary steps is related more to their *awareness* of an earlier tornado than to their actual experience with one.

Kates also found that very few people take preventive measures much in advance of hazards. Those who do make advance preparations tend not to engage in many preventive activities. They are more inclined to accept losses than to try to

Learning from Las Vegas

The architect Robert Venturi and his associates (Venturi, Brown, and Izenour, 1973) have contended that the simple lines of most modern architecture make buildings bland and uninteresting. They suggested that we can learn how to make architecture more exciting by observing the Strip in Las Vegas. The rich visual complexity of the Strip, they argue, epitomizes the vitality and contradiction that are central to contemporary experience. They see the apparent contradictions of the Las Vegas Strip as actually part of an underlying order. To them the Strip's order—unlike the rigid order of many modern urban apartments and offices that are "easy" on the eye—is vital and dynamic.

Venturi and his associates caution that designers should not translate the Las Vegas image too literally into any particular architectural work, but should strive instead to imbue their work with the sense of liveliness, visual excitement, and even playfulness that characterize the Strip. It should be pointed out that the views of Venturi and his colleagues are not shared by all designers. In fact, most of us would probably find an environment modeled after the Las Vegas Strip uncomfortable if we were to live in it for a considerable length of time. Yet the blandness of so much of today's design is also unfulfilling. A sense of visual excitement and diversity would certainly benefit many modern settings; in this sense, we can learn from Las Vegas.

The Las Vegas Strip presents a panoply of vivid, clashing visual stimulation.

© *Peter Menzel/Stock, Boston.*

reduce damages, and try to modify potential costs rather than to prevent them. Most important, Kates concludes that people are more interested in reducing the damages from a hazard than in changing their means of livelihood or their land use in order to avoid the hazard. Few people are willing to change their place of residence even in the face of severe natural hazards. In fact, Kates found that people who live in close physical proximity to a potential natural hazard are inclined to underestimate its threat to a greater extent than persons who live at an intermediate distance from it. This latter finding has been also confirmed by a study of residents' perceptions of a similar risk associated with a technological environment involving a nuclear reactor in Austria (Maderthaner, Guttman, Swaton, and Otway, 1978). People who lived very close to the reactor perceived nuclear reactors to be less hazardous than those who lived at an intermediate distance from it (see also Otway and Paher, 1976). These findings suggest that disaster planners might engage in educational programs oriented toward increasing residents' perceptions of the potential personal costs of hazards and encouraging them to explore a wider range of relatively uncostly preventive activities.

SUMMARY

The diverse stimuli that impinge on us from the environment are organized by our perceptual processes to form a coherent and integrated picture of the world. While psychologists have devoted considerable attention to the perception of isolated objects, they have tended to ignore the process by which people perceive the large-scale physical environment. Environments surround those who perceive them, and require people to move about in order to perceive all of their aspects. Environments provide such an abundance of perceptual information that all of it cannot be processed at once. Perceptual information from the environment arrives through several sensory modalities simultaneously, and at any given time we receive both central and peripheral information. Finally, environments are so complex that we cannot perceive them passively; we must actively explore, sort, and categorize the vast array of sensory inputs that confront us.

Four dimensions of environmental stimulation of particular interest to psychologists engaged in the study of environmental perception are novelty, complexity, surprisingness, and incongruity. Research in environmental perception is faced by a formidable methodological challenge, since studies of environmental perception in real-world settings cannot achieve the experimental control that is possible in the laboratory. Environmental psychologists have been able to operationalize the stimulus dimensions under study by obtaining subjective ratings from trained judges and by employing multidimensional scaling on perceptual judgments from untrained observers. Another research strategy has involved *simulation* of real-world settings that allow greater experimental control than would be possible in naturalistic settings. Environmental psychologists have employed a variety of measures to reflect the richness of perceptual responses, including questionnaires, interviews, and eye-movement records.

A chief psychological function of environmental perception is to direct and manage the activities that make up our daily lives. Environmental perception provides the basis for our knowledge about the world, and this knowledge is essential to our ability to function adaptively. Environmental perception helps us to manage our communication and social interaction with other persons, to identify important features of our everyday environment, and to enjoy a range of aesthetic experiences. One important way in which environmental perception helps to direct our environmental activity is by providing information necessary to orient ourselves in the environment.

Environmental psychologists have made the process involved in perceiving the environment more open to observation by observing how people respond to novel environments. Such research has shown that some people respond to a new environment in a *structural* way, seeing it as completely separate from themselves, while others respond in an *experiential* manner, seeing themselves as involved in and a part of the environment. Because environmental perception is so closely tied to our activities, our style of perceiving the environment will, over time, become tailored to the unique characteristics and demands of the particular environmental settings where we habitually function. Environmental psychologists have devised two research strategies to investigate some of the ways in which environmental perception differs in different types of environmental settings—cross-cultural studies and studies in artificially distorted environments.

The Gestalt theory of environmental perception contends that human perception can be understood only as a holistic process: "the whole is greater than the sum of its parts." The Gestalt theory explicates the underlying "principles of organization" that enable the individual to perceive a collection of discrete stimuli as a holistic pattern, including the principles of proximity, similarity, continuity, and closure. While Gestalt theory has made important contributions to the investigation of environmental perception at a general level, the Gestalt approach has been criticized as a specific theory of perception.

The ecological theory of environmental perception explains environmental perception in terms of the nature and properties of environmental stimulation. It proposes that all of the information that we need to perceive the environment is already contained in the pattern of stimulation that impinges on us from the environment. According to the ecological theory, environmental perception is absolutistic, and does not require intervening processes of reconstruction and interpretation on the part of the perceiver. Learning plays a role in environmental perception in that the perceiver learns to discriminate more stimulus variables in the environment and to discriminate more important stimulus variables from less important ones. The invariant functional properties of objects that are discovered through active exploration of the environment are termed *affordances*.

The probabilistic theory of environmental perception, in contrast, emphasizes the active role people assume in the perceptual process. It proposes that the sensory information that reaches us from the environment is never perfectly correlated with the real environment. The observer resolves perceptual ambiguity by coming to a *probabilistic* estimate of the true environment based on a sampling of sensory cues from a great many environmental settings. The individual can test the accuracy of these probabilistic estimates by trying out a series of actions in the environment and assessing their functional consequences. The *lens model* proposes that the perceptual process operates somewhat like a lens that captures and focuses light rays in a single plane. According to the probabilistic theory, the individual's role in the perceptual process is an especially dynamic and creative one. A person's probabilistic judgments about the environment reflect his or her own unique perspective, needs, and environmental goals.

Knowledge of the way people perceive the physical environment has been applied to the design of settings that can be clearly and efficiently perceived by their users, to the design of environments that achieve an optimal balance between visual simplicity and visual complexity, and to planning for natural disasters.

3 Environmental Cognition

A mong the most fascinating discoveries in environmental psychology has been the finding that each of us has a personal and unique "mental map" of the environment. If you were to sketch from memory a map of the neighborhood where you spent most of your time as a child, you would find that your map was uniquely different from any other person's map of the same environment. An element occupying considerable space in a prominent part of your map might be relegated to a small corner on someone else's map, or even be ignored entirely.

New York's Museum of Modern Art recently presented a remarkable photographic exhibition that concerned people's mental maps of the urban environment. The exhibition, which was prepared by a team of American and Japanese architects, focused on contrasting mental images of Tokyo, Japan. By means of retouched aerial photographs of Tokyo, the exhibition communicated some of the many ways the city is viewed mentally. To represent the notion that in a mental map some parts of the city are seen more clearly than others, a photograph showed Tokyo as a pattern of zones that varied in clarity and detail. Another photograph, in which vivid colors were superimposed on sections of a black-and-white photo of Tokyo, communicated the sense that in a mental map, various sections of the city are associated with different meanings as a function of the particular activities (working, shopping, entertainment) in which people engage in those areas.

In two decades of research, environmental psychologists have investigated such mental or "cognitive" maps of urban environments in cities throughout the world. A consistent finding has been that residents' cognitive maps of the urban environment are not true representations of the objective spatial environment. In studying cognitive maps in Paris, Stanley Milgram (1977) found that although the Seine flows through Paris in a great arc, many Parisians imagine it as a gentle curve, and some even think of it as a straight line. As we shall see, these personal distortions in cognitive maps are not the products of random error, but rather bear a consistent and meaningful relationship to the personal ways in which individuals have used the environment.

Unlike objective representations of the spatial environment that can be captured in the photographer's lens or the architect's blueprint, the subjective image of an individual's cognitive map of his or her environment reflects, in significant part, personal meaning. In the study of cognitive maps, we can thus learn about the manner in which individuals come to see, understand, and cope with the geographical environment at a personal level. We shall find, too, that by understanding people's cognitive maps of the environment, we can learn how to design environments that are more congruent with people's needs.

The Nature of Environmental Cognition

The Image of the City

As you recall from Chapter 2, environmental cognition involves the storage, organization, reconstruction, and recall of images of environmental features that are not immediately present. Psychological interest in the ways in which people form mental images of the geographical environment was initiated by a series of groundbreaking studies by Kevin Lynch, a professor of urban planning (Appleyard, Lynch, and Myer, 1964; Lynch, 1960, 1965; Lynch and Rivkin, 1959). Lynch's work, which was influenced by Kenneth Boulding's book *The Image*, laid the foundation for the psychological study of environmental cognition in two important ways. First, Lynch's investigations legitimized the scientific study of environmental images at a time when psychological investigation of cognitive processes and mental imagery was not in vogue. Second, Lynch's work provided both a framework for thinking about environmental cognition and a research methodology appropriate to the scientific study of mental maps.

Urban maps Lynch's pioneering work in the area of environmental cognition is presented comprehensively in his classic book, *The Image of the City* (1960), which details his investigations of residents' mental maps of three American cities—Boston, Los Angeles, and Jersey City. Lynch employed two different research method-ologies to investigate the mental images of these three cities. In one method, trained observers conducted a systematic field reconnaissance of the central area of each of the cities. The observers recorded the presence and characteristics of particular elements in the cityscape that might serve as focal points for residents' maps, such as a historic landmark or a city park. The observers also noted those areas that contained no prominent features and so might lend themselves less eas-ily to the formation of clear mental images. Thus the observations by the trained observers permitted the identification of those characteristics of the city environ-ment that played a positive role in facilitating the formation of an urban image, along with those features that tended to inhibit the development of a clear mental map.

Lynch's second research strategy involved lengthy interviews with residents themselves to elicit the nature of their own maps of their urban settings. During the interviews residents were asked to draw sketch maps of the city, to describe in detail various trips through the city environment (such as an individual's trip home from work), and to list those parts of the city that were most vivid and dis-tinctive in their minds. The purpose of the interviews was to ascertain the "public image" of each city, which could then serve to identify the particular features of the cityscape that were important to the formation of residents' own mental maps.

Lynch's discussion of Bostonians' images of Beacon Hill—a residential district with a distinctive early-nineteenth-century character, located near Boston's com-mercial center—is instructive (Lynch, 1960: 162):

> Beacon Hill was considered to be very distinctive, often felt to be a symbol of Boston, and often seen as from a distance. . . . More than half the subjects expressed the following as part of their image of the Hill (in roughly descending order):
> a sharp hill
> narrow, pitching streets
> the State House
> Louisburg Square and its park
> trees
> handsome old houses
> red brick
> inset doorways

Legibility   Central to Lynch's analysis of the images of major American cities is an interest in discovering the relative *legibility* of different urban environments. Lynch defines the legibility of an urban environment as the ease with which its features can be recognized and organized into a clear and unified pattern. The meaning of Lynch's notion of legibility becomes clear when we compare some im-portant differences in legibility in the three urban centers he studied.

The highest degree of legibility was found in Boston, which has a clear, pat-terned, and unified mental image. Bostonians saw their city as having very dis-tinctive districts and an easily grasped spatial structure. Readily identifiable ele-ments in the Boston image included the Boston Common, the Public Garden, and the Charles River. The core of many residents' urban image was the Boston Common, with its extensive green open space, sense of history, and central loca-

Figure 3-1
The prominent location and rich history of the Boston Common make it a dominant feature in mental maps of Boston.

Fredrik D. Bodin/Stock, Boston.

tion (Figure 3-1). Ironically, however, traversing the Common itself can be disorienting because of its unusual five-sided, asymmetrical shape.

Lynch found that both Los Angeles and Jersey City are much less legible than Boston. Although the grid pattern of downtown streets and the well-known Pershing Square provide a rudimentary structure to the image of Los Angeles, some of the city's other characteristics militate against its legibility: the decentralization of the metropolitan area, the relatively small number of well-known architectural features, the sprawling development of the downtown area, the absence of medium-sized districts. When Los Angeles residents were asked to describe the city, typical responses were: "spread-out," "formless," "without centers."

Lynch found the legibility of Jersey City to be particularly low. Jersey City suffers from both the absence of a single downtown center and a somewhat uncoordinated street system. The cityscape has almost no distinctive elements. In fact, the only feature of Jersey City's environment that residents referred to as distinguishing was the dramatic sight of the New York City skyline across the Hudson River. Residents' mental maps of Jersey City were highly fragmented and tended to have extensive blank regions.

Common map features When he compared the findings from the three cities he studied, Lynch discovered that the types of environmental features in residents' mental maps were similar in all three cities. Open areas were a common map feature, particularly open spaces characterized by natural features, such as grass, vegetation, or water. Natural greenery in the urban landscape was typically referred to with a sense of delight. The early portion of residents' descriptions of their trip from home to work was full of detailed images of flowers and trees. Many people even reported that they went out of their way in order to pass near a park or a body of water. Lynch offers an excerpt from a Los Angeles interview that reflects the characteristic delight residents took in urban greenery:

> You cross over Sunset, past a little park—I don't know what the name of it is. It's very nice, and—oh—the jacarandas are coming into bloom. One house about a block above has them. On down Canyon and all kinds of palm trees there: the high palms and the low palms; and then on down to the park.   [Lynch 1960: 44]

Another environmental feature that occurred as a common element in city images in all three urban areas was a vista or a broad view of the city from some distant point. Bostonians referred to the sweeping view of the city offered from the Charles River, while residents of Jersey City remarked on the panoramic view of New York's skyline from the Palisades. Another feature that tended to recur was a reference to social class. Bostonians envisioned Beacon Hill as divided into two distinct sides along class lines; Los Angeles residents saw Broadway as "lower class"; respondents in Jersey City viewed the Bergen section of the city as "upper class."

Map elements   Lynch identifies five types of basic elements that make up people's mental maps of the urban environment: paths, edges, districts, nodes, and landmarks (Figure 3-2). He defines these mental map elements as follows:

1. Paths. Paths are channels along which people travel, such as streets, bus routes, or railroad lines. Paths were the predominant element in the mental maps of most of the people Lynch interviewed. Bostonians' maps tended to include Boylston Street; residents of Jersey City tended to mention Hudson Boulevard; and people in Los Angeles thought of the freeways.

2. Edges. Edges are linear elements that do not function as paths. They may be barriers between different parts of the city or seams along which two parts of the city come together. Walls, edges of development, shores—all are edges. The Charles River was an important edge in Bostonians' mental maps.

3. Districts. Districts are medium to large sections of a city that are seen as having an identifiable character. Common districts in residents' mental maps of Boston were Beacon Hill and the North End.

4. Nodes. Nodes are strategic points in the city to and from which residents travel. Some nodes are formed by a crossing of important paths or by breaks in a transportation line. Others develop through concentrated use, and they may be the focal points in their districts. Common nodes are traffic rotaries, major railroad stations, and heavily used squares or plazas. Two important nodes in Lynch's maps were Pershing Square in Los Angeles and Boston's South Station.

5. Landmarks. Landmarks are points that are viewed from an external vantage point. Some landmarks, such as a golden dome or a great tower, are quite distinct and may be seen from great distances. Others are more local—a store front or sign—and are visible only in a particular area. Significant landmarks in the maps Lynch collected included the peak of the Los Angeles City Hall, the weathervane of Boston's Faneuil Hall, and the Jersey City Medical Center.

(1)

(2)

(3)

Figure 3-2

Some common elements in people's mental maps of New York City: (1) 42nd Street (a path); (2) the East River (an edge); (3) Chinatown (a district); (4) Grand Central Station (a node); and (5) the United Nations Building (a landmark).

© Leonard Speier 1981; © Lawrence Frank 1980; © I. A. Gonzalez/Black Star; © Russell A. Thomson/Taurus Photos and © Eric Kroll 1979/Taurus Photos.

(4)

(5)

Some limitations of Lynch's work   Although Lynch's research provided an important impetus to the study of environmental cognition, it has some shortcomings. Lynch's observations are based on a very small number of subjects; his interviews were conducted with just thirty residents in Boston and only fifteen residents each in Los Angeles and Jersey City. The background characteristics of the respondents Lynch selected reveal a bias toward middle-income professional people. Thus the "public" images that Lynch presents reflect the mental maps of only a small segment of each city's total population. As we shall see, the characteristics of one's background have been found to have significant influences on the ways in which one forms a mental map. In addition, there are limitations to the measurement procedures Lynch employed; we shall return to a fuller consideration of these measurement issues later in this section.

COGNITIVE MAPPING

Encouraged by Lynch's initial work and discoveries, environmental psychologists have sought to learn more about the nature of the psychological processes involved in the formation of mental maps. In a series of research studies and scholarly articles, Roger Downs, a geographer, and David Stea, a psychologist, have pursued a general answer to this question (Downs, 1970; Downs and Stea, 1973, 1977; Stea, 1969, 1974; Stea and Blaut, 1973a, 1973b; Stea and Taphanel, 1974).

The products of the mental mapping process have been variously referred to as "mental maps," "environmental images," "schemata," and "cognitive maps." Of these labels, "cognitive map" has most frequently been used to designate the product of mental mapping, and "cognitive mapping" to refer to the mapping process itself. Downs and Stea define cognitive mapping as a process that enables us to collect, organize, store, recall, and decode information about the relative location and attributes of features of the geographic environment. They note that without the ability to engage in this mental process, normal everyday behavior in the environment, such as a journey from home to school, would be impossible.

Representation   A cognitive map is an individual's organized *representation* of some part of the geographic environment: it "stands for" or "portrays" the environment. This representation is both a likeness of the spatial environment and a simplified model of it. Thus a particular person's representation of the environment is not an exact replica of the objective environment, but rather a shorthand, somewhat distorted, individually tailored version of the real world. Downs and Stea note that although there is no one-to-one correspondence between the "real" and "imagined" environments, there is a meaningful and stable relationship between the two.

Visual and nonvisual aspects of maps   Although cognitive mapping involves predominantly *visual* representations of the environment, the nature of cognitive representations is complex, and includes other sensory and motor modalities in addition to vision. In referring to Tolman's (1948) classic studies of cognitive maps in rats, Downs and Stea note that the information used by the rats was essentially tactile and olfactory. In addition to visualizing environmental settings, we typically recall the distinctive sounds and smells of a particular locale, and we may even be aware of the tactile and thermal characteristics of places.

To emphasize that cognitive mapping involves more than visual characteristics, Downs and Stea explain that congenitally blind persons also form mental images of their spatial environment. In fact, the blind employ a cognitive mapping process that is very similar to that of sighted persons, as is evident in the following excerpt of a blind person's description of his trip from a bus stop to his home:

> After descending from the bus you have to walk straight ahead a little on Kalyayevskaya Street, with the houses being on the right; . . . near the next house there are always a good many people as this is a trolley bus stop; . . . behind the house there is a vacant space; here you must walk along the fence for in rainy weather there are puddles; after a few steps to the right there are the steps which lead to my house.   [Shemyakin, 1962, quoted in Downs and Stea, 1977:69]

Downs and Stea conclude that the second element in the term "cognitive maps" is used in a figurative rather than a literal sense. Cognitive representations are not stored in the brain as pictures, like color slides or a road map, but rather as complex structures with multiple properties that may be reconstructed on demand. The question of precisely how cognitive representations of the spatial environment are constituted is quite complex. Later in this chapter, when we consider theories of environmental cognition, we shall see that experimental psychologists who have investigated human cognition have tended to fall into contrasting and often opposing camps in regard to the nature of cognitive representations. For now, we may agree that cognitive representations of the spatial environment are remarkably complex and include both visual and nonvisual information.

Active process   Downs and Stea emphasize that cognitive mapping is an active process that typically involves direct interaction with the environment. In the cognitive mapping process, "learning by doing" plays a central role. Research evidence has supported the notion that an accurate picture of the spatial environment is dependent on the motor-sensory feedback gained through an active involvement with the environment (Held and Rekosh, 1963). Similarly, Ittelson (1973) argues that in the process of environmental cognition the individual actively participates in interaction with the environment. Of course, while cognitive mapping usually involves direct interaction with the environment, cognitive maps may also be based on "mediated" or "secondary" interaction, as through books, television, and the movies.

THE SCALE OF COGNITIVE MAPS

The urban environment   The overwhelming majority of research studies on cognitive mapping have been modeled on Lynch's pioneering work and have involved the cognitive mapping of cities throughout the United States and in other parts of the world (Figure 3-3). Additional cognitive mapping studies in the United

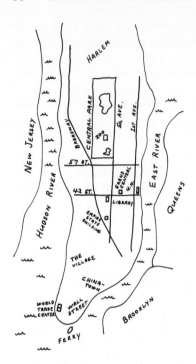

Figure 3-3
One resident's personal map of New York City.
Can you identify the particular elements referred to
in Figure 3-2?

States have been conducted in Chicago (Saarinen, 1969), New York (Milgram, 1977; Milgram, Greenwald, Kessler, McKenna, and Waters, 1972), and Los Angeles (Orleans, 1973). In Europe, cognitive mapping research has been carried out in Paris (Milgram, 1977), Rome and Milan (Francescato and Mebane, 1973), and in Amsterdam, Rotterdam, and The Hague (de Jonge, 1962). Further studies of urban images have been conducted in Ciudad Guayana, Venezuela (Appleyard, 1969, 1970, 1973), and in Mexico City, Puebla, Guanajuantó, and San Cristóbal in Mexico (Stea and Wood, in press).

Stanley Milgram's cognitive mapping research in New York City is typical of the urban mapping studies that have been carried out in the tradition of Lynch's work. In 1975 Milgram inserted an ad in *New York* magazine, asking people to send him their personal mental maps of New York City. He explained that he was not seeking objective maps in the manner of Rand McNally, but rather maps that reflected people's personal views of the city. He received 332 maps.

By examining respondents' reported background characteristics, Milgram was able to ascertain how this self-selected sample of New Yorkers compared with the city's actual population. He found that respondents lived throughout the city, were balanced over sex distribution, and were typically young adults who held minor professional positions. Their maps afforded a rare glimpse of these New Yorkers' views of their city, and of some of the fascinating ways in which individual perceptions of the city differed.

Urban districts   A small number of investigators have examined people's cognitive maps of particular districts in urban areas, such as a resident's own neighborhood, the central business district, and even an urban shopping center. Terence Lee (1968, 1973) studied neighborhood maps among a group of British housewives, and Florence Ladd (1970) gathered neighborhood sketch maps from a group of black high school students in Boston. Their findings reveal that neighborhood maps show a remarkable degree of individual variation, even among residents who live only a few houses apart.

In another study, Lee (1970) interviewed more than 200 female shoppers in England in order to study shoppers' cognitive maps of a large urban shopping center in Bristol. He found that respondents' images of the shopping center reflected a highly complex mental construct. Their mental maps included physical design features of the shopping center, such as architectural properties, pedestrian movement, and physical attractiveness, along with elements connected with retail establishments, such as the quality of services, prices, and shopping hours.

National and international environments   A small amount of cognitive mapping research has dealt with people's environmental images at a national scale and even at the scale of the entire world. In an especially interesting study, Thomas Saarinen (1973) asked high school students in the United States, Canada, Finland, and Sierra Leone to draw sketch maps of the world, labeling those places they considered interesting or important. He discovered dramatic distortions in students' world maps as a function of their personal investment in their own country. They tended to draw their own country in the center of their maps and to exaggerate its size relative to the size of other nations of the world (Figure 3-4).

Kevin Cox and Georgia Zannaras (1973) conducted a survey of undergraduate students to ascertain individual patterns of imagined similarity or dissimilarity among the states of the United States. They found that the American states were differentiated according to several identifiable cognitive categories. Some of these cognitive categories were essentially geographical (e.g., east or west of the Mississippi River), while others combined geographic factors and regional characteristics (e.g., the South vs. New England).

Figure 3-4
This Finnish student's world map places Finland in a prominent position and shows it as larger than Canada or Europe.

Copyright © 1973 by Aldine Publishing Company. Reprinted with permission from T. F. Saarinen, "Student Views of the World," in R. M. Downs and D. Stea (eds.), Image and Environment: Cognitive Mapping and Spatial Behavior, *New York: Aldine Publishing Company, p. 159.*

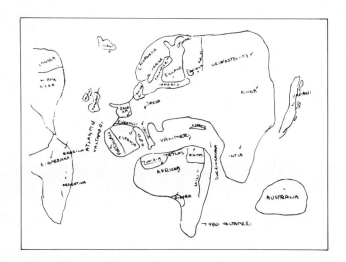

MEASURING COGNITIVE MAPS

The most common method of measuring cognitive maps has involved map drawing modeled after the *sketch* maps obtained in Lynch's (1960) initial research (see box, "Sketching Your Own Cognitive Map"). A subject is given a blank piece of paper and asked to sketch from memory a map of a particular environmental con-

——Sketching Your Own Cognitive Map——

Although the sketch-map technique does have some measurement limitations, it provides an important and especially detailed picture of people's environmental images. In fact, the sketch-map technique can offer a way for each of us to gain a fuller personal appreciation of our own cognitive representations of the environment. If you would like to explore your own environmental images through the sketch-map technique, simply follow the instructions below.

The supplies needed to make a sketch map are remarkably simple; you need only a piece of white, unlined paper (approximately 8½ inches by 11 inches) and a sharpened pencil with an eraser. On the paper, *draw a map of the town or city where you are now living.* Your map need not be an exact replica of the place, like a conventional road map. It may be a personal expression of how *you* view your town or city. Include any features of the environment that are part of your own view of the area. You may erase if necessary, and while there is no time limit, it should be possible to complete your map in about fifteen minutes.

What can you learn about your own image of the town or city where you are living from your sketch map? Keep in mind that there is no right or wrong way to draw a sketch map—your map is your own personal way of viewing the area. Did you map the whole town or city or only parts of it? Are some sections emphasized more than others? How are the parts of the area you mapped or emphasized related to the way you view or use the area?

Can you identify particular elements in your map—paths, edges, districts, nodes, landmarks? Which types of elements are most important in your map? Which elements were drawn first and which last? Are the map elements concentrated more in some parts of your map than in others? How do the types and locations of these elements reflect the way you view and use the area?

How does your sketch map differ from a conventional road map of your town or city? When you compare your sketch map with a conventional map, do you find you have altered or rearranged the location of elements, such as a path or a landmark? How do the shape and contours of your town or city (and its internal districts) in your sketch map compare to those on the road map? How have you oriented your sketch map in relation to the conventional northerly orientation of road maps? How do the ways in which your sketch map differs from a conventional road map indicate your personal view and use of the area?

Finally, what does your sketch map suggest about the environmental character of the town or city where you live? Is it an easy or a difficult environment to understand? Does it have many distinctive elements or very few? Do its internal areas have distinctive place identities or are they generally homogeneous?

text, such as the city where he or she lives. The sketch-map technique has provided an appealing research tool for two primary reasons: it offers an easy and efficient way to collect research data relevant to people's environmental images, and
it provides a vivid and qualitatively rich format for externalizing the pictorial aspects of individuals' environmental cognitions.

Unfortunately, the sketch-map technique also has some limitations, and these
limitations assume added weight in light of the overreliance on sketch maps by
researchers in this area. The most serious limitation of sketch maps involves the
potential unreliability of the measures derived from them. Gary Evans (1980) notes
that frequently the criteria employed in measuring such factors as the accuracy
and complexity of sketch maps have not been clearly stated. Thus, not only has the
reliability of such measures not been determined, but it is sometimes impossible
for later investigators to duplicate the measurement approaches used in previous
studies.

The potential unreliability of sketch-map measures presents a threat to the
internal validity of studies that employ the technique. Some investigators have also
questioned the *external validity* of sketch maps. It has been argued, for example,
that individual variations in drawing ability may threaten the validity of sketch
maps as accurate representations of people's mental images of the environment,
and that the confounding influence of drawing ability may be especially pronounced in the sketch maps of children (Golledge, 1976; Kosslyn, Heldmeyer, and
Locklear, 1977).

A final limitation of the sketch-map technique is its tendency to emphasize
the visual aspects of cognitive representations to the relative exclusion of nonvisual aspects. As we noted earlier, cognitive representations include both visual and
nonvisual properties. For this reason, it is valuable to accompany the sketch-map
technique with additional measures that can tap the nonvisual components of
cognitive maps. Some researchers have asked respondents to give *verbal descriptions*
of a particular environmental setting (Rozelle and Bazer, 1972). A person might,
for instance, be asked to write a one-paragraph description of his or her home
town. Although Lynch (1960) employed such verbal descriptions with considerable success, Stea and Downs (1970) note that subsequent investigators have
tended to underuse verbal responses and to equate cognitive maps with sketch
maps.

Although the predominant method of measuring environmental images has
involved sketch maps, additional measurement strategies have been used. Some
investigators (Lynch, 1960; Milgram, 1977) have employed a *picture-recognition* technique to measure aspects of people's urban images; subjects are asked to identify
particular urban features in a series of photographs of the urban environment (e.g.,
Fisherman's Wharf in San Francisco, Grand Central Station in New York City).
Other researchers (Cadwallader, 1979; Lundberg, 1973) have used measures of *cognitive distance* to assess aspects of the accuracy of people's mental maps; subjects are
asked to judge the distance (or relative distance) between geographic points (e.g.,
which two cities are farther apart—Chicago and Atlanta or Chicago and Dallas?).

Because the steps involved in applying measures of picture recognition and
cognitive distance have been precisely stated and communicated, the *internal validity* of these measures is higher than that of sketch maps. At the same time, how-

ever, these measures provide considerably less information than sketch maps, and the somewhat artificial nature of the tasks involved may reduce their *external validity*. The precision in measurement of the various techniques for measuring cognitive maps may be further refined through the application of advanced statistical techniques, such as *multidimensional scaling* (see Baird, 1979; Evans, Marrero, and Butler, 1981; Golledge, 1977), and secondary response measures, such as *reaction time* in assessing relative distance (see Evans and Pezdek, 1980).

PSYCHOLOGICAL FUNCTIONS OF ENVIRONMENTAL COGNITION

SPATIAL PROBLEM SOLVING

The major psychological function of environmental cognition is to enable people to solve spatial problems in their everyday physical environment. Our ability to form and use cognitive maps is essential to a wide range of daily activities—driving or walking between home and the university, finding the best stores and restaurants in the community, traveling between our own homes and those of friends, taking a holiday drive in the country. Because our cognitive mapping ability is so well developed, we are inclined to take it for granted. In fact, most of the spatial problems in our daily lives are so efficiently resolved through our ability to use cognitive maps that we are typically not consciously aware that they are problems at all. Yet without this remarkable ability to locate and find the physical and social resources we need, we would be unable to function; such everyday tasks as attending a new class across campus or visiting a friend on the other side of town would loom as major, possibly insurmountable, problems.

Adaptive value   Because cognitive mapping plays such a central role in our ability to solve spatial problems, it has a high adaptive value in human life. Cognitive maps tell us where to go to fulfill our needs and how to get there. Downs and Stea (1977) point out that cognitive mapping plays an adaptive role in our daily lives similar to the function of a navigator aboard an aircraft. They summarize this important adaptive role of cognitive mapping in our lives in the following way (1973:10):

> Given a cognitive map, the individual can formulate the basis for a strategy of environmental behavior. We view cognitive mapping as a basic component in human adaptation, and the cognitive map as a requisite both for human survival and for everyday environmental behavior. It is a coping mechanism through which the individual answers two basic questions quickly and efficiently: (1) Where certain valued things are; (2) How to get to where they are from where he is.

In discussing the adaptive value of cognitive mapping in the evolution of human beings, Stephen Kaplan (1973:64) explains that "a grounded ape of not particularly formidable proportions" had to rely on intelligence for survival. Prehistoric humans needed to anticipate events in order both to capture game and to avoid personal danger. The survival of prehistoric humans depended on their ability to identify where they were located spatially. The problem of identifying where one is located is quite complex, since the environment is uncertain and constantly changing. Yet prehistoric humans competing for survival could not afford to respond to each new situation as though they had never experienced anything like it before. Similarly, early hunters and gatherers needed knowledge of large spatial areas to find and use vital resources, while keeping track of their home territory. Thus, Kaplan concludes, an ability to make sense out of environmental uncertainty and diversity and to extract the essential recognizable characteristics from a new situation were critical to the early survival of the human species. The adaptive ability that has evolved in response to this need is cognitive mapping, which offers a complex representational framework within which to fit novel environmental experiences.

Problem solving   We have found that our ability to form cognitive maps plays an essential adaptive role in allowing us to solve spatial problems—but exactly what are spatial problems? Spatial problems involve decisions about where to go to meet our basic everyday needs. We must use "the past, in the present, to solve the future" (Downs and Stea, 1977:83). To make such decisions effectively, the individual must draw together and organize an immense number of bits of information from past experience.

Downs and Stea explain that in order to solve spatial problems, we need two types of information about the environment. First, we must know *where* necessary resources or important people are located. Second, we need to know *what* are the essential attributes or characteristics of the resources or people located in those particular places. The locational aspect of cognitive maps is complex, and includes information that relates to both the distance and the direction of essential environmental resources. While we sometimes take directional information for granted in solving the spatial problems that occur in our daily lives, we often consider how much distance stands between us and desired resources: we typically calculate the amount of time we will need to reach a university class, a movie theater, or a friend's home, and we carefully plan how far we can afford to travel on a vacation.

Cognitive distance   Downs and Stea draw an important distinction between *cognitive* distance and *perceived* distance. Perceived distance is the distance people judge to exist between themselves and a visible object, and depends on the accuracy of the individual's *perception* of the visible object. Cognitive distance, in contrast, involves the judgment of distance between oneself and an object that is not now visible. Here the accuracy of the distance estimate depends on the spatial information stored in a cognitive map.

Some researchers have examined residents' estimates of cognitive distance away from the city center (Golledge and Zannaras, 1973; Lee, 1970). David Canter and Stephen Tagg (1975) have speculated that the accuracy of distance judgments in the city depends on the clarity or sharpness of the particular city's image; it may be that the clearer the cognitive map the city allows, the more accurate will be residents' judgments of cognitive distance. Ronald Briggs (1973) has suggested that the accuracy of people's judgments of urban distance is a product of both characteris-

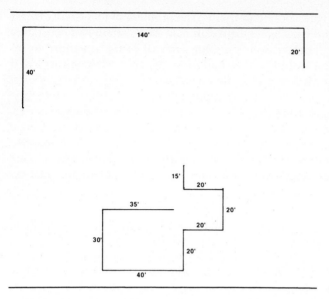

Figure 3-5

The path at the bottom, with seven right-angle turns, was judged to be significantly longer than the path at the top, with only two right-angle turns. Actually, both paths are the same length.

From E. K. Sadalla and S. G. Magel, "The Perception of Traversed Distance," Environment and Behavior, 12:65–79, © 1980 Sage Publications, Beverly Hills, with permission of the Publisher.

tics of the environment and the specific behaviors the individual engages in to obtain information about the environment. Finally, Ulf Lundberg (1973) found that individuals' estimates of distance between places located in different parts of the world improved in accuracy as a function of their emotional involvement with the places judged.

Edward Sadalla and his associates (Sadalla and Magel, 1980; Sadalla and Staplin, 1980b) conducted a series of studies to identify some of the particular features of environments that influence people's judgments of cognitive distance. In three laboratory experiments they investigated how the number of right-angle turns along a path affected individuals' estimates of the path's length. They had college students at Arizona State University walk along a variety of paths that were laid out with tape in the hallway of a building at the university. The investigators found that the students' estimate of the length of a path was positively related to the number of right-angle turns along the path. A 200-foot path with seven turns, for instance, was judged to be longer than another 200-foot path with two turns by 38 of 40 subjects (Figure 3-5). Sadalla and his colleagues also demonstrated in a laboratory experiment and in a field study at a shopping mall in Scottsdale, Arizona, that the number of intersections along a path is positively related to people's estimates of the path's length.

COMMUNICATION

Shared symbols   Another psychological function of cognitive mapping is to provide a basis for communication between people concerning the physical environment. A central feature of the city is that its residents are linked by a shared system of symbols and a common mode of communication (Strauss, 1961). Kevin Lynch (1960) contends that shared representations of the environment provide the symbols and collective memories that are essential to social communication. Lynch believes that this important communicative function of cognitive maps may be deficient in urban settings that do not present a sufficiently legible city image. He notes that although the relatively low imageability of Jersey City did not seriously block residents from navigating the city, it did impede the development of clear and commonly agreed-on urban symbols. Lynch reports the following comment by one resident of Jersey City: "This is really one of the most pitiful things about Jersey City. There isn't anything that if someone came here from a far place, that I could say, 'Oh, I want you to see this, this is so beautiful' " (p. 29).

Downs and Stea (1977) note that communication about the urban environment is typically very rich in symbolism. They add that urban symbols are evaluative, communicating both positive and negative feelings about the city. They explain (1977:92):

> Our attempts at cognitive organization are often encapsulated in *symbols* that offer a quick, shorthand method of characterizing a place. To be successful, a symbol must be immediately recognized by people as standing for a particular place. The meaning and value of a symbol goes beyond the immediate recognition of the identity of a place. It acts as a trigger to help us recall the characteristics of that place, the specific set of whatness, whereness, and whenness information that gives it a unique identity. Given the symbol, we can fill in the necessary detail.

Downs and Stea point out that most major cities are characterized by commonly recognized symbols, including nicknames and pictorial symbols (Table 3-1).

Public image   Lynch defines public images as "the common mental pictures carried by large numbers of a city's inhabitants" (1960:7). He explains that the shared public image of a city is essential to cooperative activity between people. Lynch contrasted the strength and clarity of the public images of Boston, Los Angeles, and Jersey City by comparing the relative number of distinctive elements of each city that emerged in his mapping studies.

Recognition   The clarity of a city's public image has also been investigated in terms of the number of its features that can be commonly recognized by residents. Lynch (1960) asked sixteen Boston residents to identify particular urban features in a collection of photographs of the Boston area. He found that the public image of some areas of Boston was very clear; over 90 percent of his respondents recognized Commonwealth Avenue and the Charles River. Other parts of Boston, however, had very weak public images; most respondents failed to identify features in the South End and the North End.

Stanley Milgram and his students (Milgram, 1977; Milgram, Greenwald, Kessler, McKenna, and Waters, 1972) conducted a similar recognition test in New York City. They asked 200 New Yorkers to identify the locations of 152 environmental features of New York City that were shown in a series of color slides. Milgram discovered that the public image of New York City had an uneven "psychological texture." While some parts of the city were commonly recognized, others were poorly known. For example, environmental features in Manhattan were located in the correct borough more than twice as frequently as features in the other four boroughs. In another study, Milgram (1976, 1977) identified the 50 environ-

Table 3-1.   Many cities are known by familiar nicknames and symbols.

Nicknames for cities:
  Detroit: the Motor City
  Cleveland: the Mistake by the Lake
  Chicago: the Windy City
  New York: the Big Apple, Fun City, Sin City
  Boston: Beantown
  Pittsburgh: the Steel City
  Milwaukee: the Beer Capital of the World
  Philadelphia: the City of Brotherly Love
  St. Louis: the Gateway to the West

Graphic symbols for cities:
  Seattle: the Space Needle
  St. Louis: the Gateway Arch
  Washington (D.C.): the Capitol, the Washington Monument
  San Francisco: the Golden Gate Bridge
  New York: the Empire State Building, the Statue of Liberty
  Philadelphia: the Liberty Bell
  Copenhagen: the Little Mermaid statue
  Paris: the Arc de Triomphe, the Eiffel Tower
  Athens: the Parthenon, the Acropolis
  London: Big Ben, the Tower of London
  Venice: canals, gondolas
  Sydney: the Harbor Bridge

Source: R. M. Downs and D. Stea, Maps in Minds: Reflections on Cognitive Mapping, p. 92.
Copyright © 1977 by R. M. Downs and D. Stea. Reprinted by permission of Harper & Row,
Publishers, Inc.

mental features of Paris that occurred most frequently in the mental maps of over
200 Parisians. Later in this chapter, when we discuss theories of environmental
cognition, we shall see that the differential rates at which urban features are recognized are due in part to how familiar or unfamiliar people are with those features;
people are more likely to recognize urban elements they have been exposed to
often.

PERSONAL IDENTITY

A further psychological function of environmental cognition is to provide a framework for organizing our sense of personal identity (Proshansky, 1978). Our memories, beliefs, feelings, and fantasies can be organized according to our cognitive
maps of the world. Mental maps may also serve to organize present activity, and as
a framework for the acquisition of future knowledge about the world. Lynch adds
that this important organizing function of cognitive maps grants the individual a
sense of emotional security in the world. Downs and Stea (1977) argue that a
"sense of place" is essential to our ability to understand our lives in an ordered
and holistic manner.

Personalized maps   The relationship between cognitive maps and personal identity is especially apparent in the highly personal nature of cognitive maps. Although, as we have seen, cognitive maps contain many shared elements, each map also reflects in important ways the unique and idiosyncratic view that a particular person has formed of the world. Unlike a photograph or an architect's blueprint, an individual's conception of the spatial environment is influenced by the particular meaning the setting holds for that person. In fact, each individual's cognitive map subtly reconstructs and reorganizes the physical environment into a fabric woven of personal meaning. In this sense, environmental cognition is an active and creative process in which the individual construes the spatial environment in terms of a complex array of personal feelings, attitudes, and experiences (Holahan, 1978).

Variations   Downs and Stea (1973) analyze some of the ways in which cognitive maps vary as a function of individual differences. They identify three types of variations in mental maps—incompleteness, distortion, and augmentation. *Incompleteness* occurs when an environmental feature or a surface area in the objective environment is missing from an individual's personal cognitive map. All cognitive maps depict surfaces that are discontinuous, with some spatial areas of the cityscape and some urban features left unmapped. In Lynch's maps, for instance, Bostonians failed to map one of the city's most prominent features—the John Hancock Building (Crane, 1961).

Downs and Stea define *distortion* as "the cognitive transformations of both distance and direction, such that an individual's subjective geometry deviates from the Euclidean view of the real world" (1973:19). They note Lee's (1962, 1970) finding that residents perceived facilities located downtown as closer together than equidistant facilities located away from the center of the city. And as Milgram (1977) has pointed out, although the Seine actually flows through Paris in a great arc, almost forming a half circle, most Parisians envision it as forming a much gentler curve, and sometimes even a straight line.

Finally, Downs and Stea explain that *augmentation* occurs when an individual includes in a cognitive map an environmental feature that does not exist in the corresponding objective environment. Appleyard (1970) reported that a European engineer newly arrived in Ciudad Guayana embellished his map of the city with a nonexistent railway line, because his previous engineering experience led him to expect a rail linkage between a major steel mill and an important mining area.

Unique neighborhood maps   Researchers have discovered that cognitive maps of neighborhoods vary remarkably even among people who are close neighbors. Lee (1963–64, 1973) found that the neighborhood maps of eight British subjects who lived within a hundred yards of one another in Cambridge showed almost no agreement on neighborhood boundaries. Similarly, Ladd (1970) found extensive variation in the maps drawn by black adolescent Bostonians of their Roxbury neighborhood. In fact, the neighborhood maps of two teenaged brothers who had lived in the same house for almost four years showed almost no similarity in the size, shape, and arrangement of neighborhood buildings (Figure 3-6).

Figure 3-6

Two brothers living in the same house drew very different personal maps of their neighborhood.

From F. C. Ladd, "Black Youths View Their Environment," Environment and Behavior, *2:74–99, © 1970 Sage Publications, Beverly Hills, with permission of the Publisher.*

Social class and gender   In addition to the unique variation in cognitive maps that is associated with individual differences, maps differ as a function of particular group characteristics. Environmental psychologists have found, for example, that cognitive maps vary according to social-class level. Two studies conducted in Los Angeles (Los Angeles Department of City Planning, 1971; Orleans, 1973) found that as social class increases, cognitive maps become both more accurate and more extensive. Similarly, Appleyard (1970) and Barry Goodchild (1974) found that people of higher social class tended to map broader urban areas and to include a wider network of streets. They found in addition that the cognitive maps of working-class residents tended to be oriented toward function, while those of middle-income residents tended to be oriented toward aesthetic and historical factors.

Gender has also been found by some researchers to relate to variations in cognitive maps. Appleyard (1970) reports that women tended to make more errors in their maps of Ciudad Guayana than did men. Two studies conducted in Los Angeles found that women's maps were more home-oriented, while those of men included more comprehensive images of the surrounding environment (Everitt and Cadwallader, 1972; Orleans and Schmidt, 1972).

Other studies, however, have reported no differences between men's and women's cognitive maps (Francescato and Mebane, 1973; Kozlowski and Bryant, 1977; Maurer and Baxter, 1972). Current evidence seems to indicate that overall *ability* in cognitive mapping is comparable in men and women. Those variations in mapping that have been reported as a function of sex (e.g., Appleyard, 1970; Everitt and Cadwallader, 1972; Orleans and Schmidt, 1972) are probably due to differences in environmental *behavior* that are associated with traditional sex roles. Similarly, the variations in mapping that have been found for social-class level (e.g., Appleyard, 1970; Goodchild, 1974; Los Angeles Department of City Planning, 1971; Orleans, 1973) are also probably a function of class-related differences in environmental behavior. When we consider theories of environmental cognition, we shall analyze more fully how contrasting modes of spatial behavior, such as behavior patterns associated with gender and social class, can affect cognitive mapping.

THEORETICAL PERSPECTIVES ON ENVIRONMENTAL COGNITION

HOW SPATIAL BEHAVIOR AFFECTS COGNITIVE MAPS

Although environmental psychologists have addressed considerable research attention to the content of cognitive maps, our understanding of how cognitive maps are formed and how they change over time is still incomplete. The question of how mental maps are generated and altered represents one of the most exciting areas for the development of theory in environmental cognition. Several researchers have proposed that cognitive maps are influenced by spatial behavior (Weitzer, 1980). They believe that the particular environmental features that are included in cognitive maps, the features that are ignored, and even the types of spatial distortion that appear in maps are closely related to patterns of spatial activity in the mapped areas.

Here we shall examine some of the effects of contrasting patterns of spatial behavior associated with *life-style*, environmental *familiarity*, and neighborhood *social involvement* on cognitive mapping. We will then apply this knowledge of the relationship between spatial behavior and environmental cognition to explain the variations in mapping that have been found for social class and gender. Finally, we will consider a model for predicting the recognizability of environmental features that includes the influences of spatial behavior.

Lifestyle   Milgram (1976, 1977) reports that the *selectivity* and *distortion* he found in cognitive maps of Paris were in part a function of differences in lifestyle among the Parisians he studied. Lifestyle influences the parts of a city that a resident uses habitually, becomes familiar with, and sees as personally important and meaningful. Milgram notes that a former university student began his map of Paris by first drawing the areas of the city that were related to his experiences as a student. A woman of fifty drew the neighborhood where she had lived longest in Paris in unusually minute detail, including even the direction of one-way streets in her neighborhood.

Familiarity   While familiarity includes other factors besides increased spatial behavior in an area (see Evans, 1980), we may assume that people who are more familiar with a setting, such as those who have worked or lived there for a long time, have generally engaged in more spatial behavior in that setting than have persons who are less familiar with it. Some studies have investigated how familiarity with a setting affects the *breadth* and *detail* with which the area is mapped. Saarinen (1964) found that maps of Chicago's Loop were affected by residents' familiarity with the Loop area. People who worked there mapped the Loop in considerably more detail than did people who worked in other areas of the city. Later Saarinen (1967) observed that students' maps of the University of Arizona campus differed according to the areas of the campus the students were most familiar with as a function of their academic major.

Similarly, Holahan and Mirilia Dobrowolny (1978) systematically compared the campus cognitive maps of more than 100 students at the University of Texas with their daily activity patterns in the campus environment. They found that students mapped those campus areas they used more frequently in greater detail than areas they used less frequently. Appleyard (1970) reports that the manner in which residents mapped Ciudad Guayana differed according to their familiarity with the urban area. Newcomers to the city tended to map a relatively restricted portion of the city, while older residents depicted more extensive urban images.

Additional research has examined how familiarity with an area affects the *accuracy* of cognitive maps. Gary Moore (1974), using independent judges to rate the accuracy of students' maps, found that urban areas that students were more familiar with were mapped more accurately than those with which they were unfamiliar. He explained that the maps of unfamiliar areas accurately related environmental elements *within* small, proximate clusters, but failed to reflect accurate relationships *among* the clusters themselves. Maps of familiar areas, in contrast, showed high accuracy both within and among clusters. Similarly, Appleyard (1970, 1976) reports that people who had been in a city less than six months accurately placed urban elements in correct city zones, though they tended to confuse the relative locations of the zones. Long-term residents, in comparison, accurately mapped relationships both within and among zones.

Other investigators have reported that the exact *placement* of elements in cognitive maps improves as a function of familiarity with the areas mapped. Gary Evans and his colleagues (Evans, Marrero, and Butler, 1981) examined the cognitive maps of University of California students who lived in Irvine, California, and Bordeaux, France, over a one-year period. Using a multidimensional scaling technique, they found that the exact locations of features in space were more accurate after students in both samples had lived in their respective cities for almost a year than when they had lived there less than two weeks. Interestingly, the investigators noted that after only two weeks the students were able to map the relative positions of urban features accurately, but increased familiarity was needed to fine-tune the exact location of the features in the city. Similar results were reported by Alexander Siegel and Margaret Schadler (1977), who investigated schoolchildren's model reconstructions of their classroom. They found that after six months' experience in the classroom, the exact placement of objects was significantly more accurate, although relative object placement did not change over the six months.

In summary, familiarity in a setting increases the *breadth* and *detail* with which the area is mapped. Familiarity also enhances map *accuracy*, though the effects are complex. In some studies, the *exact placement* of objects has increased with familiarity, although the *relative placement* of objects has not changed. In other studies, while the relative placement of objects *within* small clusters has not improved with increased familiarity, familiarity enhanced the relative placement of the clusters themselves.

Social involvement   Lee (1968) analyzed cognitive maps of neighborhood environments as a function of people's social relationships in the neighborhood. He suggests that an individual's neighborhood reflects a "social space" that is defined by the pattern of the individual's social activities in the neighborhood. Lee developed a social index he calls the "neighborhood quotient," which measures the level of a resident's participation and involvement in the social activities of the neighborhood. He proposes that the neighborhood quotient operates as an intervening variable in mediating the effects of neighborhood characteristics on cognitive maps.

Social class   We saw earlier that social-class level is related positively to the extensiveness of urban maps. It is probable that social-class differences in cognitive mapping are products of differential patterns of spatial behavior that are associated with social class. William Michelson (1976), referring to the work of D. Caplovitz (1963), argues that financial constraints limit the spatial mobility of poorer people, so that they cannot travel so far to shop and work as more affluent people.

Orleans (1973) analyzed class differences in cognitive mapping as a function of residents' social contacts and organizational affiliations. In general, persons from higher social-class backgrounds tended to make friends and join organizations located throughout the urban environment. In contrast, people of lower social-class backgrounds made friends and joined organizations in the immediate neighborhood. A study of slum dwellers' urban images of Patna, India, revealed that the detail and degree of coverage of individuals' cognitive maps were partially explained by the location and length of the path they traveled between home and work (Karan, Bladen, and Singh, 1980). Appleyard (1976) notes that when individuals from lower-class backgrounds have a greater opportunity for locomotion than higher-class persons, the typical pattern of mapping is reversed. Finally a related study involving ethnicity rather than social class reported that the more extensive cognitive maps of Anglo children compared to those of black and Mexican-

American youngsters are related to systematic differences between these groups in mode of transportation, geographic mobility, and the spatial extensiveness of friendship choices (Mauer and Baxter, 1972).

Sex differences   We noted earlier that men's cognitive maps are typically more comprehensive than those of women. It seems likely that these sex differences in cognitive mapping are caused by differential patterns in spatial behavior that are associated with traditional sex roles. Traditional patterns of sex-role behavior tend to keep women more involved in home-based activities and to lead men to spend more time dealing with the environment outside of the home (Gutman, 1965) (see box, "Environmental Cognition in Men and Women").

Appleyard (1970), in interpreting his findings in Ciudad Guayana, suggests that women's tendency to make more errors in their urban maps may have been due to their relative immobility in the city as compared to men. Similarly, Robert Beck and Denis Wood (1976) suggest that sex differences in cognitive mapping may be related to mode of travel, since automobile drivers produce the most accurate maps and women learn to drive later and are less likely to own cars than men. Boys and girls also differ consistently in their "home range"; that is, boys use and are familiar with a larger territory (Anderson and Trindall, 1972). Children's sketch maps of their environment reflect these underlying differences in home range.

A predictive model   Following up on earlier work by Appleyard (1969), Milgram and his associates (Milgram, Greenwald, Kessler, McKenna, and Waters, 1972) have proposed a model for predicting the recognizability of urban areas that takes account of the influences of both spatial behavior and environmental characteristics. They note that people can recognize an area only after they have been exposed to it. Thus, they argue, a recognizable area must be located centrally relative to major population flow, for even a highly distinctive area will not be recognized if it is situated in a place that residents do not use frequently. An area's recognizability will be further enhanced if it is socially or architecturally distinctive. The social distinctiveness of Chinatown and the architectural distinctiveness of Rockefeller Center in New York City play an important role in making them recognizable to New Yorkers. On the basis of these observations, Milgram and his associates propose the following formula for predicting an area's recognizability:

$$R = f(C \times D)$$

The formula states that the recognizability of an area (R) is a function of both its centrality to population flow (C) and its social or architectural distinctiveness (D).

DEVELOPMENTAL THEORIES OF ENVIRONMENTAL COGNITION

Recent evidence has indicated that children exhibit the ability to form and use cognitive maps at a considerably earlier age than used to be assumed. Furthermore, environmental psychologists have discovered that cognitive mapping ability develops in important ways throughout the childhood years. Researchers have found that the manner in which we orient ourselves in space, as well as the accu-

Environmental Cognition
in Men and Women

Although current evidence suggests that overall cognitive mapping *ability* is comparable in men and women, there are some important ways in which men's and women's *orientations* toward the environment differ. For example, Holahan and Holahan (1977, 1979), who studied Texas college students' descriptions of their living environments, found that women's cognitive maps included more personal references than men's, while men's maps were relatively more objective than women's. One male student described his living environment in this way:

> Small, one-bedroom apartment, off-campus, i.e., "efficiency," with kitchen, the usual furniture, high rent and utilities, roaches, and no view.

A female student, in contrast, responded in highly personal terms:

> I'm living in an apartment on Lakeside Road with my sister, her best friend, and my best friend. We have a cat and lots of mostly dead plants! The apartment is colorful, with lots of posters of places we'd like to visit. We try to keep it pretty clean. I'm planning on getting married, so it's got lots of stuff lying around.

Women's cognitive maps were also more socially oriented than men's. A male respondent described his environment as follows:

> In an apartment, which is the upstairs of an old house. It's small. It has one bedroom, kitchen, living area, bath, rundown.

In contrast, a woman responded in a highly social manner:

> I live in an apartment with two other girls. We get along very well; our personalities are compatible, and we enjoy each other's company. My old roommate and I did not get along well. Living with Joan and Betty is a very pleasant break from Alice. All in all, we get along great—it makes for a very happy environment.

We conclude that these sex differences in orientation toward the environment are a function of traditional sex roles, which urge men to adopt a more objective and impersonal stance in approaching the world outside the home, while women are encouraged to assume a more communal and personalized attitude in the domestic sphere. We may assume that as sex-role behavior becomes less rigidly defined, these sex-related differences in orientation toward the environment will become less pronounced.

(From C. J. Holahan, *Environment and Behavior: A Dynamic Perspective*, 1979, pp. 135–37. Reprinted by permission of Plenum Publishing Corp.)

racy and complexity of spatial images, all progress through identifiable developmental stages. In fact, as we shall see, the developmental theory of the Swiss psychologist Jean Piaget has been successfully applied to the developmental process through which people learn to generate and use cognitive representations of the spatial environment.

Developing spatial frames of reference   The established theories of cognitive development in psychology have tended to deal with small spaces that can be readily perceived rather than the large-scale spatial environments that have been the focus of cognitive mapping studies (Downs and Stea, 1973). Some environmental psychologists, however, have felt that the theory of cognitive development proposed by Jean Piaget and his colleagues (Piaget, 1954a, 1954b, 1963; Piaget and Inhelder, 1967; Piaget, Inhelder, and Szeminska, 1960) may be broadened to deal with the development of cognitive mapping ability. Piaget's theory of cognitive development emphasizes the interactive and reciprocal relationship between environmental cognition and environmental behavior throughout the developmental process.

Gary Moore and Roger Hart (Hart and Moore, 1973; Moore, 1979), in an extensive literature review, have advanced a theory of the development of cognitive mapping based in large part on Piaget's general theory of cognitive development and his more specific theory of the development of the child's ability to comprehend the geometric properties of objects in space. They propose that in developing cognitive mapping ability, the child progresses through three sequential stages involving progressively more complex *frames of reference:*

> . . . there is considerable evidence that in developing topographical representations of the large-scale environment, the child utilizes a framework or system of reference for interrelating different positions, routes, patterns of movement, and himself in the environment, and that this system of reference is the most important component of spatial representation.   [Hart and Moore, 1973:283]

Moore and Hart label the successive frames of reference through which the child develops "egocentric," "fixed," and "coordinated":

1. Egocentric. At this stage the child's frame of reference is centered on his or her own activities. Environmental features are disconnected in the child's cognitive map; the environment is fragmented.

2. Fixed. During this phase the child's mental map is oriented around fixed places in the environment that the child has explored, such as the home area. These known areas are, however, disjointed in the child's map.

3. Coordinated. In this stage the child's frame of reference assumes the characteristics of a spatial survey map. The child's mental map includes a holistic and integrated view of the spatial environment.

Three cognitive maps that reflect each of these frames of reference are shown in Figure 3-7.

Empirical evidence related to the development of spatial frames of reference is available from a large number of sources. Some studies have demonstrated that by the time young children enter school they are able to adopt perspectives other than their own in identifying environmental features shown in aerial photographs. Studies by James Blaut and his associates (Blaut, McCleary, and Blaut, 1970) in both the United States and Puerto Rico indicate that children between the ages of 5 and 7 can interpret aerial and satellite photographs. When children were shown an aerial photograph and asked to explain what they saw, they were able to identify

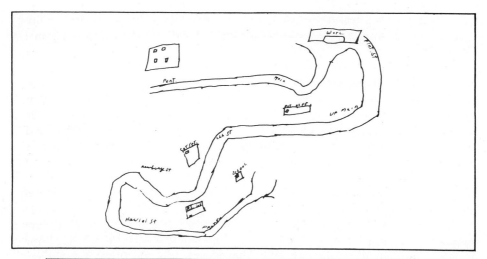

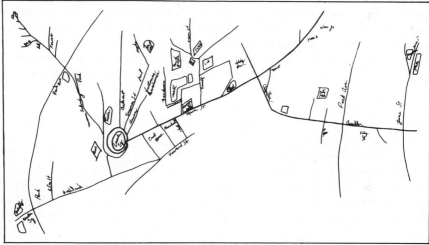

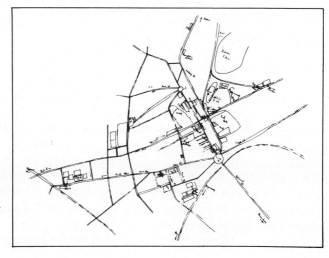

Figure 3-7

Sketch maps at three developmental levels: egocentric (*top*), fixed (*middle*), and coordinated (*bottom*).

From G. T. Moore, "Developmental Variations Between and Within Individuals in the Cognitive Representation of Large-Scale Spatial Environments," Man-Environment Systems, 1974, 4:55–57. Reprinted by permission.

such environmental features as cars, houses, roads, and trees. Children were also able to make a trace map of the aerial photograph and to use it to sketch the path between two houses.

Similarly, David Stea and James Blaut (1973), who used aerial photographs to test kindergarten children in four Puerto Rican communities, found that the most pronounced increase in correct responses occurred between kindergarten and grade 2. Children showed little improvement in the type of environmental learning measured by aerial photograph identification beyond grade 2. The authors suggest that considerable environmental learning occurs through toy play and environmental games, such as blind-man's-bluff and football. In one study they observed five-year-olds playing with environmental toys, such as buildings, cars, and trees, and found that some children used the toys to build an imaginary community as a part of their play.

Linda Acredolo (1976, 1977), using laboratory experimentation, has provided empirical support for the view that the development of spatial frames of reference progress through three distinct sequential stages. She examined the ability of three-, four-, and ten-year-olds to reorient themselves correctly in an experimental room where the child's location, as well as the location of a table, were altered while the child was blindfolded. She found that three-year-olds reoriented themselves either egocentrically or in terms of a fixed frame of reference (the altered table position), four-year-olds responded predominantly in terms of a fixed frame of reference, and ten-year-olds demonstrated a coordinated frame of reference (using the experimental room itself). Acredolo later found that the egocentric responses of younger children were reduced when landmark cues were provided in the experimental room.

Somewhat similar findings for the development of spatial frames of reference have been reported by Douglas Hardwick and his colleagues (Hardwick, McIntyre, and Pick, 1976), who employed a field experimental approach. They investigated the ability of first graders, fifth graders, and college students to sight shielded objects through a sighting tube under a variety of perspective shifts. They found that first graders responded in an egocentric manner (ignoring requested perspective shifts); fifth graders demonstrated correct sightings at a gross level but were unable to fine-tune their responses; and college students showed accurately fine-tuned sightings.

Developing cognitive representations   Environmental psychologists have also explored how the *accuracy* and *complexity* of cognitive representations of the spatial environment develop during the childhood years. Alexander Siegel and Sheldon White (1975), again applying the developmental theory of Jean Piaget, suggest that children's representations of the spatial environment progress through four sequential developmental stages. First, spatial *landmarks* are noticed and remembered. In the next stage, *paths* are learned between particular landmark pairs. Then, at the third stage, the child organizes proximate landmarks and paths into small *clusters* that are well organized internally, though poorly related to other clusters. Finally, in the fourth stage, the environmental features are correctly coordinated in an *overall framework*.

Empirical evidence pertaining to the development of cognitive representations of the spatial environment has been derived from a variety of sources. J. Herman and Alexander Siegel (1978), for example, walked children in kindergarten, the second grade, and the fifth grade through a large model of a town, and then asked them to reconstruct the town from memory. The fifth graders were significantly more accurate than younger children in reconstructing the town after an initial walk through the model. Interestingly, the accuracy of the younger children's reconstructions improved with repeated walks through the model when the model was placed in a classroom, whose walls could provide spatial cues. When the model was moved to a large gymnasium, however, where the walls were too far from the model to offer additional spatial cues, repeated experience with the model did not significantly enhance the younger children's accuracy.

Three field experiments have examined the accuracy of children's cognitive representations in real-world settings. One study demonstrated that five-year-olds were able to describe what was behind the walls of their home bedroom and kitchen more accurately than three- or four-year-olds (Pick, Acredolo, and Gronseth, 1973). A later experiment showed that eight-year-olds could locate a designated area in a school hallway more accurately than three- or four-year-olds when no landmarks were available; there were no age differences, however, when landmarks were present (Acredolo, Pick, and Olsen, 1975). Finally, researchers found that five- and six-year-olds were able to build more accurate models of a series of interconnected rooms they had navigated than were younger children; only six-year-olds could accurately infer the location of landmarks not located directly on their route (Hazen, Lockman, and Pick, 1978).

Cognitive mapping over the life span   Although environmental psychologists have accumulated a considerable body of knowledge about how cognitive mapping ability develops over the childhood years, much less is known about how it continues to develop throughout the life span. There is some weak evidence that some changes in cognitive mapping may occur during the adolescent years. Gary Moore (1973, 1974) found that "developmental levels" were apparent in the sketch maps of adolescents aged 15 to 19, but he was unable to relate these developmental levels to age significantly in the adolescent sample. Howard Andrews (1973) found that Toronto geography students in their junior year in high school demonstrated higher levels of urban knowledge than ninth-grade students in terms of the ability to identify urban landmarks and districts on an outline map. These differences, however, were observed only among students who lived close to Toronto, and may reflect group differences in urban travel behavior.

A few studies have shown developmental differences in cognitive mapping as a function of age differences among adults. Donata Francescato and William Mebane (1973), who studied cognitive maps of Italian cities, found that paths were more important than landmarks for people aged 30 and younger. For respondents over 30, in contrast, landmarks were at least as important as paths. J. Douglas Porteous (1977), who studied the cognitive maps of adults in four English towns, found that the dominant images in the maps of people under 35 tended to be newer environmental features, while the maps of respondents over 60 tended to be dominated by older, derelict, and even demolished features. Similarly, Milgram (1976, 1977) noted that the maps of youthful Parisians were more likely than those of older residents to include newer, contemporary elements as prominant images, such as the Zamanski Tower at the Faculté des Sciences and the fifty-story Maine-Montparnasse office tower. He concludes that the mental maps of older residents appear to have been internalized many years earlier, and fail to admit recent environmental changes. It is possible that the tendency of younger adults' maps to contain both more contemporary features and a greater emphasis on paths may reflect differences in the pattern and style of urban mobility that are associated with aging. Finally, there is some evidence that elderly persons process information pertaining to spatial frames of reference more slowly and less accurately than do young adults, although the developmental level at which they function is similar to that of young adults (Ohta, Walsh, and Krauss, 1977).

MODELS FROM EXPERIMENTAL PSYCHOLOGY

Information processing   Some theories that have been proposed by experimental psychologists to explain the workings of the human mind are of relevance to our understanding of cognitive mapping. Experimental psychologists have advanced a model of *information processing* (see Coombs, Dawes, and Tversky, 1970) that can help us appreciate how the mental information that forms the building blocks of a cognitive map is acquired, processed, and stored. The information-processing model can help us, for example, to understand the nature of cognitive representations of the spatial environment. John Anderson (1978) has reviewed the research from experimental psychology that has examined how cognitive representations of the environment are constituted.

Anderson explains that experimental psychologists have historically fallen into two opposing camps in regard to the nature of mental representations. Those who take the *imagery* position have argued that the properties of cognitive representations are quite spatial and maintain a rough correspondence to the physical structure of actual environmental information (see Shepard, 1975). Imagery theorists contend that visual information and verbal information are coded in quite different ways. Anderson points out that although imagery theorists do not claim that cognitive representations are literally "pictures in the head," they do believe that a *picture metaphor* can be used to describe the essential properties of mental representations.

Experimental psychologists who take the *propositional* position have proposed that mental representations are encoded in an abstract propositional form (see Plyshyn, 1973). Propositional theorists argue that the same format is used to encode both visual and verbal information. This format, they say, is essentially abstract and has explicit rules of formation (i.e., there are rules for deciding what is and what is not a well-formed proposition). Propositional theorists hold that the picture metaphor of cognitive representations is misleading, and that a propositional representation can more parsimoniously handle all of the properties ascribed to a pictorial representation.

Recently the either-or nature of the debate over cognitive representations has begun to break down. Stephen Kosslyn and James Pomerantz (1977) have suggested that cognitive representations are coded in a propositional form but, once formed, can be manipulated in a manner that is consistent with the imagery position. Anderson (1978), after critically examining both the imagery and propositional positions, concludes that it is impossible to decide between the two views on the basis of behavioral data. Evans (1980), in an extensive review of research on environmental cognition, contends that mental representations of the spatial environment are stored in both imaginal (e.g., the relative spatial positions of objects) and propositional (e.g., semantic labels for landmarks) forms.

Application to the molar environment   Herbert Leff (1978) has provided an interpretation of the model of information processing developed by George Miller and his colleagues (Miller, Galanter, and Pribram, 1960) that is appropriate to the study of how individuals form mental maps of complex environmental settings, such as a neighborhood or an entire city. Although Leff's model is not typical of traditional thinking among investigators in the information-processing area, it is especially relevant to the study of environmental cognition as it relates to molar settings. Leff contends that human information processing is an especially *active* process. He notes that an individual exerts a high level of self-direction in carrying out complex cognitive operations, such as acquiring new information, calling upon images stored in memory, or solving new environmental problems.

Drawing on the concepts and terminology of Miller and his associates, Leff explains that the individual controls the mental operations that are engaged in and the sequence for completing these operations through *plans*. Miller and his colleagues define a plan as a series of hierarchical instructions for performing specific mental operations, and note that a plan may be compared to a complex computer program. A plan is hierarchical in that it may include subplans, which in turn may include sub-subplans.

Miller and his coworkers note that a plan proceeds through a sequence of operations that may be envisioned as "*test-operate-test-exit*," or simply as a TOTE sequence. The function of the TOTE sequence is to enable the individual to complete a designated task. The "test" phase of the plan ascertains whether the desired objective has been attained. If it has not, the "operate" sequence is called into play, and then is followed by another test. When the test indicates that the goal has been achieved, the plan enters the "exit" phase of the TOTE cycle and proceeds to a further plan.

The TOTE sequence provides a model for understanding the mental operations by which environmental information is acquired, processed, organized, and directed toward managing complex environmental behaviors. Once a plan has been developed and employed successfully, it may be stored in memory as information that may be retrieved and used to solve future environmental problems. Leff has proposed a graphic model (Figure 3-8) to represent the complex process by which environmental information is received, operated on, stored, revised by environmental experience, and used in managing environmental behavior. Note that the model shows that overt behavior affords one form of feedback that in a cyclical fashion may influence the further development of cognitive processes.

Edward Sadalla and Lorin Staplin (1980a) have recently applied an information-processing model to people's judgments of cognitive distance. They propose, on the basis of earlier work by Milgram (1973), that distance estimates are influenced by the amount of information an individual has stored about the spatial

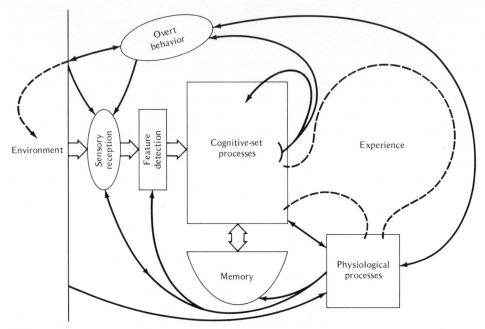

Figure 3-8
A model of environment-experience-behavior interrelations.

From H. L. Leff, Experience, Environment, and Human Potentials, *p. 91. Copyright © 1978 by Oxford University Press, Inc. Reprinted by permission.*

properties of the area judged. For instance, they predicted that a path about which more information had been stored would be estimated as longer than a path of equal length about which less information had been stored. To test this hypothesis, they asked students at Arizona State University to walk along two different paths that were marked with tape on the floor of a large room. The paths were of identical length, and each had a total of fifteen intersections (marked with additional strips of tape that intersected the paths at a 90° angle). The paths differed in that one had intersections marked with proper names that occur frequently among English-speaking populations (e.g., Smith, Thomas), while the intersections of the other were marked with low-frequency proper names (e.g., Milrow, Jillson). On the basis of the well-established finding that word frequency and word recall are positively related, the investigators predicted that the path marked with high-frequency names would lead to the storing of more information than would the path marked with low-frequency names. In accordance with their information-processing model of cognitive distance, they also anticipated that the path with high-frequency names would be judged to be longer than the other.

Consistent with their prediction, Sadalla and Staplin discovered that the students estimated the path with high-frequency names to be 46 percent longer than the path with low-frequency names. The investigators speculate that the information-processing model might explain the earlier findings that distance judgments are positively correlated with the number of right-angle turns along a path (Sadalla and Magel, 1980) and with the number of intersections along a path (Sadalla and Staplin, 1980b). They reason that if structural changes in a path, such as turns and intersections, provide information that is remembered, then the information-processing model is consistent with the finding that paths with more structural changes are estimated to be longer than those with fewer changes.

APPLICATIONS TO ENVIRONMENTAL PLANNING

How might our understanding of the manner in which people develop mental images of the environment help us to design habitats that are better fitted to human needs? Lee (1968, 1973) has explicitly addressed this question, and concludes that environmental psychologists interested in the process of environmental cognition can make a significant contribution to environmental design. He proposes that cognitively oriented psychologists may play an important role in the early stages of architectural training and as members of environmental design planning teams. Porteous (1974) has suggested that cognitive mapping studies might help to suggest new questions and to encourage new awareness in the environmental design process.

Although environmental psychologists agree that our understanding of environmental cognition offers a potentially useful contribution to architectural design, they are also aware that our knowledge in this area is at an initial phase of development. Because the process of environmental cognition is so complex and because the link between theory and design is less immediate than that between environmental behavior and design, our ability to translate research findings from cognitive mapping studies into specific design advice is still primitive. There are several ways, however, in which our knowledge of environmental cognition can be applied to practical design questions.

DESIGNING A LEGIBLE CITYSCAPE

Kevin Lynch (Lynch, 1960, 1965, 1970; Lynch and Rodwin, 1958) has been at the forefront of the effort to apply findings from research on environmental cognition to environmental design. Lynch's works have been required reading at schools of urban design for two decades, and have been highly influential in enhancing the visual awareness of design students and in offering a frame of reference for describing the visual aspects of the environment (Bell, Randall, and Roeder, 1973). Lynch argues that the planning of urban environments should be oriented toward building cities that are highly legible; that is, the city and its parts should be easy to recognize and to represent as a unified pattern of environmental elements. Appleyard and his associates, in fact, have applied Lynch's notion of legibility in planning the new city of Ciudad Guyana, Venezuela (Appleyard, 1976).

Lynch believes that just as individuals may be characterized according to their ability to form clear and accurate cognitive maps, cities may be differentiated according to their ability to facilitate the formation of legible environmental images. Cities differ in their ability to attract attention, to be recognized, and to be orga-

Figure 3-9
Florence has a legible environment that is easy to map and understand.

© *C. Ray Moore/Black Star.*

nized into unified mental images. The dramatic difference Lynch observed between the highly legible cityscape of Boston and the difficult imagibility of Jersey City provides a vivid contrast in the legibility of two American cities.

Florence   Lynch has selected Florence, Italy, as a city that reflects a uniquely high degree of legibility (Figure 3-9). Florence takes advantage of the natural geography of its region; it lies in a bowl of hills and is linked to the surrounding landscape by the Arno River, which cuts through the city. Settlements in the northern part of the city are situated on distinctive hills. The center of Florence is typified by a range of well-defined environmental features, including its stone-paved streets, its stone and stucco buildings with their distinctive Florentine eaves, and its many strong nodes. Finally, the highly distinctive Duomo, the dome of the cathedral of Santa Maria del Fiore and the symbol of Florence, is situated in the heart of the city, and can be seen from miles around.

On the basis of his analyses of cities around the world, Lynch advances several design suggestions for urban designers concerned with building more legible urban settings. Lynch points out that the basic building blocks for the design of more legible urban environments are the urban elements of paths, edges, landmarks, nodes, and districts.

Paths   Lynch addresses particular attention to the design of urban paths, pointing out that for most people they are the central element in an urban image. He proposes that paths be given some unique design quality to distinguish them from surrounding urban features. For example, a path might be set off by distinctive uses, a unique floor texture, a particular pattern of planting, or special lighting. Lynch further proposes that a path have a clear sense of direction. This might be achieved by a strong element at each end of the path or by a distinctive gradient along its course, such as a thickening of plantings, stores, or signs as a shopping node is approached.

Edges   Lynch proposes that edges have continuity of form and lateral visibility for some distance, as are typically found in the walls of a medieval city. When an edge separates two distinctive regions, the two sides of the edge may be differentiated through contrasting design materials or by a particular pattern of planting. When an edge is not a continuous line, a distinctive element at each of its ends can help to locate its line. Finally, an edge can be made more visible by increasing its use, as by opening a waterfront to recreational activity.

Landmarks   Lynch suggests that highly legible landmarks be sharply contrasted with their environmental background. Richly designed detail on a landmark will serve to attract and hold an observer's attention. Thus an intricately designed church situated among stores is likely to function as a strong landmark. Lynch adds that a landmark should be visible from a considerable distance, and ideally some unique features of its design should be visible from any direction. Finally, Lynch notes that a cluster of landmarks will effectively reinforce one another's visibility. Appleyard (1969) has provided some relevant information concerning the particular attributes of buildings that enhance their memorability. Buildings that are likely to be recalled are characterized by heavy use, symbolic significance, great height in contrast to the surrounding environment, sharp contours, and bright surfaces.

Nodes   Lynch contends that few existing urban nodes have distinctive environmental characteristics that enhance their legibility beyond the simple concentration of activity. Lynch emphasizes that a node should be environmentally distinct; it should be different from any other place. Such distinctiveness may be achieved through a continuous and unique style of lighting, planting, floor texture, or detail. A node's distinctiveness is also enhanced by an easily identifiable boundary. Finally, a well-designed node can serve to integrate the districts that surround it, as when a gradient of sycamore trees leads up to a town square that is known for its sycamores or when cobblestone streets lead into a cobblestone plaza.

Districts   Lynch stresses that legible districts must be characterized by internally homogeneous features. Such homogeneity may be achieved in many ways— through distinctively narrow streets, a particular building type or building feature, unique coloring, or specialized building material. The greater the number of types of such distinguishing elements, the higher the district's legibility. For example, the highly distinctive Boston district of Beacon Hill is typified by uniquely identifiable elements—its narrow streets, bricked surface, small row houses, recessed entrances. Lynch adds that the legibility of a district will be enhanced by the clarity and closure of its boundaries.

Qualities of legible design   Lynch identifies ten key qualities that tend to characterize legible design in each type of environmental element:

1. Singularity. The feature should be sharply contrasted and differentiated from its environmental background.

2. Simplicity. The environmental element should be clear and simple in its geometrical form and in the number of its parts.

3. Continuity. The element should have continuous edges, surface characteristics, and form.

4. Dominance. One part of the setting should dominate others in size, interest, or intensity.

5. Clarity of joint. Joints and seams should be highly visible.

6. Directional differentiation. Direction should be indicated by gradients, asymmetrical elements, or radial lines.

7. Visual scope. The range of vision should be enhanced either actually or symbolically.

8. Motion awareness. The sense of motion should be facilitated by both visual and kinesthetic cues.

Mapping the New York Subway

The popular song that tells of the disappearance of Charlie on the Boston MTA is not irrelevant to the twenty million people who ride New York's subways each year. New York's subway system is a maze of 27 train routes, covering almost 250 miles of track and 500 subway stations. The difficulty transit officials experience in attempting to help riders form a clear mental map of the subway system is reflected in the fact that the design of the New York subway map has been revised four times in the last ten years. Arline Bronzaft and her colleagues (Bronzaft, Dobrow, and O'Hanlon, 1976) recently conducted a field study to evaluate the effectiveness of the then current version of the subway guide.

To test the map's effectiveness, Bronzaft and her associates asked twenty persons who had lived in New York less than seven months to complete four subway trips through parts of the subway system with which they were unfamiliar. None of the twenty subway riders was able to plan reasonably direct routes on all four trips. In fact, more than half of the subway trips were judged by the experimenters to be unacceptable because they followed too indirect a route to their destinations. Of the unacceptable trips, a significant number were directly related to misuse of the subway guide.

On the basis of these findings, the experimenters suggested some specific design changes to clarify the subway guide. They concluded that the best way to help passengers to use the subway system efficiently would be to make available alternative types of information aids to complement the maps. Subjects reported that public announcements aboard the trains concerning direction, approaching stations, and transfer points were especially helpful.

9.   Time series. Elements should be linked in a sense of temporal sequence.

10.   Names. The recognizability of features should be increased by distinctive names, meanings, and associations.

It should be pointed out that Lynch's suggested qualities of legible urban design are essentially speculative, and have generally not been subjected to empirical examination. One noteworthy exception is the matter of names. Recent experimental evidence has indicated that the presence of names on small models of buildings *reduced* their visual recognition, although the names enhanced locational information (Pezdek and Evans, 1979).

DESIGN FOR ENVIRONMENTAL COMPLEXITY

Lynch's notion of legibility in the urban environment emphasizes the goal of simplicity and clarity in the design of urban forms. We must ask, however, whether a high degree of legibility is a sufficient planning objective for all aspects of the urban environment under all circumstances. Might complexity and even ambiguity in form be beneficial for some features of the urban environment in particular situations?

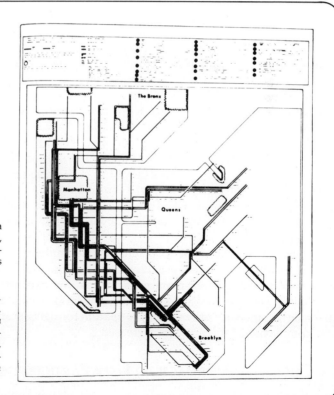

New York's subway is a challenging environment, as can be seen in this reproduction of the city's subway guide.

From A. L. Bronzaft, S. B. Dobrow, and T. J. O'Hanlon, "Spatial Orientation in a Subway System," Environment and Behavior, 8:575–95, © 1976 Sage Publications, Beverly Hills, with permission of the Publisher.

Stephen Carr (1967) has posed this question in discussing design criteria for the urban environment. He contends that urban designers should stimulate people to explore new settings and new experiences in the urban environment by designing some degree of novelty and complexity in urban forms. By exploring complex and novel environments, people expand the number of categories available to them for understanding the world and enhance their sense of competence in dealing with the environment. The architecture of Robert Venturi (1966), which we discussed in Chapter 2, offers an example of a commitment to complexity and novelty in architectural practice.

Of course, an excess of novelty and complexity in the urban environment would not be desirable, as people would be unable to comprehend or cope effectively with the city. Carr argues that a delicate balance must be achieved between complexity and order. While some degree of novelty and complexity in the urban environment is desirable, the designer must also strive to create settings that can be easily recognized, identified, and remembered. Carr's advice is congruent with the design recommendations of Rapoport and his colleagues (Rapoport and Hawkes, 1970; Rapoport and Kantor, 1967) for achieving an "optimal range" in the *perceptual* input of designed settings (see Chapter 2). We shall encounter this notion of optimal environmental complexity again in Chapter 4, where we shall learn that people prefer settings that offer an intermediate level of environmental stimulation.

Carr encourages designers to structure urban forms in such a way that their design characteristics will facilitate people's formation of mental representations of the environment. He points out that a simple and effective way to help people comprehend complex urban settings is to provide them with information aids— street maps in telephone directories, information boards at complex decision points, clear guides to public transportation systems (see box, "Mapping the New York Subway").

D. J. Bartram (1980) conducted a laboratory experiment with students at a British university to investigate the relative efficiency of various methods of presenting information about bus routes. A street map of an area of North London was used to present information on bus routes in four different formats: (1) an alphabetical list of bus stops, (2) a list of bus stops in sequential order, (3) a conventional road map of the area, and (4) a considerably simplified schematic map of the area. The students were asked to solve a variety of route-finding problems that varied in complexity from problems involving two buses to problems involving four buses.

Bartram measured the total time it took students to solve the route-finding problems with information provided in each of the four formats (Figure 3-10). The two map formats proved more efficient than the two list formats, and the schematic map was more effective than the conventional road map. The differences in the relative efficiency of the four information formats became more pronounced as the complexity of the problem increased.

A Cognitive Approach to User Participation

Another aspect of the planning process to which our knowledge of environmental cognition might be applied concerns user participation in design decisions. Ste-

Figure 3-10

Total time required to solve bus-route problems as a function of information format and task complexity.

From D. J. Bartram, "Comprehending Spatial Information: The Relative Efficiency of Different Methods of Presenting Information About Bus Routes," Journal of Applied Psychology, *1980, 65:103–10. Copyright 1980 by the American Psychological Association. Adapted by permission of the publisher and author.*

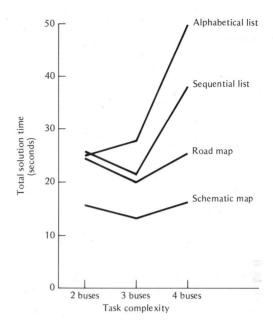

phen Kaplan (1977) has explicitly addressed the question of how our appreciation of the cognitive mapping process might affect planning for user participation in design. Kaplan points out that "experts" (designers, planners, and architects) and ordinary citizens go about solving environmental problems in quite different ways. Experts' training and experience allow them to reduce a vast quantity of environmental information to achieve efficient and highly abstract problem solutions. Ordinary citizens, while familiar with environmental problems, lack the experts' specialized training and experience, and are less capable of viewing the problem compactly or of abstracting only those key ingredients that are required to achieve an efficient solution.

Kaplan argues that designers must develop and apply collaborative techniques that can help citizens to see the problem in compact, efficient, and abstract terms. In developing a planning strategy that might achieve these objectives, Kaplan began by noting some of the cognitive assets citizens bring to environmental problem solving: a facility for nonverbal (visual-spatial) cognition, an ability to form mental representations of the environment, and an ability to imagine themselves in hypothetical situations.

Kaplan concludes that in order to tap these cognitive strengths in citizens, planning strategies should incorporate the following ingredients: visual/spatial elements, familiar features of the environment, and an "as if," exploratory stance. On the basis of these assumptions concerning users' natural cognitive talents in problem solving, Kaplan has devised a collaborative planning strategy that makes use of highly simplified physical models of the environment. Instead of the detailed and intricate models typically employed by designers, Kaplan proposes very simple models that would bring to mind only those key environmental elements that are essential to exploring and resolving the particular problem in question. For example, to explore the possible location of a new dormitory on a college campus, a simple wood block might be shifted on a highly simplified representation of the campus.

Using such simplified models in exploratory studies, Kaplan found that users were readily able to employ the models to solve environmental problems. The simplified models seemed to be easier to relate to and less confusing to users than complex architectural models. Kaplan emphasizes that the greatest benefit of simplified models is in facilitating the creative exploration of alternative design solutions around focused issues early in the design process. Of course, as we noted earlier, it is important to evaluate carefully the external validity of such simulation models.

SUMMARY

Research interest in environmental cognition was initiated by Lynch's classic investigations in Boston, Los Angeles, and Jersey City. Lynch's work legitimized the scientific study of mental maps and afforded a research methodology for the study of mental images. He identified five basic types of elements that make up people's mental maps of the urban environment: paths, edges, districts, nodes, and landmarks.

Cognitive mapping has been defined as a process that enables us to collect, organize, store, recall, and decode information about the relative location and attributes of features of the geographic environment. A cognitive map is a representation of the physical environment; it is both a likeness of the spatial environment and a simplified model of it. The nature of cognitive representations is complex, and involves sensory and motor modalities in addition to vision. Cognitive representations are stored in the brain not as pictures, but rather as complex structures with multiple properties that may be reconstructed on demand. Cognitive mapping is an active process in which "learning by doing" typically plays a central role.

The overwhelming majority of research studies in the area of cognitive mapping have investigated mental maps of urban environments. A small number of investigators have examined cognitive maps of particular districts in urban areas, such as a particular neighborhood or a central business district. A small amount of cognitive mapping research has dealt with environmental images at national and international scales. The most common method of measuring cognitive maps involves the drawing of *sketch maps*. While sketch maps are efficient and qualitatively rich, their internal and external validity need to be better established. Cognitive maps may also be measured through *verbal descriptions*, *picture recognition*, and indices of *cognitive distance*.

The chief psychological function of environmental cognition is to solve spatial problems. Cognitive maps tell us where to go to fulfill our needs and how to get there. Another psychological function of cognitive maps is to provide a basis for social communication about the physical environment. Shared environmental representations provide essential symbols and collective memories for social communication. Public images of the city are also necessary for cooperative social activity. A further psychological function of environmental cognition is to provide a basis for the individual's sense of personal identity. Cognitive maps can serve as a framework in which the individual can organize memories, beliefs, feelings, and fantasies, as well as present and future activity.

The relationship between cognitive maps and personal identity is especially apparent in the highly personalized nature of cognitive maps; mental images reflect a particular person's unique, personal, and idiosyncratic view of the world. These individual differences are reflected in variations in the completeness, distortion, and augmentation of maps. Researchers have discovered that cognitive maps of neighborhood environments vary widely among neighborhood residents.

Theoretical knowledge in the area of environmental cognition is still at an early stage of development. Researchers have suggested that cognitive maps are influenced by spatial behavior. The environmental features that are included in maps and the types of spatial distortion that appear in them are believed to be related to patterns of spatial activity in the environment. Support for this view has been derived from research studies that have examined map differences as a function of individual differences in lifestyle, familiarity with the environment, and social involvement. The variations in cognitive mapping that have been associated with social-class level and gender are probably the result of underlying differences in spatial behavior patterns.

Environmental psychologists have found that the ability to form cognitive maps develops with age. Theorists believe that children progress through three sequential *frames of reference* in forming environmental representations: egocentric, fixed, and coordinated. Investigators have proposed that the *accuracy* and *complexity* of cognitive representations proceed through four sequential developmental stages, in which the individual emphasizes successively landmarks, paths, small clusters of features, and an overall framework.

Additional theoretical knowledge about how cognitive maps are formed has been derived from the work of experimental psychologists in the area of information processing. Historical differences between information theorists who adopted an *imagery* position (i.e., cognitive representations are quite spatial and correspond roughly to the physical structure of actual environmental information) and those who took a *propositional* position (i.e., cognitive representations are encoded in an abstract propositional form) no longer arouse so much partisan feeling; cognitive representations are probably encoded in both imaginal and propositional forms.

Findings from research on environmental cognition have been applied to environmental design in the form of suggestions for the design of a more legible urban environment; that is, an environment whose parts are readily recognizable and easily represented as a unified pattern. At the same time, however, the ideal urban environment should encourage exploration through an optimal level of novelty and complexity. A recently advanced planning strategy for user participation in the design process has been based on the cognitive assets that citizens bring to environmental problem solving.

Environmental Attitudes

T he scope of environmental psychology extends beyond the research laboratory and the classroom to encompass attitudes toward the physical environment and major environmental issues. The nuclear "accident" at the Three Mile Island nuclear power plant in Pennsylvania and the ensuing public debate concerning the relative costs and benefits of nuclear power fall within the scope of environmental psychology. A recent report by the Energy Project at the Harvard Business School (Stobaugh and Yergin, 1979) emphasizes that the public and governmental controversy over Three Mile Island and the future of nuclear power reflects people's underlying attitudes as much as scientific facts.

The question of how our attitudes toward the environment are formed and changed touches many aspects of our lives. Environmental attitudes form the basis of our decisions about where we want to live and of our satisfaction or dissatisfaction with our present living environment. They shape our opinions about energy production, from our views about the Alaska oil pipeline to the prospect of solar energy. They underlie our feelings concerning the protection of the natural environment and the conservation of natural resources, from our sentiments about strip mining to concern for the California redwood. They affect many everyday decisions in our lives, such as our choice of driving, walking, or using public transportation to travel to school or work.

Because environmental attitudes are the basis of so many of our decisions, they constitute an exciting and important focus of environmental psychology. The potential benefits to society from the application of knowledge in this area are considerable. Research findings concerning environmental attitudes can be useful in the enactment of environmental law and in the preparation of environmental impact statements. The expansion of research on environmental attitudes has, in fact, coincided with the passage of legislation concerned with evaluating the potential human impact of proposed environmental alterations, such as the 1969 National Environmental Policy Act in the United States and the 1971 Town and Country Planning Act in England (Stokols, 1978). Research and knowledge may also be applied in programs to change public behavior in respect to conservation and the preservation of the natural environment, such as waste-recycling programs and efforts to eliminate littering in public places. Empirical findings concerning residential preference can be applied in decisions on the location of new suburban development, the selection of urban neighborhoods for redevelopment funds, and the determination of the ideal type of housing for residents in a particular locale.

THE NATURE OF ENVIRONMENTAL ATTITUDES

WHAT IS AN ATTITUDE?

Before we can discuss the nature of environmental attitudes, we must first agree on a definition of attitudes in general. Although the study of attitudes has occupied a central place in social psychology for many years, social psychologists are not in complete agreement on the best way to define them (see McGuire, 1969). While social psychologists agree that attitudes involve people's *feelings* toward some object or issue, some social psychologists have proposed that the definition of attitudes should also include people's *opinions* about the object or issue and their *behavior* toward it. Leonard Berkowitz (1975) has suggested that the best definition of attitudes is the simplest one, which focuses on the way people *feel* about something. Feelings and opinions do not always coincide. Someone might hold a particular opinion, but not care very deeply about it. Of course, as we noted in Chapter 2, the processes by which people cope with the environment do not operate independently, but constantly influence one another in complex ways. Thus our knowledge may influence our attitudes, and vice versa. Our knowledge about the past voting record of a congressperson, for instance, may affect our attitudes toward that person. And while attitudes do influence people's behavior, the relationship between attitudes and behavior is not a simple one.

Here we shall adopt the definition of attitudes suggested by Berkowitz: *"attitude" refers to the favorableness or unfavorableness of feelings toward some object or issue.* Thus attitudes involve *evaluative* feelings; they tell us how much we like or dislike something. Berkowitz explains, for instance, that when social psychologists measure attitudes, they ask people how good or bad they judge an object or issue to be, or how pleased or displeased they are with it. When a public opinion poll asks citizens how favorably or unfavorably they regard a particular political leader, it is measuring people's attitudes. Similarly, when a marketing firm asks customers how well they like or dislike a particular consumer product, it, too, is assessing attitudes.

ENVIRONMENTAL ATTITUDES

Environmental attitudes are people's favorable or unfavorable feelings toward some feature of the physical environment or toward an issue that pertains to the physical environment. You might ask yourself just how much you like or dislike the physical design of campus buildings and the physical layout of the campus at the college you attend. Or you might evaluate your feelings about the features of the natural environment of the state where you live. Or you might assess the favorableness or unfavorableness of your feelings toward the issue of conserving our natural resources. In each case, you are examining your environmental attitudes.

Environmental psychologists have investigated environmental attitudes in several problem areas. Researchers have studied people's level of satisfaction with their residential environment and their preferences for future or ideal environments, such as preferences for particular features of the natural environment. Investigators have also been increasingly interested in assessing people's attitudes toward environmental conservation as societal awareness of the need to preserve our natural resources has grown.

Residential satisfaction   One aspect of environmental attitudes that has received considerable attention from environmental psychologists involves the feelings of satisfaction or dissatisfaction that people have toward their residential environment. Holahan and Wilcox and their colleagues (Holahan and Wilcox, 1978, 1979; Holahan, Wilcox, Burnham, and Culler, 1978) investigated residential satisfaction among students who lived in a high-rise university dormitory. Only the attitudes of freshmen were assessed, because freshmen were randomly assigned to the various dormitory settings, while more advanced students tended to choose their own living environments. The investigators found that the residents of the high-rise dormitory were much more dissatisfied with their living environment than were students who lived in low-rise dormitories on the same campus, and students who lived on floors 9 through 14 were more dissatisfied than those who lived on lower floors. Respondents on the high floors expressed more dissatisfaction with the level of social participation among students in the dormitory and with the ease of meeting people and making friends. The investigators point out that friendship patterns in university residential settings tend to develop around spaces that are used frequently. The higher floors were less accessible to students than lower floors, and so proved less conducive to the formation of social bonds among residents. We shall learn more about the ways physical design can influence the formation of friendships in Chapter 10.

Of course, the factors that determine residential satisfaction are complex, and we should not conclude that all forms of high-rise living are unsatisfying to all of their occupants. In fact, one study of nonstudent residents of high-rise and low-rise housing (Francescato, Weidemann, Anderson, and Chenoweth, 1975) found that overall satisfaction in the two types of environment was comparable. Particular areas of satisfaction and dissatisfaction, however, differed between the two settings. While high-rise residents were more satisfied with recreational facilities than low-rise occupants, for example, they were less satisfied with a lack of privacy from their neighbors than were low-rise occupants.

Scenic preference   Environmental psychologists have also investigated people's preferences for features of the natural environment (Figure 4-1). Rachel and Stephen Kaplan and their associates at the University of Michigan (Herzog, Kaplan, and Kaplan, 1976; Kaplan, 1977a, 1977b; Kaplan, Kaplan, and Wendt, 1972) found that people prefer scenes of nature to scenes of urban features, and that within each of these areas people prefer more complex scenes to less complex ones. They also found that people prefer familiar features in the environment to unfamiliar ones.

   Other investigators have tended to agree that people prefer "naturalness" in

Figure 4-1
Environmental psychologists have studied people's preferences for natural scenic beauty.

© L. Lorusso/The Picture Cube.

the environment (Daniel, Wheeler, Boster, and Best, 1973) and natural scenic beauty (Calvin, Dearinger, and Curtin, 1972). People's preferences for outdoor recreational environments were negatively associated with both urban development and the presence of large numbers of people (Carls, 1973). Similarly, Gary Evans and Kenneth Wood (1980) show that individuals' evaluations of the roadside environment are negatively affected by signs of human intrusion and roadside development (see box, "Preferences for Scenery in the Roadside Environment"). The particular physical features in natural landscapes that people prefer include patterns of forest and grassland, water, and natural vegetation (Brush and Shafer, 1975; Zube, Pitt, and Anderson, 1975).

Ervin Zube and his associates (Zube, 1974, 1976; Zube, Brush, and Fabos, 1975; Zube, Pitt, and Anderson, 1975) have been particularly interested in learning how individual differences, such as personal background or professional training, affect people's evaluations of scenic quality. They have found that people's ratings of the scenic quality of natural landscapes generally show a high degree of consensus. One exception to this finding occurs in ratings of scenery by inner-city res-

Preferences for Scenery in the Roadside Environment

Gary Evans and Kenneth Wood (1980) asked university students to assess the scenic quality of the roadside environment along a country road that wound through rolling hills and canyons in Southern California. In order to evaluate students' preferences for potential developments along the roadside as well as the present roadside setting, the investigators devised a photographic simulation technique. Three groups of students were asked to view one of three slide presentations. One series of 100 slides showed the present roadside environment. A second series of 100 slides included some that showed "sympathetic" developments that had occurred along similar roads, such as curbing, lighting, moderate agricultural development, and some rustic roadside fencing. The third series of 100 slides included some that depicted "unsympathetic" developments that had occurred along similar roads, such as housing, planted greenery, and metal fencing.

Evans and Wood then asked each group of students to rate their overall visual impressions of the roadside environment they had viewed by means of a series of bipolar adjectives (e.g., pleasant–unpleasant) on a five-point scale. The students preferred the present roadside environment over any form of development, whether sympathetic or unsympathetic. People felt that as intrusions by humans in the roadside setting increased, the environment became proportionately more "worthless," "useless," "cluttered," "unpleasant," "ugly," and "drab."

The researchers also asked the students in each group to rate the overall scenic quality of the roadside environment on an item labeled "high–low scenic quality." They found that both sympathetic and unsympathetic development were of equal importance in reducing the students' evaluations of the roadside's scenic quality. Students' preferences for the different roadside environ-

idents, apparently because of their more limited personal experience with natural landscapes. These investigators also observed more variance in evaluations of scenic quality when people were asked to rate scenes that included some human influences or urban development, such as houses, stores, and automobiles. Nonprofessionals tended to evaluate influences from people or towns more favorably than did professionals in the environmental design fields. This was especially true for features of the built environment with which nonprofessionals were familiar or which added to their convenience and comfort.

Attitudes toward conservation   Researchers have attempted to identify personal characteristics that are associated with favorable attitudes toward environmental conservation. John Pierce (1979) reports that such personal values as "a comfortable life" and "responsibility" were associated with favorable attitudes toward preservation of water resources among owners of waterfront property in Washington State. Wohlwill (1979) found that support for preservation of the natural beauty of the California coastline was positively related to social values reflecting

ments as revealed in the "like–dislike" item also showed that sympathetic and unsympathetic roadside developments were essentially equally disliked. The authors point out that the external validity of this study is limited by the fact that only one country road was studied; a replication of these findings with additional scenic roads would enhance the study's external validity.

Mean ratings of pairs of bipolar adjectives describing a natural roadside environment and two types of development.

Adjective pair	Unmodified scenic corridor	Sympathetic development	Unsympathetic, urbanized development
Valuable-worthless	3.8	3.4	3.2
Useful-useless	3.5	3.0	3.1
Uncluttered–cluttered	4.2	3.8	3.6
Pleasant-unpleasant	4.5	4.1	3.8
Beautiful-ugly	4.3	4.1	3.8
Vivid–drab	3.7	3.2	3.2
High scenic quality– low scenic quality	4.2	3.5	3.4
Like–dislike	4.3	3.7	3.6

Source: Adapted from G. W. Evans and K. W. Wood, "Assessment of Environmental Aesthetics in Scenic Highway Corridors," Environment and Behavior, 12:255-73, © 1980 Sage Publications, Beverly Hills, with permission of the Publisher.

liberal views oriented toward social change. Other studies have also reported that favorable attitudes toward environmental conservation are associated with a liberal social philosophy (Buttell and Flinn, 1978b; Dunlap, 1975; Koenig, 1975; Tognacci, Weigel, Wideen, and Vernon, 1972).

While some early work also reported that positive attitudes toward conservation are positively related to such demographic variables as education and income (Buttel and Flinn, 1974; Devall, 1970; McEvoy, 1972; Morrison, Hornback, and Warner, 1972; Rosenbaum, 1973), later research (Buttel and Flinn, 1978a; Wohlwill, 1979) has suggested that proconservation attitudes are represented across a wide spectrum of society. Of course, individuals who have a vested interest in industries that would be affected by pollution-control policies often oppose the enactment of legislation to protect the environment. Phillip Althoff and William Greig (1974) found that feedlot operators and manufacturing executives in Kansas preferred less stringent pollution-control regulations than did elected officials or Environmental Protection Agency officials in the state.

PERCEIVED ENVIRONMENTAL QUALITY INDICES

In recent years environmental psychologists have devoted considerable attention to devising techniques to measure environmental attitudes. The major type of measurement techniques that have been developed in this area have been labeled *perceived environmental quality indices,* or PEQIs. A PEQI affords a quantitative measure of the *quality* of a particular physical setting, such as a housing development or a recreational area, as it is *subjectively* experienced by a particular group of people. Craik and Zube (1976a) explain that the subjective, observer-based information afforded by a PEQI can complement the objective data supplied by physical indices of environmental quality. For example, officials interested in assessing the environmental quality of a scenic lake might employ both a physical index of chemical pollutants and a PEQI reflecting how the lake's environmental quality is experienced by people who use it for boating, fishing, or swimming. While physical indices have typically stressed the negative aspect of environmental quality, such as pollution levels, PEQIs usually assess both pleasing and unpleasing aspects of environmental quality.

Although PEQIs provide a subjective appraisal of environmental quality, they may show a positive correlation with objective indices of environmental quality. Mary Barker (1976) explains that in public opinion surveys conducted in several cities, citizens' expressed concern and dissatisfaction with air quality increased with the level of objectively measured pollutants in the air. As air quality increasingly affected health and visibility in St. Louis, people expressed increasingly greater awareness of a pollution problem and perceived poor air quality as more of a nuisance. It should be pointed out, however, that public recognition of air pollution is generally specific to particulate matter in the air and other highly visible sources of pollution; gaseous air pollutants usually receive less recognition. In a similar vein, Robert Coughlin (1976) reports that ordinary citizens are able to recognize the objective condition of water quality. In a study of citizens' perceptions of pollution levels in small streams, subjective judgments of "clean," "healthy," and "polluted" were correlated with a large number of objectively measured chemical characteristics.

Types of PEQIs   Some researchers have hoped that it would be possible to develop a standard form of PEQI that could be used to assess people's perceptions of all types of environments. As Robert Bechtel (1976) has cautioned, however, because surrounding conditions as well as the activities in which people engage are likely to affect environmental judgments, it is probably more realistic to envision a range of PEQIS, each tailored to a particular type of environment and environmental condition. Several authors have identified a distinction between two major types of PEQIS, one based on *preferential judgments* and the other involving *comparative appraisals* (Brush, 1976; Craik and McKechnie, 1974; Craik and Zube, 1976a; Daniel, 1976; Marans, 1976).

PEQIS based on preferential judgments express an entirely personal evaluation of environmental quality based on a particular person's environmental standards. Craik and Zube (1976a:16) note that a PEQI based on preferential judgments might ask, "On a completely personal basis, how much do you like your neighborhood?" In contrast, PEQIS based on comparative appraisals assess the relative quality of a particular environmental setting in relation to some identified standard of comparison. A PEQI based on comparative appraisals might ask, "Setting aside your personal feelings now, how do you compare your neighborhood with other neighborhoods in the Bay Area?"

Craik and Zube point out that while preferential judgments more fully reflect observer characteristics, comparative appraisals yield a greater consensus among observers and better agreement between experts and nonexperts. Craik (1972a, 1972b) notes that in a landscape evaluation in England (Fines, 1968, 1969), the subjective variation in assessments by a group of surveyors was reduced by asking them to disregard personal preferences and by supplying illustrative examples of previous judgments on a comparison standard. Most important, this distinction between preferential judgments and comparative appraisals can help environmental psychologists to tailor PEQIS to particular research needs. When the research concern involves an effort to identify the preferences of a *unique subgroup* of environmental users, such as ethnic residents of an urban ghetto or middle-income professionals in a suburban subdivision, the preferential judgment approach might be used to tap personal views and perspectives. In contrast, when the researcher seeks to ascertain the *general* preferences of a *broadly defined* sample of environmental users, the comparative appraisal format might be employed to provide more highly predictive and valid preferences of a "standard" user group.

Developing PEQI measures   The initial step in developing a PEQI is to determine the best way to present the environmental stimulus that is to be judged. An environmental stimulus can often be presented more simply and less expensively in simulation than directly (Craik and Zube, 1976b). PEQI assessments have often been based on people's responses to photographs, color slides, drawings, scale models, or verbal descriptions of the environment (Daniel, 1976; Marans, 1976). As we saw in our discussion of simulation techniques in Chapter 2, the issue of external validity is an important concern when environmental simulations are used, and researchers who employ them must systematically ascertain whether their findings can be generalized to real-world settings (Danford and Willems, 1975; Lowenthal, 1972; McKechnie, 1977). Sometimes, of course, as in the case of judgments of water pollution, where very subtle visual changes cannot be easily identified through simulation, observers are asked to make on-site judgments in field settings (Coughlin, 1976). Many additional steps in developing a PEQI parallel

Table 4-1. The 66 bipolar adjectives that make up the Environmental Description Scale.

Adequate size–inadequate size	Huge–tiny
Appealing–unappealing	Impressive–unimpressive
Attractive–unattractive	Inviting–repelling
Beautiful–ugly	Large–small
Bright–dull	Light–dark
Bright colors–muted colors	Modern–old-fashioned
Cheerful–gloomy	Multiple purpose–single purpose
Clean–dirty	Neat–messy
Colorful–drab	New–old
Comfortable–uncomfortable	Orderly–chaotic
Comfortable temperature–uncomfortable temperature	Organized–disorganized
	Ornate–plain
Complex–simple	Pleasant–unpleasant
Contemporary–traditional	Pleasant odor–unpleasant odor
Convenient–inconvenient	Private–public
Diffuse lighting–direct lighting	Quiet–noisy
Distinctive–ordinary	Roomy–cramped
Drafty–stuffy	Soft lighting–harsh lighting
Efficient–inefficient	Sparkling–dingy
Elegant–unadorned	Stylish–unstylish
Empty–full	Tasteful–tasteless
Expensive–cheap	Tidy–untidy
Fashionable–unfashionable	Uncluttered–cluttered
Flashy colors–subdued colors	Uncrowded–crowded
Free space–restricted space	Unusual–usual
Fresh odor–stale odor	Useful–useless
Functional–nonfunctional	Warm–cool
Gay–dreary	Well balanced–poorly balanced
Good acoustics–poor acoustics	Well kept–run down
Good colors–bad colors	Well organized–poorly organized
Good lighting–poor lighting	Well planned–poorly planned
Good lines–bad lines	Well scaled–poorly scaled
Good temperature–bad temperature	Wide–narrow
Good ventilation–poor ventilation	

Source: Adapted from J. V. Kasmar, "The Development of a Usable Lexicon of Environmental Descriptors," Environment and Behavior, 2:153–69, © 1970 Sage Publications, Beverly Hills, with permission of the Publisher.

steps essential to the development of any psychological index of subjective judgments. For example, people's responses to several items must be combined in an overall numerical scale value or index, and the index's reliability (repeatability of findings) and validity (actual measurement of what is claimed) need to be established (Craik and Zube, 1976b; Daniel, 1976).

Some measures of perceived environmental quality have employed a *semantic differential* technique as a response mode (see, for example, Lowenthal and Riel, 1972). The semantic differential technique (developed by Osgood, Suci, and Tannenbaum, 1957) asks respondents to evaluate a concept (such as an environmental setting) on a scale consisting of polar-opposite adjectives. People might be asked, for instance, to evaluate a recreational setting as "good" or "bad," "comfortable" or "uncomfortable," "pleasant" or "unpleasant." Evans and Wood's (1980) study of preferences for roadside environments, for example, used semantic differential ratings to assess people's feelings about potential developments along the roadside. Joyce Kasmar (1970) developed a series of 66 bipolar adjectives, which she termed the Environmental Description Scale, that could be used by nonprofessionals to describe and evaluate architectural spaces. The items that make up the Environmental Description Scale are shown in Table 4-1. You may use the Environmental Description Scale to evaluate your current living environment by circling the adjective in each bipolar pair that best describes the setting where you live.

PSYCHOLOGICAL FUNCTIONS OF ENVIRONMENTAL ATTITUDES

Environmental attitudes serve an important psychological function in our lives in helping us to make decisions about a wide range of behaviors. For example, when students decide where to live while they attend college, they rely on their attitudes about different types of neighborhoods, about high-rise versus low-rise housing, and about college dormitory settings. Of course, other factors also play a part in an individual's choice, such as the amount of rent he or she can afford (Ankele and Sommer, 1973). Here we shall consider the role of environmental attitudes in helping us to make two types of decision: choosing a personal living environment and making decisions about the public environment, as when a person decides to join a campaign against littering.

RESIDENTIAL CHOICE

One important function of environmental attitudes is to help us to select the residential environment where we wish to live. People's satisfaction or dissatisfaction with their current residential environments, as well as their preferences in regard to an ideal living environment, play a central role in shaping both their decisions about moving and their choices of new residential settings. Of course, residential satisfaction and residential choice are complexly interrelated. While satisfaction affects choice, choice can also influence satisfaction, as when the choice of a new home initiates a new pattern of satisfaction and dissatisfaction in the new setting.

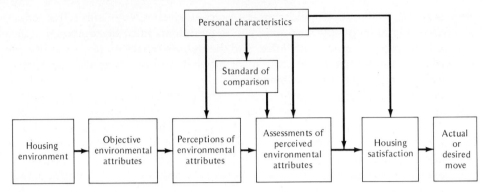

Figure 4-2
A conceptual model for understanding housing satisfaction and the desire to move.

Adapted from R. W. Marans, "Perceived Quality of Residential Environments: Some Methodological Issues," in K. H. Craik and E. H. Zube (eds.), Perceiving Environmental Quality: Research and Applications, *1976, p. 145. Reprinted by permission of Plenum Publishing Corp.*

Robert Marans and Willard Rodgers (Marans, 1976; Marans and Rodgers, 1975) propose a graphic model to explain how environmental attitudes (residential satisfaction or dissatisfaction) are translated into environmental behavior (residential mobility). A simplified adaptation of their model (Figure 4-2) shows how objective characteristics of the housing environment interact with the resident's personal characteristics (age, income, tastes, and so on) and standard of comparison to affect the resident's assessment of the housing environment. This personal assessment of the housing environment in comparison with one's standard of an ideal housing environment is the basis of an individual's feeling of satisfaction with his or her housing environment, and ultimately of the decision to move to a more acceptable housing environment. The individual's feelings of satisfaction with the immediate housing environment also interact with feelings of satisfaction with the surrounding neighborhood and with the broader community to affect the final decision to move (see box, "Residential Choice and Rapid Transit").

When Herbert Gans (1967) assessed residential satisfaction among residents of the new suburban housing development of Levittown, New Jersey, he found that despite a widespread assumption that suburban life is unsatisfying and boring, most Levittowners were quite pleased with the quality of their lives. Gans argues that it is more correct to speak of suburban happiness than suburban anomie. The suburban residents he spoke with reported feeling less depressed, bored, and lonely after their move to suburbia than they had felt before. Gans concludes that the small amount of dissatisfaction he found in Levittown related not to the suburban environment itself, but to the feeling of being left out of activities and relationships in the community.

William Michelson (1976a) conducted a five-year longitudinal study of the relationship between people's residential preferences and their residential choices in Toronto. He found that people who lived in single-family homes in the suburbs chose their homes because they liked the characteristics of the neighborhood, felt it was an ideal place to raise children, and believed that a single-family house was "status enhancing." Some suburbanites, especially women who were not employed outside the home, were somewhat dissatisfied with the remoteness of their suburban homes from recreational and shopping facilities. Overall, however, people who lived in single-family suburban homes were more satisfied with their housing environment than people who lived in multifamily, high-rise housing in urban settings.

Michelson found that most high-rise residents looked forward eventually to owning their own single-family home in a suburb or small town. The high-rise dwellers said they would prefer a single-family home in order to have better control over their premises, to increase their feelings of indoor and outdoor privacy, and to enhance their economic security. Michelson adds that when financial conditions permitted them to do so, many high-rise dwellers did in fact move to single-family homes.

Other research studies have found similarly that most people's residential preference is to own a detached, single-family home, typically in a suburban setting (Foote, Abu-Lughod, Foley, and Winnick, 1970; Onobokun, 1976). Low-income black respondents (Rent and Rent, 1978) and college students (Hinshaw and Allott, 1972) have been found to share this preference. People's preferences for suburban housing are also associated with a desire for safety and reduced traffic (Appleyard and Lintell, 1972; Kasl and Harburg, 1972) and the "right kind" of neighbors (Lamanna, 1964). Studies that have directly compared urban and suburban dwellers' satisfaction with their homes and communities have tended to report that residential satisfaction is higher among the suburbanites (Marans and Rodgers, 1975).

A word of caution is in order here. It is probably unwise to assume that a single-family home in the suburbs is the ideal living arrangement for everyone. Claude Fischer (1976) points out that suburban residents' expressions of satisfaction with their residential setting probably reflect, in part, *individual characteristics* of suburbanites that influenced their original decision to move to suburbia. Suburbanites characteristically are of higher financial status and are more involved in raising children than city dwellers, and they are more likely to be white. Certain groups, such as adolescents, the elderly, and members of ethnic minorities, may find the suburban environment boring and lonely. And one more word of caution: suburbanites may report that they are satisfied with their current living environment because a feeling of satisfaction is *consistent* with the great effort many of them have expended to acquire their homes and the large monthly mortgage payments they must make to keep them. Later we shall see how *attitude consistency theories* explain how attitudes are acquired and changed.

PROTECTING THE ENVIRONMENT

Another important function of environmental attitudes is to help us to make decisions involving the use and care of the physical environment. A decision to walk a few extra steps to deposit a piece of paper in a trash receptacle reflects underlying positive feelings about a clean outdoor environment. A decision to support the

———Residential Choice and Rapid Transit ———

Can the presence of a rapid transit system affect residential choice? Will people move near a mass transit system in order to take advantage of its travel convenience, or will they relocate away from such a system to avoid the associated noise and commuter crowds? Mark Baldassare and his colleagues (Baldassare, Knight, and Swan, 1979) pursued answers to these questions in evaluating the effects of the new Bay Area Rapid Transit (BART) system on residential choice in the San Francisco metropolitan area. The investigators suggested that if BART affected residential choice, the effects might be expected to be different in respect to residences near BART *stations* than in respect to those situated near *aerial lines* between stations. They reasoned that a residence near a BART station provided easy access to the system, and that this convenience might outweigh the costs of related noise and congestion. Homes near aerial lines between stations, in contrast, would be exposed to the costs of the system (train noise, television disruption, some reduction in privacy) without the easy access to the system available to people who lived near stations.

To ascertain how BART affected residential choice, Baldassare and his associates surveyed a sample of over 700 San Francisco Bay Area residents who lived near the BART system concerning their housing decisions. People's decisions to move away (item A in the table) were quite similiar at the station and aerial sites, and were only slightly affected by the presence of BART; they were more strongly related to such factors as overall neighborhood quality and the respondents' stage in the life cycle.

When BART was stipulated as a possible reason for moving among those respondents who were considering moving (item B), however, twice as many residents at the aerial locations as people at the station sites responded that BART

Sierra Club, to join an environmentalist organization, or simply to sign a petition calling for protection and conservation of natural resources is influenced by an appreciation of natural scenic beauty and a concern about careless disregard for our natural resources. The public media campaigns we have witnessed in recent years concerned with environmental education have been oriented toward changing people's environmental attitudes and eventually their behavior in regard to the natural environment.

Some investigators (Maloney and Ward, 1973; Maloney, Ward, and Braucht, 1975) have developed measures of environmental attitudes related to ecological concerns, such as air pollution from automotive exhaust, the use of recyclable food containers, and the level of pesticides in consumer food products. These investigators believe that the ecological crisis that threatens our world (see Commoner, 1971; Ward and Dubos, 1972) is caused by people's maladaptive behavior in respect to the physical environment. They contend that only when people's attitudes toward land use, waste disposal, and consumption are improved will more positive and adaptive behaviors emerge.

Russell Weigel and his associates (Weigel and Newman, 1976; Weigel and

was one of the reasons they might move. BART's effects on people's decisions to move *to* the area are even more apparent. Many more residents at the station locations responded that they chose their own homes because of BART (item C) or that they knew other people who had chosen their homes for BART (item D) than did residents near the aerial lines. The investigators conclude that the differential costs and benefits of a rapid transit system do influence people's reported residential choice patterns near that system, but that the transit system has a greater influence on decisions to move *to* the area than on decisions to move away from it.

Percentage of residents at BART station and aerial locations who answered yes to questions concerning BART's effect on housing decisions.

Question	Locations near stations	Locations near aerial lines
A. Considered moving away in last few years?	31%	33%
B. Was BART at least part of the reason? (asked of those considering moving)	21	42
C. Chose home because of BART?	18	5
D. Know others who chose home for BART?	17	3

Source: Adapted from M. Baldassare, R. Knight, and S. Swan, "Urban Service and Environmental Stressors: The Impact of the Bay Area Rapid Transit System on Residential Mobility," Environment and Behavior, 11:435–50, © 1979 Sage Publications, Beverly Hills, with permission of the publisher.

Weigel, 1978) conducted a series of studies to learn how people's environmental attitudes affect a variety of ecologically oriented behaviors. First they developed an attitude measure to assess people's attitudes toward the environment. The attitude survey, called the Environmental Concern Scale, measures respondents' attitudes toward a variety of ecological issues, such as pollution, conservation of natural resources, and wild-life preservation (see box, "Assessing Your Own Environmental Attitudes").

Weigel and his coworkers then developed a behavioral index to assess the degree to which the environmental attitudes measured by the Environmental Concern Scale actually predict people's behavior. A research confederate (a trained assistant of the investigators) took three environmental petitions to 44 residents of a New England town who had completed the Environmental Concern Scale three months earlier. The petitions expressed opposition to oil drilling off the New England coast, opposition to the construction of nuclear power plants, and support for stringent laws against the removal of pollution-control devices from their automobile exhaust systems. Six weeks after the subjects had been presented with the petitions, a second confederate solicited their participation in a

Assessing Your Own
Environmental Attitudes

   The Environmental Concern Scale, developed by Russell Weigel and his associates, measures people's attitudes toward a variety of ecological issues.* The scale permits people to express favorable or unfavorable attitudes toward ecological concerns, such as environmental pollution and conservation of natural resources. Not surprisingly, active members of the Sierra Club were found to have more favorable attitudes toward ecological concerns than a randomly selected sample of residents of a medium-sized New England town. If you would like to conduct a self-assessment of your own environmental attitudes, answer the sixteen questions that make up the Environmental Concern Scale. After reading each statement, simply check one of the boxes to indicate whether you "generally agree" or "generally disagree" with that statement. Then count the number of checks you have made in column 1 and column 2. Checks in column 1 indicate favorable attitudes toward ecological concerns, while checks in column 2 reflect unfavorable attitudes. Most people's scores show that their attitudes are neither 100 percent favorable nor 100 percent unfavorable. What does your own score indicate? Are your attitudes more favorable or more unfavorable toward ecological concerns?

Environmental Concern Scale	Column 1	Column 2
1. The federal government will have to introduce harsh measures to halt pollution, since few people will regulate themselves.	Generally agree ☐	Generally disagree ☐
2. We should not worry about killing too many game animals because in the long run things will balance out.	Generally disagree ☐	Generally agree ☐
3. I'd be willing to make personal sacrifices for the sake of slowing down pollution even though the immediate results may not seem significant.	Generally agree ☐	Generally disagree ☐
4. Pollution is *not* personally affecting my life.	Generally disagree ☐	Generally agree ☐
5. The benefits of modern consumer products are more important than the pollution that results from their production and use.	Generally disagree ☐	Generally agree ☐
6. We must prevent any type of animal from becoming extinct, even if it means sacrificing some things for ourselves.	Generally agree ☐	Generally disagree ☐

* The reliability and validity data for the scale were based on responses on a five-point continuum; responses with the abbreviated response format used here should be interpreted more cautiously.

7. Courses focusing on the conservation of natural resources should be taught in the public schools.

Generally agree ☐　　Generally disagree ☐

8. Although there is continual contamination of our lakes, streams, and air, nature's purifying processes soon return them to normal.

Generally disagree ☐　　Generally agree ☐

9. Because the government has such good inspection and control agencies, it's very unlikely that pollution due to energy production will become excessive.

Generally disagree ☐　　Generally agree ☐

10. The government should provide each citizen with a list of agencies and organizations to which citizens could report grievances concerning pollution.

Generally agree ☐　　Generally disagree ☐

11. Predators such as hawks, crows, skunks, and coyotes which prey on farmers' grain crops and poultry should be eliminated.

Generally disagree ☐　　Generally agree ☐

12. The currently active antipollution organizations are really more interested in disrupting society than they are in fighting pollution.

Generally disagree ☐　　Generally agree ☐

13. Even if public transportation were more efficient than it is, I would prefer to drive my car to work.

Generally disagree ☐　　Generally agree ☐

14. Industry is trying its best to develop effective pollution technology.

Generally disagree ☐　　Generally agree ☐

15. If asked, I would contribute time, money, or both to an organization like the Sierra Club that works to improve the quality of the environment.

Generally agree ☐　　Generally disagree ☐

16. I would be willing to accept an increase in my expenses of $100 next year to promote the wise use of natural resources.

Generally agree ☐　　Generally disagree ☐

Column 1 total ―――　　Column 2 total ―――

Source: Adapted from R. Weigel and J. Weigel, "Environmental Concern: The Development of a Measure," Environment and Behavior, 10:3-7, © 1978 Sage Publications, Beverly Hills, with permission of the Publisher.

roadside litter-pickup program that was being conducted in nearby areas. Subjects were also asked if they would recruit a friend for the program. Finally, eight weeks after the second contact, a third confederate approached the subjects to ask if they would take part in a recycling program. People who assented agreed to bundle their newspapers, collect their bottles in containers, and leave these recyclable materials outside, where they would be picked up once a week for eight weeks.

To measure how well the Environmental Concern Scale actually predicted people's behavior toward the environment, Weigel and his associates compared respondents' scores on the attitude scale with their actual behaviors in response to the requests from the three experimental confederates (Table 4-2). The comprehensive behavior index, which included the variety of environmental behaviors subjects were asked to engage in, was highly correlated with environmental attitude scores. Notice, however, that the correlations between the attitude survey and each of the *single* environmental behaviors requested of respondents were more modest than the correlation with the comprehensive behavior index. As we shall see when we discuss theoretical issues related to environmental attitudes, this dis-

Table 4-2. Correlations between subjects' scores on the Environmental Concern Scale and their actual environmental behaviors.

Single behaviors	r^a	Categories of behavior	r^b	Behavior index	r^b
Offshore oil	.41				
Nuclear power	.36	Petitioning behavior			
Auto exhaust	.39	scale	.50		
Circulate petitions	.27				
Individual participation	.34	Litter pickup scale	.36		
Recruit friend	.22				
				Comprehensive behavior index	.62
Week 1	.34				
Week 2	.57				
Week 3	.34				
Week 4	.33	Recycling behavior			
Week 5	.12	scale	.39		
Week 6	.20				
Week 7	.20				
Week 8	.34				

[a] Point-biserial correlations.
[b] Pearson product-moment correlations.

Source: R. H. Weigel and L. S. Newman, "Increasing Attitude-Behavior Correspondence by Broadening the Scope of the Behavioral Measure," Journal of Personality and Social Psychology, 1976, 33:793–802. Copyright 1976 by the American Psychological Association. Reprinted by permission of the publisher and author.

tinction between the prediction of a class of behaviors and the prediction of single behaviors is an important one.

THEORETICAL PERSPECTIVES ON ENVIRONMENTAL ATTITUDES

HOW ENVIRONMENTAL ATTITUDES ARE LEARNED

Considerable theoretical development has occurred in regard to the ways in which environmental attitudes are learned and change over time. You might speculate as to how some of your own environmental attitudes, such as those measured by the Environmental Concern Scale, were formed. Environmental psychologists have posed similar questions. In their attempts to answer them they have been able to rely on an important body of theoretical work concerned with attitude development and attitude change. This social-psychological work has in turn been based on more general models of learning that were developed by experimental psychologists.

Classical conditioning   The classical conditioning model of learning proposes that when a neutral stimulus—an object, an issue, even another person—is repeatedly paired with an experience that is pleasant or unpleasant, the previously neutral stimulus will come to arouse the pleasant or unpleasant feelings with which it has been associated. Let us see how the classical conditioning model would explain a tourist's negative attitude toward a city where he happened to become ill on a cross-country vacation (Figure 4-3). The stimulus that is initially without affect (the new city) is referred to as the *conditioned stimulus*, while the stimulus that is initially associated with either positive or negative affect (illness) is called the *unconditioned stimulus*. The *unconditioned response* is the automatic response (negative feelings associated with being ill) that is initially elicited by the unconditioned stimulus. When the conditioned stimulus is experienced along with or immediately after the unconditioned stimulus, the conditioned stimulus will itself come to elicit the same response (a negative attitude toward the new city)—now called a *conditioned response*—that was initially elicited only by the unconditioned stimulus.

Social psychologists have successfully applied the classical conditioning model of learning to theories of attitude formation and change (Byrnes and Clore, 1970; Zanna, Kiesler, and Pilconis, 1970). Continued research by social psychologists has also succeeded in identifying some of the parameters of the classical conditioning model that affect the strength of learned attitudes. Researchers have found, for example, that the stronger the affect associated with the unconditioned stimulus, the more intense the learned attitude (Lott, Bright, Weinstein, and Lott, 1970; Staats, Minke, Martin, and Higa, 1972). It has also been shown that the closer the temporal association between the conditioned stimulus and the unconditioned stimulus, the stronger the learned attitude (Lott and Lott, 1968). Finally, investigators have found that the more often the conditioned stimulus is paired with the unconditioned stimulus, the more intense the learned attitude (Staats, Staats, and Heard, 1960). On the basis of this research, we would expect the tour-

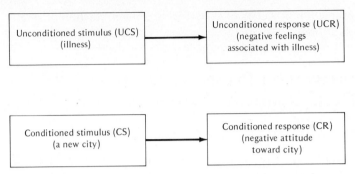

Figure 4-3
The classical conditioning model of attitude formation.

ist's attitude toward the city where he became ill to be most negative when the illness was especially uncomfortable, when he became ill immediately after arriving in the city, and when the illness recurred on revisiting the city.

Instrumental conditioning   The instrumental conditioning model of learning proposes that responses that are rewarded will be strengthened and maintained, while those that are punished will be weakened and eliminated. According to the instrumental conditioning model of learning, which is sometimes called *operant* conditioning, the individual acts on or "operates" on the environment. Those acts that are followed by favorable consequences are likely to be repeated, while those that are followed by unfavorable consequences are unlikely to be repeated. For example, a commuter who forms a car pool and is rewarded with a special car-pool lane on the freeway and free crossing at a toll bridge or tunnel is likely to develop a positive attitude toward car-pooling and to maintain car-pooling behavior. An event that tends to increase the probability that earlier behavior will occur again is called a *positive reinforcer*; an event that tends to decrease the probability that the behavior will occur again is termed a *negative reinforcer*. The special car-pool lane and the tollfree passage are positive reinforcers.

Social psychologists have successfully applied the instrumental conditioning model of learning to theories of attitude formation and change (Kiesler, Collins, and Miller, 1969; Insko and Cialdini, 1969). They have also succeeded in identifying some of the specific parameters of the learning situation that are of importance in the instrumental learning of attitudes. For example, Carl Hovland and his colleagues (Hovland, Janis, and Kelley, 1953) propose that for attitude change to occur, three social processes must intervene between the stimulus and response conditions. Before a stimulus (an appeal to commuters to car-pool) can lead to attitude change, it must be noticed or *attended* to (a commuter listens to a radio message encouraging car-pooling). Next, the individual must *comprehend* or understand the message (the commuter recognizes that some rewards—a special freeway lane and tollfree crossing—are available to car-poolers). Finally, the individual must *accept* the message (the commuter decides that car-pooling would be beneficial to himself or herself in traveling to work more quickly and inexpensively). The acceptance of the message is dependent on *incentives;* that is, the individual must perceive that the new attitude will be more personally rewarding than the old one. Other social psychologists (Bandura, 1974; Berger, 1962) have shown that behavior change, such as the development of new attitudes, can come about through *vicarious* reinforcement; we often adopt new attitudes because we observe how other people who model these attitudes are reinforced. For example, a commuter may decide to join a car pool after observing other car-poolers enjoying a special freeway lane and tollfree passage.

How Attitudes Affect Behavior

Implicit in much of our discussion of environmental attitudes has been an assumption that people's *attitudes* toward the environment will affect their *behavior* in regard to the environment. We have assumed that residents' satisfaction with their living environment influences their decision to move or to stay where they are. Similarly, we have assumed that individuals' attitudes toward protecting the natural environment will affect the way they treat it. Here we shall examine an important and rather complex body of research that has investigated the link between people's attitudes and their behavior.

Attitude consistency theories   Attitude consistency theories provide a general framework for examining the relationship between attitudes and behavior (see Kiesler, Collins, and Miller, 1969). In general, attitude consistency theories examine the degree of consistency among the various attitudes a person holds and between those attitudes and the individual's behaviors. Consistency theories hold that inconsistency among attitudes or between attitudes and behavior produces discomfort and tension, and that people will attempt to eliminate such discomfort by changing either their attitudes or their behavior.

An extensive body of social-psychological literature has dealt with Leon Festinger's (1957) theory of *cognitive dissonance.* Festinger explains that when a person possesses two cognitions (broadly defined as attitudes, beliefs, or knowledge about one's behavior) one of which is opposite to or inconsistent with the other, a state of dissonance (involving psychological tension and discomfort) exists, and the individual is motivated to reduce or eliminate it. Dissonance would exist, for example, if an individual held a positive attitude about a clean environment and simultaneously knew that he or she habitually littered the environment. According to cognitive dissonance theory, the individual would be motivated to reduce the dissonance by altering either the environmental attitude (e.g., deciding littering is not really a serious environmental problem) or the environmental behavior (e.g., deciding to deposit litter in trash receptacles).

One aspect of the research involving attitude consistency theories has, in fact, focused on the consistency between people's attitudes and their behavior. Since most efforts by environmental psychologists to change people's attitudes (e.g., attitudes toward energy consumption) are designed ultimately to change their behavior toward the environment, the question of attitude-behavior consistency is especially important.

Attitude-behavior consistency   For many years social psychologists assumed that the effects of people's attitudes on their behavior were quite direct (see Krech, Crutchfield, and Ballachey, 1962; Rosenberg and Hovland, 1960). They assumed that people who expressed strong positive religious attitudes, for example, would attend religious services regularly. Yet when Allan Wicker (1969a) reviewed more than thirty studies concerned with the relationship between attitudes and behavior, he discovered that the effects of attitudes on specific behaviors were not very strong. More recent research by social psychologists has demonstrated that attitudes do affect behavior (Kahle and Berman, 1979), but that those effects will be evident only when both attitudes and behavior are measured at equivalent levels of generality or specificity (Ajzen and Fishbein, 1973, 1977; Davidson and Jaccard, 1979).

Martin Fishbein (1973) explains that two people who hold equally favorable attitudes toward an issue may engage in different behaviors in regard to it. One person very concerned about the environment, for instance, may participate regularly in the activities of the Sierra Club, while another never attends Sierra Club functions but carefully avoids personal actions that might damage the natural environment. Thus, while the two behaviors taken together reveal a favorable environmental attitude, the attitude does not predict very well the *specific* behavior that a person will engage in. Although a *general* attitude does not strongly predict any *specific* behavior, it does predict a *general* class of behaviors related to the attitude. Fishbein and Ajzen (1974) show that a measure of general attitudes toward religion can predict a general class of 70 religious behaviors, such as donating money to a religious cause and praying before and after meals.

Richard Bagozzi and Robert Burnkrant (1979), in a reanalysis of Fishbein and Ajzen's data, agree that attitudes do predict a class of multiple behaviors, but add that the attitudes and behaviors must be scaled at the same level of specificity. In both the Fishbein and Ajzen and the Bagozzi and Burnkrant studies, while the attitude did predict a *class* of behaviors, it did not predict well any *single* behavior. Recently Mark Zanna and his associates (Zanna, Olson, and Fazio, 1980) have shown that the attitude-behavior relationship is also influenced by individual differences in the persons studied. They found, for example, that the relationship between religious attitudes and corresponding religious behaviors was strongest in individuals who both reported little variability in their past religious behaviors and tended to infer their attitudes from those past behaviors.

Clearly, research findings by social psychologists concerning the correspondence between attitudes and behavior are complex. In general, when we apply these findings to the domain of environmental attitudes and behavior, we may expect that environmental attitudes will predict environmental behaviors when both attitudes and behaviors are scaled at the same level of specificity. For example, we may predict that a general attitude toward the environment (e.g., a positive attitude toward conservation) will predict a general class of environmental behavior (e.g., a variety of behaviors that involve protection of the natural environment). This theoretical prediction is consistent with the finding we discussed earlier, that the Environmental Concern Scale predicts a comprehensive index of environmental behaviors much more strongly than it predicts any single environmental behavior.

PREDICTING AESTHETIC PREFERENCE

Theorists have asked: "What are the particular features of an environmental set-ting that cause it to be viewed as beautiful or attractive?" In Chapter 2 we saw that environmental psychologists have identified the dimensions of environmental stimulation that are of importance in investigations of the stimulus properties of the environment. Recall the work of Donald Berlyne (1960) and Joachim Wohlwill (1966) in identifying the stimulus properties of novelty, complexity, surprising-ness, and incongruity. This area of work has been extended to serve as the basis of a theory of aesthetic preference. Because this theoretical work on aesthetic prefer-ence is related to the *complexity* of environmental stimuli, it can also help us to un-derstand the findings (discussed in Chapters 2 and 3) that good environmental de-sign involves an optimal level of environmental complexity.

Berlyne has conducted a classic series of psychological studies (Berlyne, 1960, 1967, 1972, 1974) dealing with a theory of aesthetic preference. He proposes that in general people's aesthetic preferences are related to the complexity of a stimulus, and that stimuli at an intermediate level of complexity will be judged most pleas-ant and attractive. As the complexity of a stimulus increases, people view it as in-creasingly pleasant—up to a point. Beyond a certain optimal level of complexity, however, the stimulus is seen as less pleasant. The relationship between stimulus complexity and attraction may thus be seen as an inverted U. A wide body of ex-perimental evidence that has investigated optimum levels of stimulation has con-curred in describing the effects of varied levels of stimulation as an inverted U-shaped curve (Day, 1967; Fiske and Maddi, 1961; Hebb, 1955; Leuba, 1955; Walker, 1964).

These general theories of aesthetic preference have been applied in the devel-opment of a theory of environmental aesthetics. Wohlwill (1966, 1968a, 1968b, 1970, 1976) has concluded after a series of studies that evaluative responses to the built environment are most positive at intermediate levels of complexity. He notes, however, that while the inverted U-shaped curve describes evaluative or preferen-tial judgments, it does not describe the effects of stimulus complexity on the ex-ploratory behavior people show toward the stimulus. In fact, exploratory activity increases monotonically as the complexity of the stimulus is increased. This dis-tinction between the effects of stimulus complexity on evaluative judgments and on exploratory activity is consistent with earlier findings reported by experimental psychologists.

Further research that bears on the development of a theory of environmental aesthetics has been conducted by Albert Mehrabian and James Russell (Mehra-bian and Russell, 1974; Russell and Mehrabian, 1976, 1978). They measured peo-ple's approach toward a range of built and natural environments, using ratings on a series of color slides. They found that the inverted U-shaped relationship be-tween approach and the *arousal level of an environment* (the average level of arousal, ranging from sleep to frenzied excitement, reported by a representative sample of subjects) was evident only when the degree of pleasantness or unpleasantness of the environment was in a neutral range. An even stronger influence on approach was a direct relationship between approach and the degree of pleasantness of a

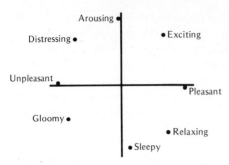

Figure 4-4

Circular ordering of eight terms to describe the emotional quality of environments.

From J. A. Russell and G. Pratt, "A Description of the Affective Quality Attributed to Environments," Journal of Personality and Social Psychology, 1980, 38:311–22. Copyright 1980 by the American Psychological Association. Reprinted by permission of the publisher and author.

setting; people expressed a desire to approach pleasant settings and to avoid unpleasant ones.

James Russell and Geraldine Pratt (1980), continuing the line of research begun by Mehrabian and Russell, have proposed a theoretical model that defines and organizes the terms people use to describe the emotional quality of environments. For example, if we describe a rural setting as "relaxing" and an urban context as "exciting," the model can help us to understand the definition of each of these qualities of the environment and also how these two qualities are related.

Russell and Pratt's model organizes eight basic terms to describe the emotional quality of environments in a two-dimensional space (Figure 4-4). The space is bipolar—each affective descriptor is located 180 degrees from its opposite descriptor. For example, "pleasant" and "unpleasant" are bipolar opposites, as are "arousing" and "sleepy." Thus, when we say a rural environment is "relaxing," the model helps us to understand that "relaxing" is the opposite of "distressing" and involves in equal combination aspects of both "pleasant" and "sleepy." Notice also that the model helps us to understand that "relaxing" and "exciting" are not opposite descriptors, since both are "pleasant" ways to describe the environment.

APPLICATIONS TO ENVIRONMENTAL PLANNING

Knowledge about people's environmental attitudes can play an important part in environmental planning and design. Craik and Zube (1976a, 1976b) discuss some of the ways indices of perceived environmental quality might be applied in the planning process. The National Environmental Policy Act of 1969 directed the executive branch of government to monitor changes in environmental quality over time and to prepare a statement on the impacts of actions that would significantly alter the quality of the human environment. Craik and Zube emphasize that measures of people's perceptions of environmental quality can play a key role in this effort.

Craik and Zube detail some of the useful purposes that assessments of perceived environmental quality might serve. First, information about how citizens perceive environmental quality can help to formulate and clarify environmental policy goals. Feedback about people's perceptions of changes in environmental quality as a result of environmental protection programs can help in the monitoring and refining of ongoing programs and in the formulation of new ones. Planners might use the perceptions of prospective users of proposed environmental projects concerning the projects' likely human impacts in conjunction with objective indices of projected environmental impacts in assessing potential project costs and benefits and in evaluating the need for new environmental projects.

Craik and Zube provide a table (Table 4-3) that summarizes some of the specific applications of PEQIS across a range of environmental domains. In general, information about people's environmental attitudes is applicable in decisions concerning the *built* environment, such as residences, institutions, and transportation facilities, and the *natural* environment, such as scenic areas, lakes, and rivers. Craik and Zube note that before the Bay Area Rapid Transit system was put into operation, an extensive study (Appleyard and Carp, 1974) was undertaken to assess some of the system's possible impacts on people's perceptions of the residential environment along the transit route. And a study of people's perceptions of landscape quality at 56 sites in the Connecticut River Valley (Zube, Pitt, and Anderson, 1975) served to identify some of the physical characteristics of the region that were associated with perceived landscape quality.

Table 4-3. Potential applications of perceived environmental quality indices (PEQIS) across eight environmental domains.

Applications	Built environments	Natural environments (scenery)	Recreational environments	Residential environments	Institutional environments	Air	Water	Sound
Monitoring:								
public information program	X	X				X	X	X
gauging social equity	X	X	X	X	X	X	X	X
Environmental impact assessments:								
alternative plan analysis			X	X	X	X	X	X
Post-hoc evaluations:								
construction project			X	X	X			
pollution-control programs						X	X	X
other environmental planning and management programs	X	X	X					

Source: K. H. Craik and E. H. Zube, "Summary and Research Strategies," in K. H. Craik and E. H. Zube (eds.) Perceiving Environmental Quality: Research and Applications, 1976, p. 285. Reprinted by permission of Plenum Publishing Corp.

User Preference in Residential Design

Environmental attitudes reflecting user preferences and satisfactions can be applied in design decisions involving residential environments. For example, knowledge about the design elements that area residents see as satisfying or dissatisfying and the features they prefer can guide designers in their selection of housing types and residential design features. Systematically collected information on housing preferences might also serve as a valuable community social indicator of relevance to a wide range of physical and social planning issues (Hempel and Tucker, 1979).

Measurement techniques are now available that might be applied by architects and planners concerned with assessing user preference and satisfaction. Frances Carp and her associates (Carp, Zawadski, and Shokron, 1976) developed an index of residential satisfaction based on a sample of more than 2,500 San Francisco Bay Area residents which they felt planners could use in making residential design decisions. A scale is also available (Canter, Sanchez-Robles, and Watts, 1974) that is appropriate in the study of housing satisfaction among residential groups of different cultural backgrounds.

Although our emphasis here is on the *process* of applying user preferences and satisfactions in residential design decisions, it may be useful to note some examples of specific design suggestions that have emerged from studies of people's attitudes toward residential environments. Claire Cooper (1971) found that residents of the St. Francis Square moderate-income housing project in San Francisco emphasized the importance of available outdoor areas, especially areas with trees. Franklin Becker (1976) also found particular concern with outdoor space among residents of multifamily housing in New York City; they wanted separate play areas for younger and older children, and they wanted some novel play equipment. In a study that employed scale models of rooms (Baird, Cassidy, and Kurr, 1978), it was found that college students preferred ceiling heights that were approximately two feet (0.61 m) higher than those they normally encountered. Michael Cunningham (1977) reports the intriguing finding that people preferred a large open space in an apartment, such as the living room, to be located toward the right rather than the left of the floor plan. Jeanette Brandt and Nancy Chapman (1980) found that college students who were allowed to modify their dormitory rooms by adding design changes, carpeting, and decorations reported more satisfaction with their living environment than did students who lived in a traditional dormitory on the same campus which permitted very little room modification (and see box, "How Students Evaluate Faculty Offices").

Holahan and Wilcox (Holahan, 1977a; Holahan and Wilcox, 1977) applied a measure of user satisfaction in planning a design change in a university residence hall. They began by surveying the student residents of a high-rise dormitory to learn how satisfied or dissatisfied they were with their residential environment. They found that students were especially dissatisfied with the dormitory's lack of privacy and with the somewhat low level of social contact that was typical in the setting. A particular area of dissatisfaction was the communal dining area, which students felt was too institutional in appearance and uncomfortable for socializing.

Armed with this information, the investigators met with student representatives and residence hall staff to plan design improvements in the cafeteria. They

How Students Evaluate Faculty Offices

Can college students tell their professors something about how to decorate their offices? According to David Campbell (1979), they can. Campbell asked 251 students at the University of Kansas to evaluate a series of color slides that depicted a faculty office that varied systematically in its interior design. The office was varied in three ways. First, it varied in terms of the presence or absence of living things; in one arrangement the office had four potted plants and two aquariums with fish, while in another it had no living things. Second, the presence or absence of art objects was varied; one arrangement included four wall posters and a macramé hanging, while another showed no art objects. Finally, in one arrangement the office was tidy and neat, while in another it was messy and cluttered.

Students' evaluations of the faculty office differed significantly in response to these interior design variations. They rated the office as more comfortable, more inviting, and more welcoming (items A, B, and C) when plants and aquariums were present than when they were absent, when posters and a macramé hanging were present, and when the office was tidy rather than messy. They also felt that the professor's interests were more similar to their own (item D) and that the professor seemed more welcoming of visitors and less busy and rushed (items E and F) when living things and art objects were present and when the office was tidy. Campbell concludes that professors who want to make students feel comfortable and welcome in their offices would do well to get some plants and fish, hang some posters and artwork, and try to keep the office free of disorderly piles of books, notes, and scraps of paper.

Mean ratings for three interior design variations, by impression conveyed (scale: 1 to 9).

	Design variations					
Impression	Living things	No living things	Art objects	No art objects	Tidy	Messy
A. Comfort of visitor	5.95	4.60	5.68	4.86	6.10	4.45
B. Invitingness of office	5.88	4.33	5.59	4.65	6.14	4.10
C. How welcome visitor feels	6.02	4.68	5.75	4.95	6.20	4.51
D. Similarity of professor-visitor interests	5.76	4.21	5.44	4.53	5.62	4.35
E. How much professor welcomes visitors	6.18	4.95	5.98	5.16	6.32	4.81
F. Professor busy and rushed	5.84	5.68	5.87	5.76	4.47	7.07

Source: Adapted from D. E. Campbell, "Interior Office Design and Visitor Response," Journal of Applied Psychology, 1979, 64:648–53. Copyright 1979 by the American Psychological Association. Adapted by permission of the publisher and author.

decided to add attractive and colorful partitions to part of the previously open cafeteria to reduce the institutional scale of the setting, to increase a sense of privacy,
and to facilitate social contact among students. To evaluate the effectiveness of the
design changes, they surveyed students seated in the partitioned and unparti-
tioned areas of the cafeteria during the semester after the design changes had been
completed. They found that students seated in the new partitioned arrangements
were significantly more satisfied with the cafeteria environment than were students in the nonpartitioned area.

USER ATTITUDES IN NATURAL RESOURCE DEVELOPMENT

Craik (1972a, 1972b) has argued for the application of information concerning user
preferences and evaluations in natural resource management. He points out that
user attitudes are relevant to decisions involving the preservation of natural landscapes, the development of scenic and recreational sites, and the routing of roads
and power lines through open country. Zube and his associates (Zube, 1974, 1976;
Zube, Brush, and Fabos, 1975; Zube, Pitt, and Anderson, 1975) have argued that
our natural landscape constitutes an important scenic resource, and have developed a descriptive format to help people of different disciplinary and personal
backgrounds to communicate about the quality of the natural landscape. Other
investigators (Lucas, 1966; Smith and Alderdice, 1979) have also emphasized the
relevance of user-oriented research to wilderness policy decisions, such as decisions on zoning, the distribution of recreational sites, and environmental preservation and service provision in national parks.

Several researchers have applied user-based evaluations to planning decisions
involving outdoor recreation settings (Figure 4-5). L. D. Gustke and R. W. Hodgson (1980), observing the behavior of people walking along an interpretive trail in
a Michigan park, found that park visitors spent most time at and most enjoyed the
part of the trail just after a change in the surrounding environment (passing from a
field into a wooded area). They suggest that interpretive messages will be best
learned and remembered if they are located immediately after such environmental
transitions.

Peter Dorfman (1979) surveyed over 700 campers concerning their satisfactions and preferences in camping environments. He found that camping satisfaction was strongly related to the absence of crowding, annoying neighboring
campers, and pollution. While crowding and pollution can be dealt with through
effective recreation management, solutions to these problems necessarily involve
some financial costs. The level of regulations capable of controlling annoying
camping behavior is difficult to achieve because, although campers do not want to
be annoyed by other inconsiderate campers, neither do they wish their own wilderness experience to be infringed on by regulations.

Some investigators (Litton, 1972; Shafer, 1969) have carried out essentially
measurement-oriented studies involving the development and testing of assessment techniques that can be used in managing and developing natural settings,
such as national forests. Other investigators have applied users' evaluations to
planning decisions involving the roadside environment in natural settings. Rachel
Kaplan (1977) describes a study of users' preferences for roadside landscape fea-

Figure 4-5
Environmental psychologists have applied users' evaluations to the planning and management of outdoor recreational environments.

© *Lawrence Frank 1981.*

tures in Michigan's Upper Peninsula. Using a photographic presentation technique, she found that people most preferred woodland scenes (especially relatively open and spacious woodland) and least preferred flat farmland. Interestingly, however, open fields beyond a roadside screen of trees were also highly preferred. Kaplan suggests that a relatively modest improvement in the roadside environment can substantially improve scenic quality.

Evans and Wood (1980) suggest that the simulation technique they devised (see box, pp. 94–95) offers a simple and inexpensive way to present decision makers and citizen advisory groups with realistic simulations of proposed changes in the roadside environment. Gary Winkel and his colleagues (Winkel, Malek, and Thiel, 1969) developed a similar simulation strategy, involving the retouching of photographs of the roadside environment, which permits the systematic removal of particular features (such as utility poles and billboards) to measure people's assessments of possible alterations in the roadside setting.

User attitudes are also of relevance in decisions involving the programming and development of natural water resources. In fact, new standards developed by the Special Task Force (1970) of the United States Water Resources Council stipulate that proposed water projects be estimated in terms of positive and negative impacts on environmental quality and social well-being in addition to traditional concerns with national economic impacts (Craik, 1972b; Holahan and Kovalic, 1975). Information on area resident's preferences and evaluations concerning potential project impacts could play an important role in meeting the requirements of the new standards. User-based assessments might include perceptions of water quality among groups who use the water for different recreational purposes, such as swimming, boating, and fishing (Ditton and Goodale, 1973), and area residents' evaluations of the aesthetic and contemplative benefits of the water resource (Coughlin, 1976).

Environmental Attitudes and Conservation Programming

Energy conservation   A recent report from the Energy Project at Harvard Business School (Stobaugh and Yergin, 1979) concludes that America's future energy needs will have to be met in part through energy conservation. The report contends that a serious commitment to energy conservation in the United States could reduce energy needs by from 30 to 40 percent—the equivalent of more than all of America's imported oil. The report explains that a key component of an effective energy-conservation program is a change in people's attitudes.

Knowledge from environmental psychology related to environmental attitudes could play an important part in an effort to bring about attitude change consistent with energy conservation. A first step in changing attitudes toward the conservation of energy might involve educational programs oriented toward increasing people's awareness of energy concerns and needs (see O'Riordan, 1976). Of course, as we saw earlier, the relationship between attitudes and behavior is complex, and environmental education programs may not immediately result in new environmental behaviors (Cone and Hayes, 1977; Lingwood, 1971). Yet current research knowledge (Weigel and Newman, 1976; Weigel and Weigel, 1978) leads us to expect that when environmental education programs are tailored to a level of specificity in attitude change that is consistent with the desired behavioral change, they can make a meaningful contribution to energy-conservation programs.

Research projects by environmental psychologists have identified some features of energy-conservation programs that can help to increase their effectiveness. Studies have demonstrated, for example, that the provision of financial rewards to people who cooperate increases the effectiveness of programs to conserve electrical energy in the home (Hayes and Cone, 1977b; Kohlenberg, Phillips, and Proctor, 1976), to reduce driving mileage among college students (Foxx and Hake, 1977), and to increase bus ridership on a college campus (Everett, Hayward, and Meyers, 1974). In addition, we can decrease fuel-oil consumption by rewarding people with positive social feedback (Seaver and Patterson, 1976).

A special problem area involving energy conservation is that of "master-metered apartments"; that is, apartment complexes where electricity is metered at a single point for all tenants rather than for individual apartments. Studies have shown that energy consumption in master-metered complexes can be as much as 25 percent more than in individually metered apartments (McClelland, 1980). James Walker (1979) studied the effects of cash rewards offered to tenants in a master-metered apartment complex in Texas for specific energy-conserving behaviors. Apartments were checked at random over ten weeks, and residents who were found to have set their thermostats lower than normal on a cold day or higher than normal on a hot day were awarded $5. When electricity use in the apartment complex before and after the experiment was compared with electricity use in a similar complex that did not offer awards, energy consumption was found to have been reduced considerably in the experimental complex.

In another study involving financial awards in master-metered apartments, Lou McClelland and Stuart Cook (1980) conducted an energy conservation contest among four groups of apartments (heated and cooled by natural gas) in a family housing complex at the University of Colorado. During each two-week period over twelve weeks the apartment group that consumed the least energy was awarded $80 to use in any way it determined. Natural gas consumption over the twelve weeks was reduced by 6.6 percent, although the level of energy savings tended to decrease over the period of the study.

Other investigators have demonstrated that residential energy consumption can be significantly reduced if residents are given specific feedback about their energy use even when financial awards are not provided. Clive Seligman and John Darley (1977) found that homeowners who were given feedback about their energy consumption several times each week for a month reduced their electricity consumption by 10.5 percent compared to a control group. Similarly, Lawrence Becker (1978) found that providing New Jersey homeowners with feedback about their electricity consumption (when combined with the setting of specific energy-conservation goals) resulted in a 13 percent reduction in electricity use in comparison to a control group. Finally, Becker and Seligman (1978) show that a unique signaling device placed in the homes of New Jersey homeowners to signal an "energy-conservation opportunity" helped to reduce electricity costs by 15.7 percent compared to a control group. The signaling device consisted of a blue wall light (connected to both the air conditioner and an outside thermostat) that lit up when the outside temperature dropped below 68° and the air conditioner was turned on.

Carl Hummel and his associates (Hummel, Levitt, and Loomis, 1978) have shown that knowledge of people's environmental attitudes might be relevant to the formulation of energy-conservation legislation. They interviewed residents of Fort Collins, Colorado, during and after the 1973 gasoline shortage to ascertain how they felt about proposed solutions to the energy crisis that would alter their lifestyle. The investigators found that as the energy crisis became clearly established, people became more inclined to favor mandatory rather than voluntary energy-conservation programs—but they preferred mandatory policies directed against "extra" energy consumption rather than across-the-board changes. For example, respondents favored policies that would tax people whose cars had poor mileage ratings. These attitudes were consistent with energy-conservation legislation being considered in Congress, which was oriented toward the establishment of mandatory constraints on "extra" energy consumption.

Controlling littering   Studies have shown that other environmental problems, such as littering and environmentally destructive behavior, can be reduced through reinforcement techniques (Figure 4-6). One study in Tennessee (McNees, Schnelle, Gendrich, Thomas, and Beagle, 1979) found that reinforcement consisting of free snacks at McDonald's encouraged a litter-hunt program among schoolchildren that reduced litter in a local community by over 30 percent. A similar study (Clark, Burgess, and Hendee, 1972) found that rewarding children with for-

Figure 4-6
Psychologists have found that reinforcement techniques can reduce littering in public settings.

© *Peter Menzel/Stock, Boston.*

est ranger badges and Smokey the Bear insignia facilitated a successful litter cleanup in a campground. We may hope that such reinforcement programs for youngsters generate positive attitudes toward the environment that endure after the immediate reinforcement study has been completed.

Finally, field studies have shown that environmental cues, such as strategically placed signs, can help to elicit appropriate environmental attitudes and positive environmental behaviors. Research has demonstrated that signs discouraging littering can significantly reduce littering (Baltes and Hayward, 1976; Geller, Witmer, and Orebaugh, 1976), as can the presence of colorful trash cans (Finnie, 1973). The presence of signs discouraging walking on the grass was found to decrease such behavior significantly in a small university park (Hayes and Cone, 1977a). Interestingly, some recent studies have shown that the presence of litter itself acts as a cue for further littering behavior. Susan Reiter and William Samuel (1980) found that littering in a California parking garage was significantly greater when the garage floor was already littered than when it was clean—whether or not antilitter signs were present in the area. Similarly, Robert Krauss and his colleagues (Krauss, Freedman, and Whitcup, 1978) demonstrated in both a controlled laboratory study and a field observational study in New York City that the rate of littering is positively correlated with the amount of litter already present.

Cost-effectiveness   Before concluding our discussion of the application of research findings to the reduction of environmentally wasteful and destructive behaviors, we should consider the issue of the *cost-effectiveness* of the various programs we have mentioned. At what point does a cash-award program for energy conservation become more expensive than the electricity it saves? When does a program that offers financial rewards for litter cleanup cost more than simply increasing litter-collection services? Clearly, the question of cost-effectiveness is important. For research findings from environmental psychology to be effectively applied to environmental problems, they must be economically feasible as well as scientifically sound. Each program will need to be evaluated individually before we can know whether any particular program's benefits outweigh its costs. While the economic feasibility of environmental programs is essential, noneconomic benefits should be included in the evaluation, such as the benefits to community pride and aesthetic enjoyment derived from a litter cleanup campaign. On the basis of environmental psychology research in this area, we may advance some suggestions relevant to the devising of cost-effective environmental programs.

First, when energy becomes costly enough, the financial awards offered for energy conservation actually pay for themselves. The monthly electricity savings in the master-metered apartment complex that Walker (1979) studied in Texas were more than enough to pay for the monthly cash awards as well as the labor costs for apartment inspectors. Second, as Daniel Stokols (1978) pointed out after reviewing a large number of studies in this area, there is considerable evidence that inexpensive programs can effectively promote conservation efforts, such as programs based on feedback without financial awards or those based on periodic rather than continuous reinforcement of environmentally positive behaviors.

The evidence that litter leads to more litter suggests that the least costly method of controlling littering may consist of quick, regular litter cleanups that can be carried out with a minimum of effort (Reiter and Samuel, 1980). Finally, although the long-term behavioral effects of environmental education programs still need to be assessed, the potential benefits of school-based environmental education programs should not be minimized. One recent study (Kushler, 1980) involving an energy-education program in more than 100 Michigan high schools concluded that conservation-related classroom instruction positively affected both the environmental attitudes and the self-reported conservation behaviors of high school students.

SUMMARY

Environmental attitudes are favorable or unfavorable feelings toward some feature of the physical environment or toward an issue that pertains to the physical environment. Environmental psychologists have investigated environmental attitudes in several areas, including residential satisfaction, scenic preference, and attitudes toward environmental conservation. The major techniques that have been developed to measure environmental attitudes have been labeled *perceived environmental quality indices*, or PEQIS. A PEQI affords a quantitative measure of the quality of a particular physical setting, such as a housing development or a recreation area, as it is subjectively experienced by a particular group of people. PEQIS based on *preferential judgments* express an entirely personal evaluation of environmental quality in terms of a particular person's environmental standards. PEQIs based on *comparative appraisals,* in contrast, assess the relative quality of a particular environmental setting in relation to some identified standard of comparison.

Environmental attitudes serve an important psychological function in our lives by helping us to make decisions that involve a wide range of behaviors in regard to the environment. One important function of environmental attitudes is to help us to select a residential environment. The level of our satisfaction with our current residence and our preferences in living environments influence our decision to move to a new residential environment or stay where we are. Another important function of environmental attitudes is to help us to make decisions on the use and care of the physical environment, such as a decision to avoid littering or to support a conservation program.

Social psychologists have applied the classical conditioning model of learning to theories of attitude formation and change. This model proposes that attitudes toward an object or issue (conditioned stimulus) are learned when that object or issue is repeatedly paired with another experience that is pleasant or unpleasant (unconditioned stimulus). The intensity of the learned attitude increases as a function of (1) the level of affect associated with the unconditioned stimulus, (2) the shortness of the delay before the conditioned stimulus follows the unconditioned stimulus, and (3) the proportion of times the conditioned stimulus is paired with the unconditioned stimulus.

Social psychologists have also applied the instrumental conditioning model of learning to theories of attitude formation and change. This model proposes that attitudes that are rewarded will be strengthened and maintained, while those that are punished will be weakened and eliminated. For attitude change to occur, (1) the stimulus must be attended to by the individual, (2) the individual must comprehend the message, and (3) the individual must accept the message.

Further theoretical work by social psychologists has indicated that the effects of attitudes on behavior will be evident only when both attitudes and behavior are measured at equivalent levels of generality or specificity. Thus a general attitude will predict a general class of related behaviors.

Research on aesthetic preferences indicates that the relationship between stimulus complexity and evaluative or preferential judgments may be described as an inverted U-shaped curve. People's evaluative responses to the built environment are most positive when the environment is at an intermediate level of complexity. Knowledge of environmental attitudes has been applied in residential design, natural resource development, and conservation programming.

Performance in Learning and Work Environments

I t may be surprising to hear that the physical environment can influence a student's success at college. Yet it is true that the physical settings where students attend class, study, and take examinations can affect their academic performance. Similarly, the physical design of primary and secondary schools can influence children's learning and school performance. And the physical settings where people work, whether professional, clerical, or industrial, can shape their performance on the job.

The study of how the physical environments where we learn and work affect our performance represents an important area of investigation in environmental psychology. Researchers have found that people's performance in learning and work environments can be affected by a range of environmental factors, including the level of lighting, the quality of sound insulation, and the layout of designed spaces. The effects of the physical environment on performance can be substantial, influencing productivity, efficiency, accuracy, fatigue, and boredom. Investigators have found, for example, that open-space classrooms and offices can

123

impede the effectiveness of learning some tasks and can sometimes be detrimental to work performance.

Such research findings are especially relevant to environmental planning. Administrators and planners are committed to discovering ways to enhance people's performance in schools, colleges, offices, and factories. Yet, remarkably, long-established solutions to design needs in these vital human settings have often been accepted and reapplied uncritically. Conversely, innovations in design in learning and work environments, such as open-space classrooms and open-plan offices, have been adopted without sufficient evaluation of their effects on performance.

THE NATURE OF ENVIRONMENTAL PERFORMANCE

Because some of the same environmental variables (i.e., noise and temperature) will be examined in this chapter and the next, which discusses environmental stress, it will be helpful to distinguish between the orientations of these two chapters. This chapter explores how features of the physical environment can affect human *performance*, and focuses specifically on learning and work environments. We shall consider ways the environment can positively affect performance as well as impede it. In considering environment and performance, we shall draw on research in the field of human factors psychology as well as environmental psychology.

Chapter 6 will examine how features of the physical environment can be a source of *stress*, affecting people's physiological functioning, social relationships, and emotional well-being. Our major emphasis will be on the urban environment, although we shall also consider residential and institutional environments for the elderly. Our focus will be exclusively on some of the negative ways the environment can affect people's lives. In discussing environmental stress, we shall draw on research in the health sciences and social psychology as well as environmental psychology.

IDENTIFYING THE ENVIRONMENTAL VARIABLES

An initial task for environmental psychologists who study performance in learning and work environments has been to identify the particular environmental variables that are relevant to human performance. The efforts of environmental psychologists in this area have been aided by earlier work by researchers in the field of *human factors psychology*, (called *ergonomics* by the British). Human factors psychologists apply information about human behavior to the design of work-related products to enhance the effectiveness with which people can use them (McCormick, 1976). Human factors psychologists are interested in some work-related products that go beyond the investigative interests of environmental psychologists, such as machines and protective clothing. They are also interested in learning how the *molar physical environment* in which people work can affect their job performance, and here their interests are shared by environmental psychologists. Two important journals in the human factors area are *Human Factors* and *Ergonomics*.

Ambient environment   Light, sound, and temperature are aspects of what is called the *ambient* environment; that is, the environment that "surrounds" us in any particular setting. In discussing the ambient environment, we are not concentrating on *things* in the environment, but rather on the *properties* or characteristics of the physical environment. For example, we may be interested in knowing how much light a ceiling can reflect, to what degree a carpeted floor can absorb sound, or how well a wall area can insulate against heat and cold.

A dramatic example of how features of the ambient environment can affect human performance is provided by a grade school I recently consulted with in Texas. Increasing enrollment in the school had forced school administators to seek out additional classroom space in the already overcrowded schoolhouse. To meet the growing demand for space, a large storage area near the school's gymnasium was converted into a fourth-grade classroom. It failed miserably as an educational environment. Student performance dropped precipitously and the classroom teacher had numerous job-related complaints. An examination of the classroom setting revealed that the room's ambient environment presented a variety of threats to effective teaching and learning.

The lighting, which had been ample for a storage area, was inadequate for classroom learning; many students complained that they could not read the blackboards. The small air conditioner that had been added to the storage area during the conversion was unable to cool the classroom sufficiently during occasional hot spells, and the teacher reported that students became especially restless at such times. Finally, the new classroom's location near the school gymnasium presented a major source of distracting noise during the regular gym periods. Unfortunately, while the improvised classroom did afford much-needed additional space, its ambient environment was particularly inappropriate for the learning process.

Layout   Human factors and environmental psychologists concerned with human performance have also investigated the *layout* of designed settings. Here the psychologist is interested in how the layout of spaces in which several interrelated activities take place affects the ease and efficiency with which the overall job or learning task is carried out. A major goal is to arrange the layout of such spaces so as to maximize the performance of the total set of interrelated activities carried out in the setting (Bennett, 1977).

The layout of designed settings has two aspects—the *location* of environmental features and their *arrangement*. Ernest McCormick (1976) explains that the location of environmental features involves the decision of where to place features within a general area, such as the placement of computerized learning equipment in a school setting or the location of a conference room in an organizational context. In arranging features one must determine how to relate a group of environmental features to one another in a particular area—the placement of teachers' supplies, arts and crafts materials, and individual and group work spaces in an open-space classroom, say, or the arrangement of interrelated pieces of assembly-line equipment in an industrial setting.

An example of how the layout of environmental features can affect job performance may be seen in a personnel office I consulted with in California. The personnel office was located in a large general hospital that was attempting to adjust to serious budget limitations. In order to conserve building costs and make maximum use of available space, the administrators had moved the personnel office to a large, open area in one wing of the hospital. The desks of an administrative assistant, two job counselors, and two secretaries were placed together in the open area, along with a row of chairs that served as a waiting area, without intervening walls or partitions. All personnel staff members complained that they could not do their work effectively in the open area.

The job counselors found themselves conducting personal interviews over the clatter of the secretaries' typewriters, within earshot of both their co-workers and waiting visitors. The administrative assistant complained of constant interruptions from visitors and staff, who were forced to pass by her desk when they entered and left the personnel area. The secretaries found themselves constantly intruding into other workers' space to use the file cabinets, which were awkwardly arranged throughout the area as (ineffective) boundary markers. Although the new open arrangement succeeded in reducing building costs, the organizational costs of the spatial layout in terms of reduced job efficiency and lowered staff morale offset the initial construction gains.

Defining the Performance Variables

McCormick (1976) explains that performance in work and learning environments involves behaviors that are oriented toward accomplishing some objective. Performance covers a wide spectrum of responses, from essentially physical activities (loading produce) to psychomotor behavior (operating a telephone switchboard) to strictly mental activities (studying for an exam). Fred Steele (1973) divides human performance into three categories of instrumental tasks: physical activities, which occur *outside* people; mental activities, which take place *within* people; and interactional activities, which occur *between* people.

Physical activities   Steele points out that such physical activities as operating machines, lifting objects, painting, hammering, sawing, and rolling dough can be facilitated or impeded by the physical environments in which they occur. He describes a New England company that operated for almost fifty years with no heating in the plant. In order to keep warm during the winter, the workmen engaged in only those tasks that required enough physical movement to raise their body heat. More stationary tasks were simply left until spring. Steele also describes an artist who moved his studio to a large barn when he realized that the scale of his earlier work had been determined by the size of his small studio rather than by his own creativity.

Mental activities   The mental activities that occur within people include such things as thinking, reading, concentrating, and remembering. Steele emphasizes that the performance of these mental tasks is also affected by the nature of the settings within which they take place. Our mental activities are hindered in settings where we are subject to interruptions that we cannot control. In open-space offices, the lack of a door or other visual screening when a work area is located near a heavy traffic flow can negatively affect the performance of mental tasks. Steele adds, however, that an optimal level of stimulation can be helpful to mental activities, and that work settings that are completely isolated from personal contact and visual stimulation can be detrimental to people's creative imagination.

Interactional activities   Interactional activities—those that take place between people—include such things as group planning meetings, "brainstorming" sessions, selling services or merchandise to customers, and verifying or sharing information with another person. The emphasis here is on interaction that takes place as a *means* toward the accomplishment of some task, rather than on the interaction as an end in itself. These joint tasks, too, are subject to influences from the physical settings

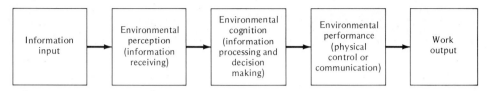

Figure 5-1
A holistic model of environmental performance.

Adapted from E. J. McCormick, Human Factors in Engineering and Design, *p. 9. Copyright © 1976. Adapted by permission of the publisher, McGraw-Hill.*

where they occur. Many organizations have only small conference rooms, which permit only small groups of people to come together. When a problem is addressed that could better be handled by a large group, the size limitation of the meeting room can stand in the way of effective problem solving. Steele notes too that group meetings that are held in areas through which other people must pass are subject to constant intrusions and distractions that can impede group decision making.

A holistic model of performance   In discussing the nature of people's performance in learning and work environments, we must keep in mind that performance does not operate in isolation from the other psychological processes we have considered. Environmental performance always occurs in combination and in constant interaction with other psychological processes in the environment, such as environmental perception and environmental cognition. Ernest McCormick (1976) has presented a holistic model of environmental performance that reflects the interaction between performance and other psychological processes in the environment (Figure 5-1).

As the model shows, environmental performance is one link in a chain of related psychological processes. These related processes are initiated when some type of *information* from the environment impinges on us. We receive the information through one or more of our senses, in the process of environmental perception. Next we encode and organize the perceptual information through the process of environmental cognition, which results in a decision to act or not to act on the information. The operations that occur on the basis of that decision constitute the process of environmental performance. The performance may be a physical action that controls some aspect of the environment, a communication delivered to another person or ensuing mental activity.

An example of how environmental performance occurs as one part of a holistic process may be seen in the work of a safety engineer in a nuclear power plant. The process is initiated when a serious malfunction in the plant's cooling system is shown by a red indicator (information input). The engineer sees the indicator (environmental perception), and realizing that the problem may be dangerous, he decides to close down the reactor (environmental cognition). He then throws the appropriate control switches and phones other plant personnel to inform them of the emergency situation (environmental performance). The importance of designing work environments so as to facilitate human performance is underscored by the Three Mile Island nuclear reactor accident in Pennsylvania, where, according to a presidential task force, the seriousness of the accident was compounded by an initial series of interrelated machine and operator errors.

Making a Behavior Map

By making a behavior map of an environmental setting, the environmental psychologist can study the relationship between specific types of behavior and particular locations in the setting. After observational categories have been established, the environmental psychologist constructs data sheets on which behavioral data can be recorded. The data sheets include observational categories, a list of subareas in the environmental setting, and observation intervals.

A typical behavior-mapping data sheet, which was developed for observations in a hospital setting, is shown here. The data sheet shows that observer Pam conducted observations on Ward 4 at 10:00 A.M. on February 17, 1978. During the first observa-

A behavior-mapping data sheet.

Date: 2/17/78   **Time:** 10:00 A.M.

Observation interval / Location	Dayroom	Bedroom	Corridor	Dayroom
Observation category:				
Lying awake	1	2		1
Sleeping		1		
Sitting alone	5			4
Writing	1			1
Personal hygiene		1		
Reading	2			2
Standing	2		3	
Pacing	3		5	5
Eating				
Games	2			2
Talking	2			3

Source: C. J. Holahan, "Action Research in the Built Environment," in R. H. Price and P. E. Politser (eds.), Evaluation and Action in the Social Environment. New York: Academic Press, 1981, p. 97. Reprinted by permission.

tion interval, she observed a total of thirty patients on the ward. Of these patients, eighteen were in the dayroom, four were in the bedroom, and eight were in the corridor. We can see that the dayroom was not an especially active setting. At observation interval 1, five patients in the dayroom were sitting alone, three were pacing, two were standing, and one was lying awake. There was, however, some dayroom activity. We can see that at interval 1, two patients in the dayroom were talking, two were playing games, two were reading, and one was writing. By comparing observation intervals 2 and 3 with interval 1, we can see subtle changes in ward behavior over this time period.

Ward: *4*   Observer: *Pam*

2		*3*		
Bedroom	Corridor	Dayroom	Bedroom	Corridor
2		1	2	
1			1	
		5		
1				
		2	1	
	2			3
	6	4		5
		2		
		4		

RESEARCH MEASURES FOR ASSESSING PERFORMANCE

Behavior mapping   One way environmental psychologists have assessed performance in learning and work environments has been by directly observing people's behavior in these settings. William Ittelson and his associates (Ittelson, Rivlin, and Proshansky, 1976) have developed a research procedure called *behavior mapping* to use in studying people's behavior in a variety of settings. A behavior map of an environmental setting consists of a record of the number of individuals engaged in each of several predetermined behavior types in each subarea of the environment. In preliminary observation sessions, a list of behavior categories is developed which covers most of the behavioral variance in the setting being studied. In addition to scoring type of behavior, the observer records each subject's specific location in the environment at each observation interval. Observations thus cover all physical spaces in the selected environments at each interval. Observations are recorded on data sheets designed for quick and easy use by observers (see box, "Making a Behavior Map").

Essential to the behavior-mapping procedure is the training of a team of observers to conduct behavioral observations in a standardized and reliable manner. First, the team, which typically consists of four to six members, is acquainted with the list of observational categories, and any confusion or ambiguity concerning the meaning of the categories is resolved. Next the observers practice using behavior-mapping data sheets, working together in pairs. The level of agreement between team members is constantly monitored, and any disagreements or discrepancies are clarified. The training is considered complete when agreement between team members reaches 90 percent—typically after ten hours or so of practice. Research has shown that when this training procedure is used, the behavior-mapping technique maintains a high level of reliability. People's behavior in designed settings may also be assessed by means of Roger Barker's (1968) Behavior Setting Survey (see Chapter 1). Robert Bechtel (1977) discusses in detail how the Behavior Setting Survey may be used to assess people's behavior in a variety of designed settings. It should be pointed out that while the behavior mapping and Behavior Setting Survey methods are especially useful in examining people's performance in learning and work environments, they have also been applied in studying a wide range of other behavior problems in environmental psychology.

Unobtrusive measures   One limitation of measurement strategies involving the direct observation of behavior, such as behavior mapping and the Behavior Setting Survey, is that the presence of observers may itself cause people to alter their performance. When people's behavior is influenced by their awareness that they are being observed as part of a research study, they are said to "react" to the observations, and the research measure employed is termed "reactive." A classic example of reactive research methods that altered job performance can be seen in the Hawthorne illumination studies (Roethlisberger and Dickson, 1939) at a Western Electric plant during the 1920s (for an alternative interpretation of the Hawthorne findings, see Parsons, 1974).

The investigators were interested in learning if an increase in lighting would improve workers' performance at assembling relays and testing telephone equipment. They found that as illumination was increased in one part of the plant, workers' productivity in that area also increased. Later, however, the researchers discovered that as lighting was *decreased* in that part of the factory, productivity continued to *increase*. The investigators concluded that the observed increases in productivity, rather than being a function of the actual level of lighting, resulted from the workers' sense of satisfaction at being singled out for the special attention of the research study.

Eugene Webb and his colleagues (Webb, Campbell, Schwartz, and Sechrest, 1966) have proposed the use of *unobtrusive measures* that will not affect or alter the behavior under observation. Unobtrusive measures are measures taken to exploit the physical evidence that is a natural product of human behavior, and which can provide an index of the type and level of behavior that produced it. The fictional detective Sherlock Holmes characteristically used such physical traces (a boot print, a cigar ash, lipstick on a glass) to infer the nature of past behavior.

Webb and his associates categorize such physical measures according to two broad types—*erosion* measures and *accretion* measures. Erosion measures are produced by the degree of selective wear on some physical material. The rate at which vinyl tiles had to be replaced at the Museum of Science and Industry in Chicago was used to indicate the relative popularity of exhibits at the museum (Duncan, 1963). Tiles around an exhibit of hatching chicks needed to be replaced every six weeks, while tiles in other areas of the museum lasted for years without having to be replaced.

Accretion measures are produced by the selective deposit or buildup of some physical material. Holahan (1976) measured the amount of graffiti on walls and doors that accumulated on a newly remodeled psychiatric ward as one index of staff and patients' feelings about the ward. Another easily available source of unobtrusive measures is the archival records systematically collected by organizations, governmental bodies, and newspapers. For instance, records of employee turnover, absenteeism, and sick leave may provide information relevant to staff morale and organizational effectiveness. Of course, in using unobtrusive measures the researcher should verify that maintenance practices and construction materials are uniform across the areas studied.

Laboratory measures   While behavior mapping, the Behavior Setting Survey, and unobtrusive measures are especially appropriate for use to assess people's performance in real-world settings, a variety of additional performance measures has been used to assess human performance in controlled settings. These measures have typically been highly refined and precise indices of particular aspects of human performance, such as measures of the accuracy of visual perception or performance on precise mechanical tasks. Some laboratory measures of performance have consisted of focused tasks, such as coding, proofreading, or puzzle solutions, that tap motivation and persistence as well as accuracy and efficiency. The precise definition of laboratory measures of performance has strengthened the internal validity of the studies that employ them, although the artificial nature of such tasks has limited their external validity as indices of performance on real learning and job tasks. In the next section we shall see that much of the research that has examined the effects of the ambient environment on performance has employed highly refined laboratory measures of performance.

Questionnaire measures   In addition to the behaviorally based measures of performance we have discussed above, questionnaire self-report measures have been used in some studies of human performance. Typically, self-report measures have been used in conjunction with behavioral measures, so that individuals' perceptions of their own job performance and indices of their feelings, attitudes, and motivations may be related to actual performance. Laboratory-based research on the effects of heat and cold on performance has often included questionnaire measures of how comfortable or uncomfortable subjects feel at various temperature levels. Similarly, field studies of the effects of contrasting office designs on job performance have often included questionnaire measures of employees' satisfaction or dissatisfaction with the spatial arrangement, as well as their perceptions of whether their job performance has been enhanced.

Converging measures   No single measurement strategy is inherently superior to all others. Ideally, investigators should strive to employ a combination of research measures, so that the inherent weaknesses of one measurement approach may be balanced by the natural strengths of another. An investigation of illumination and productivity in an industrial setting might employ a variety of "converging" research measures. First, a laboratory measure of performance in a controlled experimental setting might be employed to provide a fine-tuned index of how specific variations in illumination level affect particular behaviors (e.g., vigilance and psychomotor performance). Next, the researcher might conduct a naturalistic field study in a factory setting in order to compare recorded differences in productivity, sick days, and damaged items (unobtrusive measures) among areas in the factory that differ naturally in lighting. Next, workers' job performance might be directly observed (behavior mapping) in each of the areas. Finally, in a field experimental approach, lighting changes might be introduced in different parts of the factory and the effects on performance and productivity further observed. When these varied measures are employed together, the researcher can compare findings gained through increased experimental control with those acquired under less artificial conditions.

ENVIRONMENTAL EFFECTS ON PERFORMANCE

THE EFFECTS OF LIGHT

Light can affect visual work in two ways (Boyce, 1975). First, light can *directly* affect the difficulty of a visual task by altering the operating conditions of the visual system. For example, the level of lighting available for a task and the interference of visual glare can directly alter our ability to carry out a visual task. Second, light can *indirectly* affect performance by creating work conditions that are uncomfortable, distracting, or tiring. While the most important effects of light on performance are direct effects, the major effects of sound and temperature on performance (as we shall see) are indirect.

Illumination   In order to read a book easily, to write a letter or paper efficiently, or to carry out a detailed task, we need an adequate level of lighting. The environmental psychologist defines these problems as the effects of *illumination* on performance. When we place a light meter near a light source, such as a desk lamp, we are measuring the level of illumination of that light source. Illumination is measured in *footcandles* (or *decalux;* 1 footcandle = 1.076 decalux). As the light meter is moved farther away from the light source, the number of footcandles measured decreases. In order to read typical handwriting without difficulty, we require approximately 100 footcandles of illumination (Bennett, 1977).

The direct effects of illumination on visual performance have been well established through a series of studies carried out predominantly in experimental settings (Beutell, 1934; Boyce, 1973; Fry, 1962; Graham, 1965; Simonson and Brozek, 1948; Weston, 1945). P. R. Boyce (1975), in summarizing the results of these studies, notes that as illumination increases, visual acuity increases; that is, with more light we are able to discriminate and recognize smaller visual details. The effect of a particular change in illumination is greater on more difficult visual tasks than on less difficult ones. Corwin Bennett (1977) points out that greater illumination also allows us to perform visual tasks more quickly and more accurately. As the overall level of illumination increases, however, the performance gains from equal increments in illumination decrease. At very high levels of illumination, increases in illumination can actually decrease performance by suppressing some information cues, such as visual gradients (Logan and Berger, 1961; Stevens and Foxell, 1955).

Glare   Glare occurs when a source of light that is brighter than the general level of illumination to which our eyes are adapted is located close to the direction in which we are looking (Bennett, 1977). A common example of glare's interference with our ability to see clearly occurs when we are faced with an oncoming car using its high-beam headlights at night.

The direct effects of glare on visual performance are termed *disability glare.* Boyce (1975) and McCormick (1976) have reviewed research (carried out as early as the 1920s) that has shown that disability glare is detrimental to performance. The negative effects of disability glare increase as the glare approaches the line of sight. *Discomfort glare* occurs when the glare causes discomfort to the performer, but does not directly affect psychophysical measures of performance (Boyce, 1975). Boyce (1975) speculates that the discomfort caused by glare may indirectly affect the quality of visual performance, but points out that more research is needed on this question.

Color   A great deal of popular and professional speculation has focused on the effects of various colors on people's performance (see Birren, 1965). In fact, however, very little empirical knowledge is available on this topic. It has been shown that great contrast in colors can directly affect performance on visual tasks that involve color discrimination (Eastman, 1968). There is some evidence that color may affect people's mood and level of arousal (see Mehrabian and Russell, 1974) and their attitudes (see Blum and Naylor, 1968); we may speculate that such differences in mood, arousal, and attitudes may indirectly affect task performance. Some colors are associated with particular moods; while red is seen as "exciting" and "stimulating," blue is viewed as "secure" and "soothing" (Wexner, 1954). In addition, variations in color affect physiological reactions, such as blood pressure and rate of respiration (Acking and Küller, 1972). Finally, red induces higher levels of arousal than green, as reflected in higher galvanic skin responses (Wilson, 1966). These possible indirect effects of color on performance are probably related to cultural differences in the meaning and conventional uses of various colors.

THE EFFECTS OF NOISE

One of the most heavily investigated environmental influences on human behavior in recent years has been noise. Here we shall discuss the effects of noise on the types of performance that typically occur in learning and work environments. In Chapter 6 we shall consider the stressful effects of noise on people's health, social relationships, and emotional well-being.

What is noise? In general, we may define noise as sound that the listener does not want to hear. Not all noise is loud; a conversation close to your study carrel in the library can be as unwelcome as a pneumatic drill outside the window. While a major component of the definition of noise consists of this psychological quality of being unwanted, it is important to appreciate the physical aspects of noise as well.

The physical measure of sound intensity is the *decibel,* or dBA. Leo Beranek (1966) has provided a graphic representation of the typical decibel levels of the sounds produced by twelve common environmental sources (Figure 5-2). When you examine the figure, you should keep in mind that sound intensity increases exponentially: a sound of 10 decibels is twice as loud as a sound of 1 decibel, a sound of 20 decibels is four times louder than one of 1 decibel, and a 100-decibel sound is 1,000 times louder than a 1-decibel sound (Beranek, 1966).

Although most of us can readily point to personal experiences in which noise made the performance of some task more difficult, such as studying for an exam or writing a term paper, the empirical findings in the area are quite complex. In a recent review of the effects of noise on performance, Sheldon Cohen and Neil Weinstein (1981) point out that despite a vast body of research studies on noise and performance, we are unable to predict with much confidence how noise will affect performance in a given situation.

No negative effects   Many studies of the effects of noise on performance have been conducted in controlled laboratory settings. Typically, these studies have involved relatively simple tasks, such as adding or comparing lists of numbers. Reviews of these studies have concluded that laboratory-produced noise does not adversely affect performance on relatively simple mental and psychomotor tasks (Broadbent, 1957; Glass and Singer, 1972a; Kryter, 1970). Cohen and Weinstein note that laboratory tasks that

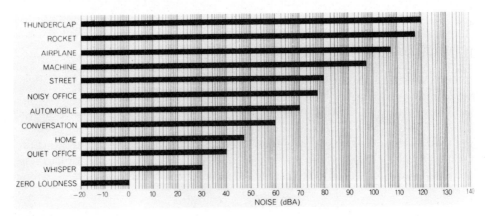

Figure 5-2
Intensities of sounds produced by common environmental sources.

have not been adversely affected by noise are typified by the following characteristics: responses are required at definite times, responses are cued by a clear warning, stimuli are easily visible, and responses involve repetitive dexterous and coordinated movements. Under some circumstances noise may actually enhance performance, as when it helps subjects to pay attention or to stay awake (Corcoran, 1962; Warner, 1969).

Negative effects under some conditions   Evidence from a wide range of studies, however, has made it clear that under some circumstances, noise does adversely affect performance. In general, the variables that determine whether noise will adversely affect performance may be grouped in three major categories: the type of *task* performed, the particular *characteristics* of the noise employed, and the *timing* of the performance measures.

Type of task   Investigators have reported that noise adversely affects performance on complex tasks (Bogges and Simon, 1968; Eschenbrenner, 1971) and on tasks that require a high level of concentration and vigilance (Broadbent, 1954, 1958, 1971; Jerison, 1959; Woodhead, 1964). Complex tasks have been defined as those that involve more than one signal to attend to, rapid target stimuli, or an irregular signal whose occurrence cannot be predicted. Performance on tasks that involve the handling of a large amount of information is also detrimentally affected by noise (Finkelman and Glass, 1970; Glass and Singer, 1972b; Hockey, 1970a, 1970b; Woodhead, 1966). When subjects are required to carry out several tasks simultaneously under noisy conditions, they may maintain adequate performance on one task at the expense of others.

Noise characteristics   Performance is generally more adversely affected by intermittent noise than by noise that is continuous, and these negative effects are greatest when the intermittent noise occurs at aperiodic or irregular intervals (Broadbent, 1957; Eschenbrenner, 1971; Sanders, 1961; Theologus, Wheaton, and Fleishman, 1974). In an interesting field correlational study, Arline Bronzaft and Dennis McCarthy (1975) examined the association between exposure to the noise of elevated trains (aperiodic intervals) and reading ability in children in a New York City grade school. The reading scores of children in classes on the side of the school closest to the train noise were lower than those of students in classes located farther from the noise. Evidence also indicates that performance is more disrupted by noise that consists of intelligible speech than by other types of noise (Acton, 1970; Obszewski, Rotton, and Soler, 1976).

Timing   Some of the most interesting recent findings concerning the effects of noise on performance have related to the time at which performance is measured. Recent studies have shown that the negative effects of noise on performance become more pronounced as a function of length of exposure to the noise (Hartley, 1973; Hartley and Adams, 1974). Investigators have also discovered that some important performance decrements from noise occur as *aftereffects* that become apparent after the subject has left the noisy situation (Glass, Singer, and Friedman, 1969; Sherrod, Hage, Halpern, and Moore, 1977) (see box, "Apartment Noise and Children's Reading Ability"). People who had been exposed to aperiodic noise were found to make more errors on a proofreading task and to give up more easily their attempts to solve difficult (actually unsolvable) puzzles *after* the cessation of the noise than a no-noise control group (Glass, Singer, and Friedman, 1969).

Apartment Noise and Children's Reading Ability

Sheldon Cohen and his colleagues (Cohen, Glass, and Singer, 1973) conducted a fascinating field correlational study to learn how noise in New York City apartment buildings was related to the reading ability of children who lived in the buildings. The apartment buildings studied were four thirty-two-story towers constructed on bridges spanning a major interstate highway. The investigators measured noise level within the apartment buildings, and demonstrated that the lower the floor of a building (the closer to highway traffic), the greater the noise level. The authors also noted that because traffic noise is *aperiodic*, it is especially likely to cause performance deficits. On the basis of this information, the researchers hypothesized that floor level of residence in the apartment buildings would be positively related to reading ability .

To test their predictions, Cohen and his colleagues assessed the reading ability and the auditory discrimination of fifty-four schoolchildren who lived in the apartments. They discovered that children who lived on the lower floors did indeed show greater impairment in both reading achievement and auditory discrimination than children who lived on higher floors. The association between noise level and reading deficits was indirect; noise directly affected auditory discrimination, and auditory discrimination, in turn, mediated the effects of noise on reading achievement. The accompanying table makes clear that floor level accounted for the major portion of variance in auditory discrimination, and was followed in importance by father's educational level, number of children in the family, and grade level.

Cohen and his associates emphasize that the deficits in auditory discrimination and reading ability appeared as *aftereffects* of prolonged exposure to noise. The detrimental effects of traffic noise on performance were greater for children who had lived in the buildings for a longer period of time. In an effort to adapt to their noisy environment, children apparently learn to screen out acoustic cues; when speech-relevant cues are among the signals screened out, the child fails to develop some of the discriminative abilities that play a role in learning to read.

Percentage of total variance in auditory discrimination accounted for by various independent variables.

Independent variable	Percentage of variance accounted for
Floor level	19%
Father's education	12
Number of children in family	10
Grade level	6
Mother's education	0

Source: S. Cohen, D. C. Glass, and J. E. Singer, "Apartment Noise, Auditory Discrimination, and Reading Ability in Children," Journal of Experimental Social Psychology, 1973, 9:407–22. Reprinted by permission

An important variable associated with the aftereffects of noise is the degree of *control* subjects can exert over the noise. Several investigators (Glass, Reim, and Singer, 1971; Glass and Singer, 1972a; Sherrod, Hage, Halpern, and Moore, 1977) have reported that the negative aftereffects of noise on performance decrease when subjects are able to control the noise. We shall discuss the important variable of control more fully in Chapter 6.

THE EFFECTS OF TEMPERATURE

A considerable body of research has examined the question of how people's performance in learning and work environments is affected by variations in room temperature. This issue is especially interesting in light of federal energy-conservation guidelines in the late 1970s that stipulated cooling and heating levels in public buildings. Most of us can point to times when working or studying has been uncomfortable because the work area was either too warm or too cold. The question of whether such uncomfortable temperatures adversely affect our performance, however, is quite complex. As we shall see, most studies have shown that temperature variations within the ranges we are likely to encounter indoors do not directly bring about performance decrements. It is possible, however, that the discomfort induced by an overly hot or cold environment may indirectly reduce the quality of our work.

Thermal comfort   Researchers who have studied *thermal comfort* have been interested in determining the temperature range within which people feel comfortable, and in identifying the points at which temperatures become uncomfortably warm or uncomfortably cold. The results of one such study show that people reported feeling most comfortable at 79° F (26° C). However, subjects felt only "slightly warm" or "slightly cool" across quite a wide range of temperatures, from a low of 68° F (20° C) to a high of 80° F (30° C) (Rohles, 1971).

Other factors should be kept in mind when we consider thermal comfort. While people do vary in the temperature range they find comfortable, many commonly assumed group differences in thermal comfort, such as those related to sex and age, have not been substantiated by empirical research (Griffiths, 1975). Controlled research studies have shown that men and women are generally similar in terms of thermal comfort (McNall, Jaax, Rohles, Nevins, and Springer, 1967; Fanger, 1972), and that thermal comfort ranges do not differ between college-age and elderly subjects (Fanger, 1972). Other physical variables in addition to air temperature do, however, affect thermal comfort, including humidity, air movement, physical activity level, and amount of clothing worn (McCormick, 1976). It seems likely that temperatures that are uncomfortable may, with extended exposure, indirectly influence people's performance by inducing fatigue, boredom, and irritability.

Heat and performance   Although most research studies on the effects of heat on performance have been conducted under highly controlled laboratory conditions, their results are exceedingly complex. A considerable number of studies indicate that heat adversely affects performance, some show no performance effects from heat, and still others reveal that heat may actually improve performance (Griffiths, 1975).

Many studies conducted over the past twenty-five years have shown that high temperatures can have a detrimental effect on the performance of a great variety of tasks. Heat adversely affects the performance of physical work (Leithead and Lind, 1964; Mackworth, 1961; Wyndham, 1969), a variety of psychomotor and vigilance tasks (Azer, McNall, and Leung, 1972; Colquhoun and Goldman, 1972; Teichner and Wehrkamp, 1954; Wyon, 1974), industrial work (Tichauer, 1962), and school-work (Lofstedt, Ryd, and Wyon, 1969).

Other studies have shown, however, that the effects of heat on performance are more complex than they may at first appear. Some studies have shown that performance may be increased under low levels of heat (Wilkinson, Fox, Goldsmith, Hampton, and Lewis, 1964) or during the period of initial exposure to heat (Poulton and Kerslake, 1965). One study revealed that, while heat negatively affects complex tasks, it does not hamper simple tasks (Griffiths and Boyce, 1971).

In summary, while heat does adversely affect the performance of a great variety of tasks under some conditions, those effects are influenced by other factors, including the complexity of the task, the level of heat, and the length of exposure to high temperatures (see Pepler, 1963). It should be noted that the temperature levels that have consistently been shown to reduce performance are generally much higher than those people are typically faced with in the built environment, and considerably above the temperature range people report as comfortable (Wyon, 1974).

Cold and performance   Much less research has been done on the effects of cold temperatures on performance than on the effects of heat. Again, most such work has been conducted in controlled laboratory settings. Such studies have generally shown that people's performance on a variety of psychomotor tasks is adversely affected by cold (Fox, 1967; Poulton, 1970; Lockhart and Kiess, 1971; Teichner and Wehrkamp, 1954). Performance decrements on manual tasks due to cold are essentially a function of hand-skin temperature (Fox, 1967). Subjects have been shown to perform manual tasks as well in cold conditions as in comfortable temperatures when their hands are warmed by infrared heaters (Lockhard and Kiess, 1971). The temperature range at which cold consistently reduces performance is well below the range of temperatures that people find comfortable.

THE EFFECTS OF LAYOUT

Although environmental psychologists and design professionals share an intuitive belief that a poor spatial layout in educational and occupational environments can adversely affect human performance, empirical evidence concerning the effects of layout on performance is scarce (McCormick, 1976). An appreciation of the way the layout of designed spaces can affect performance may be derived from principles of spatial arrangement for facilitating performance suggested by McCormick (1976). While McCormick has proposed these principles as guides for design decisions, they also provide a framework for understanding how spatial layout and human performance are interrelated.

McCormick explains that environmental features that are of particular *importance* to an organization's functioning (e.g., a teletype in a newsroom) or subject to frequent use (e.g., computer terminals in a university science department) should be located where they are optimally accessible to the persons who will use them. A group of features that will *function together* as part of a pattern of interrelated activities (e.g., the book sign-out, book return, and circulation desk in a college library) or *function sequentially* (e.g., consecutive components on an industrial assembly line) should be arranged together in a spatial grouping that is appropriate to the associated activity pattern. Similarly, environmental features associated with *competing* or interfering activities (e.g., a coffee-break area in a quiet library reading room) should not be placed close to the areas devoted to the activities with which they will compete.

Fred Steele (1973), an organizational consultant in Boston, offers an example of how improper layout can impede an organization's functioning. He describes an organization that was troubled about hostility between two of its units. The friction between the two units was discussed at an off-site conference center, and all parties agreed to work toward improving intergroup relations in the future. Soon after the participants returned to the job environment, however, these good intentions evaporated and conflict between the units recurred. An analysis of the organization's spatial layout revealed that the two units were located on different floors of a high-rise building, and that the daily contacts between group members that were essential to a positive working relationship were impossible to maintain. The physical separation between the two units played a key part in creating social distance and eventual hostility between the units despite people's otherwise good intentions to work cooperatively.

Robert Bechtel (1977) describes how a poorly laid-out office in a research laboratory in Alaska hampered the effective functioning of the office. The major problem was the location of two desks (top of Figure 5-3) that involved conflicting and mutually disruptive job functions. One desk was used by the division secretary and the other by the noncommissioned officer in command (NCOIC), who handled the sign-out of vehicles. Using the Behavior Setting Survey, Bechtel discovered that the two work sites had overlapping spatial boundaries, and that job functions carried out by one individual inevitably disrupted the work and invaded the privacy of the other. The functional difficulties of the space were further compounded by its location in a major traffic pattern and the fact that it housed the coffee urn used for laboratory coffee breaks.

Bechtel proposed a rearrangement of the laboratory layout (bottom of Figure 5-3) that was designed to resolve the functional difficulties in the office. In the new layout, the NCOIC was moved to a separate, partitioned office that allowed increased privacy, along with easy access to the coffee urn and the mail area. At the same time, the new layout was designed to enhance the job performance of the division secretary, who now had an office of her own which contained the telecopier, filing, and visitor features that constituted her main duties. Bechtel points out that additional layout changes in the new spatial arrangement, such as the relocation of the division chief and administrative assistant, were also designed to enhance the overall job performance of those individuals.

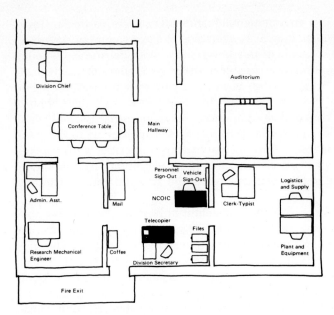

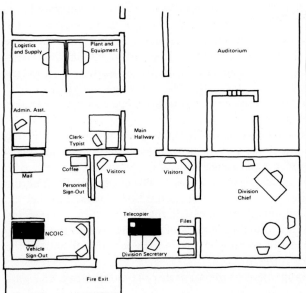

Figure 5-3

The conflicting activities of the division secretary and noncommissioned officer in command (NCOIC) (shaded features) in the spatial layout at the top are more effectively handled in the revised layout at the bottom.

Adapted from R. B. Bechtel, Enclosing Behavior, pp. 41, 43. Copyright © 1977 by Dowden, Hutchinson & Ross, Inc., Stroudsburg, Pa. Reprinted by permission of the publisher.

THEORETICAL PERSPECTIVES ON ENVIRONMENT AND PERFORMANCE

As we have seen, the effects of the physical environment on people's performance in learning and work settings are quite complex. Here we shall consider some of the theoretical perspectives advanced by environmental psychologists to explain this complex relationship. The dominant theory in this area goes beyond the immediate physical properties of the environment to postulate the role of *arousal* in mediating the effects of environment on performance. The arousal theory is relevant to all three environmental variables we have discussed—light, sound, and temperature. It also explains why seemingly poor environmental conditions do not always adversely affect performance, and may sometimes even improve it.

An innovative, though less widely held, theory, which restricts itself to the immediate physical properties of the environment, explains the effects of noise on performance in terms of *auditory masking.* A more recent theoretical model attempts to integrate the arousal and auditory masking theories.

AROUSAL THEORY

Arousal theory is relevant to a broad range of human behavior and experience, not merely to the effects of environment on performance (see Berlyn, 1960). It can help to explain people's aesthetic judgments, for example, as we saw in Chapter 3.

According to arousal theory, a wide range of environmental situations and events, from watching the team score a touchdown in the big game to being called on to deliver an impromptu speech, can cause an individual to experience a general state of emotional excitement. At a physiological level, emotional excitement is accompanied by such physical reactions as a rapid heart rate, an increased rate of respiration, and sweating. The feelings associated with arousal may be pleasant or uncomfortable, depending on whether the related environmental experience is positive or negative and on whether the level of arousal becomes uncomfortably high.

Arousal theory proposes that such environmental stimuli as noise, temperature, and even color raise an individual's level of arousal. This increase in arousal in turn mediates the reported effects of environmental variables on human performance. At the present time, arousal theory is the most widely accepted model for explaining the effects of noise (Broadbent, 1971), heat (Provins, 1966), and cold (Fox, 1967) on performance. It is an especially valuable model because it is able to explain both the adverse and the positive effects of environment on performance.

Yerkes-Dodson Law   Arousal theorists have explained the complex effects of the environment on performance in terms of a basic principle in experimental psychology established soon after the turn of the century, known as the *Yerkes-Dodson Law* (Yerkes and Dodson, 1908). According to the Yerkes-Dodson Law, the relationship between level of arousal and human performance is represented as an inverted U-shaped function. Performance will be maximal with an intermediate level of arousal, and will gradually fall off as arousal either increases or decreases. The law states further that the effects of arousal on performance interact with the complexity of the task at hand. The optimum level of arousal is lower for complex tasks than for simple ones.

These relationships between arousal level and task performance are shown graphically in Figure 5-4. Thus, for example, a moderate level of arousal might adversely affect performance on a complex task while facilitating performance on a simple one. As you may recall, we also encountered an inverted-U relationship between level of environmental stimulation and optimum psychological response

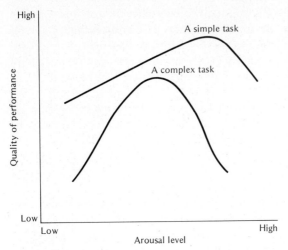

Figure 5-4

Quality of performance as a function of arousal level and task complexity.

From D. Kahneman, Attention and Effort, 1973, p. 34. Reprinted by permission of the publisher, Prentice-Hall, Inc.

in our discussion of perception, attitudes, and cognition. Clearly, the implications of the Yerkes-Dodson Law in regard to people's behavior and experience in the physical world are considerable.

Noise and arousal   Investigators have used arousal theory to explain the effects of noise on performance. Donald Broadbent (1971) argues that noise is arousal-producing, and that the level of the noise (and the accompanying arousal) will determine whether the noise enhances or impedes task performance. Central to Broadbent's position is a belief that heightened arousal leads to a narrowing of an individual's attention. He suggests that on tasks that require only a restricted range of cues, moderate noise can enhance task performance by focusing attention on essential task cues and away from competing cues. Consistent with this view, E. Christopher Poulton (1977) suggests that the arousal caused by moderate levels of noise can benefit performance in terms of vigilance, rate of work, and reaction time. Noise-generated arousal can also facilitate the performance of subjects who have been deprived of sleep (Corcoran, 1962; Wilkinson, 1963). Researchers believe that an optimal level of arousal caused by moderate noise can help to make an individual more alert and attentive, and that these effects can benefit performance on relatively simple tasks that require mental alertness (Eschenbrenner, 1971).

When noise is intense, however, it leads to a state of *overarousal* that adversely affects task performance by causing the individual to ignore task-relevant cues (Broadbent, 1971, 1978). Prolonged exposure to moderate noise may also result in a state of overarousal that is detrimental to task performance (Hartley and Adams, 1974). As the Yerkes-Dodson Law suggests, we would expect the negative performance effects of noise-generated arousal to be more evident on complex tasks than on simple ones. On more complex tasks, where incidental information is important to adequate performance, the overarousal caused by intense noise leads the individual to screen out sources of information essential to the task (Broadbent, 1971, 1978; Kahneman, 1973).

Temperature and arousal   Researchers have also applied arousal theory to the effects of temperature on performance. Moderate heat (and possibly moderate cold) can facilitate task performance by generating an optimal level of arousal (Poulton, 1976a). When the arousal caused by heat is greater than the optimum level of arousal for the task at hand, however, it adversely affects performance (Provins, 1966). Overarousal caused by very cold temperatures is also detrimental to performance (Fox, 1967). Consistent with the Yerkes-Dodson Law, temperature-generated arousal is more detrimental on complex tasks than on simple ones (Griffiths and Boyce, 1971). The relationship between heat and arousal is complex, however (Poulton and Edwards, 1974). Arousal tends to increase while temperature is rising and under moderate heat, but tends to decrease under conditions of constant mild warmth.

A Theory of Auditory Masking

Imagine that you and a friend are conversing while standing on a station platform, waiting for the next train to arrive. Your friend turns to you, and just as she begins to speak, the train roars into the station. You see your friend's lips moving, but can hear nothing over the noise of the arriving train. Environmental psychologists describe such an event as *auditory masking*, because the target auditory signal (the verbal statement) is hidden or "masked" by the extraneous noise (the arriving train) (see Lightfoot and Morrill, 1949). While our discussion here will be limited to auditory masking, masking can also occur in the visual realm when a visual target is hidden or masked by extraneous visual signals or visual "noise" (see Howell and Briggs, 1959).

E. Christopher Poulton (1976a, 1977, 1978) has proposed that auditory masking offers the best explanation for the results of those research studies that have reported that noise (particularly continuous noise) causes decrements in human performance. While the role of auditory masking in noise research was recognized as early as World War II by the late S. S. Stevens (see Kryter, 1950), it has generally been overlooked by contemporary researchers. In presenting his argument for the central role of auditory masking in noise research, Poulton reviews more than thirty research studies that have found deterioration in performance due to noise, and reports that auditory masking offers a plausible explanation of the performance decrements found in all of them.

Poulton contends that auditory masking occurred in two ways in these earlier studies. First, the noise masked the sounds the subjects made as they responded to the task. In some studies an audible click informed the subject that the response had been recorded, while in others an audible click let the subject know that the control had been pressed hard enough. Such auditory feedback would facilitate the performance of subjects in the no-noise control condition (who could hear the click), but would not be available to subjects in the noise condition (for whom the noise would mask the click). This difference in the availability of response sounds could explain the finding that performance was better in the control conditions than in the experimental conditions.

Second, Poulton proposes that in the earlier studies noise masked the subjects' inner speech. Just as noise can mask spoken words, it can mask the inner speech that subjects engage in while performing experimental tasks that have a verbal component. In some earlier noise studies, subjects could improve their performance by mentally rehearsing the complicated instructions for the experiment or the series of numerical values that had to be remembered to respond accurately. Such inner speech could proceed unhindered in the no-noise control condition, but would be hampered by auditory masking in the noise condition. In effect, subjects in the noise condition "could not hear themselves think." Again, Poulton proposes that this difference in the ability to engage in inner speech could explain the performance difference between the experimental and control conditions.

A Unifying Model

Poulton (1979) has advanced a theoretical model that attempts to integrate the arousal and auditory masking theories in a single framework. He argues that the positive effects of noise on performance are due to arousal, the negative effects to masking, and the failure to find effects to the fact that arousal and masking cancel each other. Poulton's unified model is shown in Figure 5-5.

The function at the top of Figure 5-5 reflects the effects of masking on performance. Here there is an immediate decrement in performance as soon as the noise is turned on, and a return to the normal level of performance as soon as the noise is terminated. The middle function in the figure shows the effects of arousal on performance. In this case, arousal initially benefits performance, but the arousal decreases as the subject adapts to the noise. When the noise is stopped, arousal drops below the resting level and gradually returns to normal. Poulton notes that this postnoise drop in arousal may cause the negative aftereffects reported in some noise studies.

The function at the bottom of Figure 5-5 reflects the combined effects of masking and arousal. When noise facilitates performance, the improvements generally occur soon after the noise is turned on. When noise hurts performance, the

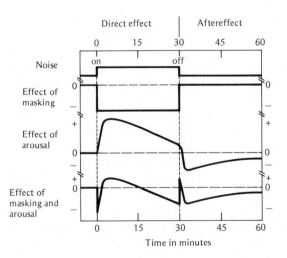

Figure 5-5

The combined effect on performance of auditory masking and arousal produced by continuous noise.

From E. C. Poulton, "Composite Model for Human Performance in Continuous Noise," Psychological Review, 1979, 86:361–75. Copyright 1979 by the American Psychological Association. Reprinted by permission of the publisher and author.

negative effects typically appear later, and may persist after the noise has stopped. Poulton suggests that the composite function at the bottom of the figure is compatible with a wide body of research on the effects of noise on performance.

Many issues remain unresolved in this important and complex area of study. Broadbent (1978) contends that while auditory masking can help us to understand some of the effects of noise on performance, auditory masking has been overapplied to findings in this area. Dylan Jones and his colleagues (Jones, Smith, and Broadbent, 1979) failed to find indications of auditory masking in a series of recent studies. And while arousal theory may explain many of the complex relationships between environment and performance, Poulton (1979) has argued that stronger evidence that the hypothesized physiological correlates to arousal do in fact occur is essential to the theory. While there is evidence that noise increases skin conductance (Glass and Snyder, 1972, 1973), blood pressure (Jonsson and Hansson, 1977), and adrenalin secretion (Frankenhaeuser and Lundberg, 1977), more research in this area is needed, as is evidence that such physiological changes are associated with corresponding changes in performance. Moreover, some recent studies (see Smith and Broadbent, 1980) have failed to support the position that moderate arousal will enhance performance on a simple task by causing the individual to use more salient cues.

APPLICATIONS TO ENVIRONMENTAL PLANNING

Research knowledge about how the physical settings where people learn and work affect their performance is of considerable practical value. The quality of school environments will exert a lasting influence on students' future progress. The character of the work settings where we spend a major portion of our adult lives plays a significant part in shaping the overall quality of our lives.

FORMULATING OVERALL PLANNING GOALS

Knowledge about environment and performance suggests some overall goals that might guide the planning of learning and work environments. Fred Steele (1973) discusses the importance of helping environmental users and decision makers to develop a high level of *environmental competence*. One ingredient of environmental competence is the ability to be aware of the important influences of the physical environment. Environmental psychologists might play an important role in developing workshops and educational programs oriented toward broadening and deepening people's awareness of the influences of learning and work environments (see Sommer, 1972). Another aspect of environmental competence is the ability to change the physical environment to meet human needs and objectives more effectively.

A second planning goal is the development of *empirically based evaluations* of the effectiveness of learning and work environments (see Friedmann, Zimring, and Zube, 1978). Michael Brill (1971) presents a general evaluation model for assessing the functional effectiveness of designed settings (Figure 5-6). The design process begins with a statement of the objectives of the setting to be designed. These ob-

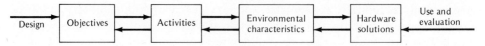

Figure 5-6
A general evaluation model for designed settings.

From M. Brill, "Evaluating Buildings on a Performance Basis," in C. Burnette, J. Lang, and D. Vaschon (eds.), Architecture for Human Behavior: Collected Papers from a Mini-Conference, 1971, p. 42. *Reprinted by permission of the Philadelphia Chapter/American Institute of Architects.*

jectives require the execution of a series of related activities. Each activity, in turn, necessitates supportive environmental characteristics, which can be achieved through particular hardware solutions.

For example, a library reading room must facilitate such activities as reading, thinking, and recollection over extended periods of time. The environmental characteristics required for such activities include adequate illumination, quiet, and thermal comfort. A variety of hardware solutions is available to support these desired environmental characteristics, such as lighting fixtures, windows, acoustic tiles, carpet, and a central air conditioning system. Brill emphasizes that the most important component of the model is *environmental characteristics*, which derive directly from performance criteria and link them to appropriate hardware solutions. The arrows in the model pointing from left to right reflect the design perspective, while those that point from right to left indicate the evaluative orientation.

APPLICATION TO PARTICULAR DESIGN PROBLEMS

Research findings on the relationship between the environment and performance have been applied to the formulation of specific design standards for some aspects of the ambient environment in buildings, such as illumination and temperature levels. Our knowledge in this area is also relevant to planning decisions involving new and challenging design concepts, such as open-space classrooms and open-plan office designs.

Illumination standards   An extensive series of research studies by H. R. Blackwell and his colleagues (Blackwell, 1959, 1961, 1964, 1972a, 1972b; Blackwell and Blackwell, 1968; Blackwell and Smith, 1970) has provided the empirical basis for illumination standards in the United States. The research procedure they developed to assess the effects of illumination on performance, called the *visibility method*, permits a holistic view of human task performance, as it focuses on such features as surface texture and the arrangement of task components as well as illumination level (Boyce, 1975).

The procedure may be used to establish illumination standards for any task in a variety of environmental settings. It has been used to determine the appropriate level of illumination needed to carry out work tasks in offices, banks, operating rooms, machine shops, and warehouses. The illumination standards developed through this procedure are based on the specific types of tasks performed in each type of setting, and are tailored to the unique task demands of various subareas in each environment. In the case of a bank environment, for example, only 50 foot-candles are needed for the performance of the type of tasks that are carried out in the lobby area, while 150 footcandles are required for the type of tasks that occur in the teller's area (Bennett, 1977).

Open-space classrooms   In addition to aiding in establishing specific design standards, research findings concerning environment and performance can serve as a basis for evaluating the functional effectiveness of innovative design plans in the educational and business worlds. A design innovation that has recently had a pronounced effect on educational planning is the open-space classroom (Figure 5-7). While the open-space classroom offers promise of an exciting and beneficial contribution to the educational scene, the environmental psychologist demands that the effects of the new design be carefully and systematically evaluated in actual use.

The open-space classroom is, in effect, a school without walls; it replaces the traditional row-and-column seating arrangement of the conventional schoolroom with a large, unbroken space that can hold from three to five regular classes of students and their teachers (EFL, 1965). A central goal of the open-space design is to promote a high degree of flexibility in the classroom; without permanent barriers, classroom space may be easily rearranged to fit changing educational needs

Figure 5-7
The open-space classroom is an exciting design innovation that has greatly affected educational planning.

(Gump, 1975). Another aim of the open-space design is to encourage interaction among teachers and between teachers and students (EFL, 1965). The adoption of open-space classroom design has often been accompanied by more flexible patterns of teaching and classroom organization.

Although systematic evaluations of the educational effects of the open-space classroom are few, some initial findings can help us evaluate its effectiveness. Leanne Rivlin, Marilyn Rothenberg, and their associates, using a field correlational design, systematically evaluated two public elementary schools in New York City that used the open-space design (Rivlin and Rothenberg, 1976; Rivlin, Rothenberg, Justa, Wallis, and Wheeler, 1974; Rothenberg and Rivlin, 1975). They employed a variety of research methods, including behavior mapping, interviews with teachers and students, and a scale model of the classroom. They found that the actual performance of the open-space classrooms was often at variance with the stated educational philosophy. For instance, the use of space in the classroom was quite uneven. In one classroom, almost half of the persons they observed were grouped together in less than 10 percent of the total room space. In another classroom, the teacher spent 72 percent of her time in one sector of the room, and was never observed in fully one-half of the room. Children spent most of their time at individual work, such as writing, rather than the group projects that the open-space design is believed to encourage. Finally, teachers and students alike expressed a need for more quiet and private spaces.

Other investigators have also reported that noise can be a problem in open-space classrooms. Noise in open-space classrooms can be a source of distraction, and noise from overheard social conversations is more distracting than noise related to schoolwork (Brunetti, 1971, 1972). The degree of distraction depends on the particular activity that people are engaged in; noise is more distracting to people in a study session than to those performing a laboratory task.

Similarly, Robert Stebbins (1973), who conducted a field correlational study, reports that open-space design can lead to disorderly and disruptive classroom behavior. He studied schools in Kingston, Jamaica, where an open design had been employed not as an educational goal, but to facilitate ventilation and to economize on construction. When he compared the Kingston schools to traditionally designed schools in Canada, he found that the open design was associated with increased inattention caused by external distractions, excessive talking, and mischievousness. After another field correlational study, Paul Gump and Lawrence Good (1976) report no overall differences between students in open-space and traditional schools in the use of small learning groups, in the time spent at particular tasks, or in student leadership positions. They did find, however, that the open-space arrangement was associated with the use of more spaces in the classroom for learning tasks, and (for primary grades only) with more teacher-led activities.

A field correlational study carried out by Joshua Burns (1972) took advantage of a naturally occurring environmental change in a high school. A science unit in the school was redesigned so that a science resource center was converted to an open-space design, while two science labs maintained their traditional design. Burns systematically compared students' behavior and attitudes in the differing environmental conditions and discovered that more science-related activities occurred in the traditional labs, while greater social interaction characterized the open resource center. Students reported experiencing more distractions in the open arrangement, though actual noise levels were similar in the two settings. Lighting and thermal conditions were better in the traditional than in the open areas.

It appears that while the open-space classroom design has been associated with some changes in spatial behavior in some studies (e.g., the use of classroom spaces for learning tasks and increased social interaction), the effects of the open design on overall educational performance remain to be demonstrated. The open-classroom arrangement has also commonly been found to result in greater visual and auditory distractions and lack of privacy.

Several investigators have suggested ways in which open-space classrooms might be made to function more effectively. Some authors, for example (Krovetz, 1977; Rothenberg and Rivlin, 1975), have argued for varied types of space in classroom design rather than exclusively open-space settings. In this way, alternative types of classroom space (some open and others more structured) could be fitted to the demands of particular tasks and the special needs of children who differ in age and personality.

Some investigators (EFL, 1965; Walsh, 1975) have also suggested solutions to the noise problem that can arise in open-space classrooms. The most common suggestion is that the classroom floor be carpeted. Carpeting removes a great deal of noise at its source: pattering feet, chairs being moved, articles being dropped. Acoustic ceiling tiles can reduce noise even further. In many open-space settings where both ceiling tiles and carpets have been used, however, teachers have complained that they have trouble making themselves heard. The best solution in most cases is probably the use of carpeting along with an ordinary, sound-reflective ceiling. In addition, work groups should be separated by a sufficient distance so that the space between groups can operate as a sound barrier. Finally, activities that are potentially disruptive, such as those involving necessary noise or movement, should be scheduled so as not to conflict with adjacent activities.

A field study conducted by Gary Evans and Barbara Lovell (1979) in a California high school that had recently been remodeled offers an empirically based evaluation of potential design improvements in open classrooms. The investigators used a quasi-experimental design (see Cook and Campbell, 1976) that included prechange and postchange measures in both the remodeled high school and an unchanged control high school. In the remodeled school, sound-absorbant partitions had been installed in a previously open-space classroom in order to reduce noise, redirect traffic away from class areas, more clearly demarcate class area boundaries, and increase privacy (Figure 5-8). The control setting consisted of an open-space classroom in another high school that was unchanged.

Evans and Lovell discovered that the design modifications to the open-space classroom had beneficial educational effects. Verbal and nonverbal interruptions in the classroom dropped sharply after the remodeling, and students asked more work-related questions. In the control classroom, in contrast, interruptions rose somewhat and questions about course content decreased slightly over the same time period. Evans and Lovell conclude that the design changes enhanced the educational process in the remodeled classroom. On the basis of their findings and those of an earlier study by Carol Weinstein (1977) (see box "Remodeling an

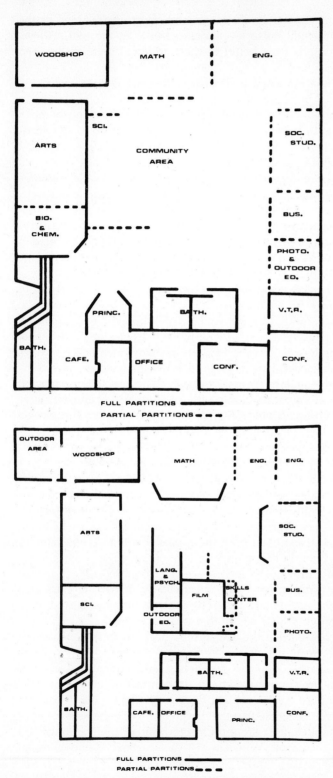

Figure 5-8

This open-space class-room (*top*) was remodeled to include sound-absorbent partitions (*bottom*) designed to enhance class functioning and increase privacy.

From G. W. Evans and B. Lovell, "Design Modification in an Open-Plan School," Journal of Educational Psychology, *1979, 71:41–49. Copyright 1979 by the American Psychological Association. Reprinted by permission of the publisher and author.*

──Remodeling an Open-Space Classroom──

Carol Weinstein (1977) carried out an intriguing field experiment in which she attempted to improve the educational effectiveness of an open-space classroom in Massachusetts through a variety of design changes. She began by conducting an initial behavior mapping in the open-space setting, which served twenty-five second- and third-grade children. The results of the initial behavior mapping revealed a variety of educational problems in the classroom. Students were not evenly distributed across the classroom space, but tended to crowd into some areas and to ignore others. The room's science and game areas were greatly underused, and the small amount of student activity that did take place in those areas was limited to only a few categories of behavior.

Weinstein planned and carried out design changes that were intended to distribute students more evenly over the classroom space and to broaden the types of behavior that occurred in the underused parts of the classroom. She added a platform table, stools, shelving for storage, individual study carrels, and a cardboard "house" for a quiet retreat, and rearranged existing furnishings to make better use of the available space. She then repeated the behavior mapping to evaluate the impact of the design changes.

Weinstein's findings from the prechange and postchange behavior mappings are summarized in the accompanying table. Before the design changes there was a considerable imbalance between the students' use of space and the space actually available; after the changes, the students made greater use of the available space. Notice especially that the science and game areas, which were greatly underused before the remodeling, got considerably more use after the room changes. In addition, Weinstein points out that the range of behaviors that occurred in the previously underused areas was significantly greater after the remodeling effort. These results demonstrate the effectiveness of small-scale design changes as a means of enhancing the educational effectiveness of open-space classrooms.

Percentage of students observed in areas of an open classroom and percentage of total space devoted to each area before and after design changes.

Area	Prechange observations		Postchange observations	
	Percentage of students	Percentage of total space	Percentage of students	Percentage of total space
Reading	28.6%	25.0%	27.3%	25.0%
Mathematics	22.2	12.0	15.8	13.0
Art	14.8	13.0	16.5	14.0
Science	10.9	15.0	15.0	17.0
File	9.8	15.0	3.7	11.0
Games	8.0	13.0	17.3	14.0
Corner	4.3	7.0	4.3	6.0

Source: Adapted from C. S. Weinstein, "Modifying Student Behavior in an Open Classroom Through Changes in the Physical Design," American Educational Research Journal, 1977, 14:249–62. Copyright 1979, American Educational Research Association, Washington, D.C.

Open-Space Classroom"), they offer the following design recommendations to improve the educational effectiveness of open-space classrooms: the reduction of visual and auditory distractions, clearer demarcation of activity boundaries, increased accessibility of class materials, provision of adequate work space and privacy, and provision of some territories where students can establish a sense of personal control.

Open-plan offices   Our knowledge of environment and performance can also help us to evaluate design effectiveness in open-plan offices (Figure 5-9). The open-plan office (sometimes called the landscape office, or *Bürolandschaft*) was originated in Germany during the 1960s (Brookes, 1972; Zanardelli, 1969). It consists of a large, open work area, with no floor-to-ceiling partitions. The arrangement of desks, work spaces, and low, movable barriers is planned to reflect the pattern of work flow and the organizational process of the particular office setting. The design is intended to enhance work flow, improve work-related communication, and permit flexibility and freedom in the performance of tasks (see Kubzansky, Salter, and Porter, 1980).

Although more research will be needed before we can adequately determine the performance effectiveness of the open-plan office (Parsons, 1976), some preliminary evaluative findings are available. One study (Brookes, 1972; Brookes and Kaplan, 1972) used a semantic differential measure to assess the reactions of em-

Figure 5-9
This open-plan office includes several work spaces in a large open area.

© *Leif Skoogfors 1978/Woodfin Camp & Assoc.*

ployees of a large retail company before and after their move from a conventional to an open-plan office. Results demonstrated that workers did not see the new open-plan setting as more functional or more efficient than the standard office environment. Employees perceived more noise and more visual distractions in the open-plan arrangement; overheard conversations were especially distracting. On the positive side, respondents felt the new office was more aesthetically attractive, and that it facilitated social contacts. Overall, the investigators concluded that the open-plan office looked better than the standard design, but performed less well.

Another study (Nemecek and Grandjean, 1973) surveyed employees in fifteen open-plan offices in Switzerland. Again respondents noted that noise and disruptions in the open arrangement interfered with their ability to concentrate on their work. Over one-third of those surveyed reported being very disturbed by noise (especially conversations), while only one-fifth of respondents said they were not disturbed by noise. Managerial personnel reported that they experienced difficulty keeping their confidential conversations from being overheard. There were some positive findings, however; 63 percent of surveyed employees felt that they could complete their work more efficiently and with less effort in the open-plan design. Another researcher (Manning, 1965), who surveyed more than 2,000 employees of a large insurance company in Great Britain, found that almost half of the workers in an open-plan setting expressed a preference for a smaller, partitioned area.

It appears that under some circumstances the open-plan office design can improve work performance, but that problems associated with noise and lack of privacy require attention. Noise might be reduced by the use of sound-absorptive surfaces (carpets, acoustic ceiling tiles, draperies, plants); machine noise could be reduced by proper maintenance and lubrication procedures, and by the use of rubber mountings and cushions; the office layout could be arranged so that tasks involving noise and movement did not interfere with surrounding activities (Bennett, 1977; McCormick, 1976). In addition, some alternative spaces should be available for private meetings, group discussions, and unavoidably noisy tasks.

SUMMARY

The features of the molar physical environment that have been most fully investigated by psychologists interested in the effects of the environment on people's performance are light, sound, and temperature. These environmental variables are aspects of what is called the *ambient* environment. An additional aspect of the molar environment that has been studied in terms of human performance is the *layout* of designed settings. Layout involves both the location and the arrangement of environmental features. Human performance ranges over a wide spectrum of responses, from essentially physical activities to psychomotor behaviors to strictly mental activities. Performance also includes interaction between people for the accomplishment of some task. Environmental performance always occurs in combination and in constant interaction with other psychological processes in the environment, such as environmental perception and environmental cognition. Research measures for the study of performance include behavior mapping, the Behavior Setting Survey, unobtrusive measures, and laboratory measures.

The direct effects of illumination on visual performance have been well established through a series of studies carried out predominantly in experimental settings; as illumination increases, visual acuity increases and visual tasks can be performed more quickly and accurately. *Disability glare* is detrimental to performance; *discomfort glare* causes discomfort to the performer, but does not directly affect psychophysical measures of performance. Color can directly affect performance on visual tasks that involve color discrimination, and it seems possible that the effects of color on people's mood and level of arousal may indirectly affect task performance.

Laboratory-produced noise does not adversely affect performance on relatively simple mental and psychomotor tasks. Under some circumstances, noise may actually enhance performance, as when it helps subjects to pay attention or to stay awake. Noise does adversely affect performance in some circumstances, however. Performance on tasks that require a high level of concentration and vigilance, are complex, or involve the handling of a large amount of information is adversely affected by noise. Noise is also more likely to affect performance adversely when it is aperiodic or consists of intelligible speech. The negative effects of noise on performance become increasingly evident as the time of exposure to the noise lengthens, and some important performance decrements due to noise occur as aftereffects.

High temperatures can adversely affect the performance of physical work, a variety of psychomotor and vigilance tasks, industrial work, and schoolwork. Performance may be increased, however, at low levels of heat or during the period of initial exposure to heat. Cold temperatures can also negatively affect performance on a variety of psychomotor tasks; the adverse effects of cold on manual tasks are essentially a function of hand-skin temperature. It should be noted that research on temperature and performance has been conducted predominantly under highly controlled laboratory conditions, and that the adverse effects generally occur when temperatures are much higher or lower than those typically faced in the built environment.

There is little empirical evidence concerning the effects of layout on performance. The location of environmental features should be determined by the importance and frequency of their use, while their arrangement should be based on their joint functioning or sequential functioning. Initial evidence has shown that layout can affect the quality of communication and the degree of interference and distraction in designed settings.

Arousal theory contends that the effects of the environment on performance are mediated by an increase in level of arousal, with concomitant increases in heart rate, respiration, and sweating. At the present time, arousal theory is the model most widely used to explain the performance effects of noise, heat, and cold. According to arousal theory, performance will be maximal with an intermediate level of arousal, and will gradually fall off as arousal either increases or falls. The optimum level of arousal for task performance is lower for complex tasks than for simple ones.

The theory of *auditory masking* proposes that the adverse effects of noise on performance occur because the target auditory signal is hidden or "masked" by extraneous noise. It is suggested that auditory masking occurred in two ways in studies that have shown performance decrements due to noise. First, the noise may have masked the sounds the subjects made as they responded to the task. Second, the noise may have masked the subjects' inner speech. A recently proposed model attempts to integrate in a single framework both the arousal and auditory masking theories. This model suggests that the positive effects of noise on performance are due to arousal, the negative effects to masking, and the failure to find effects to the fact that arousal and masking cancel each other.

Knowledge about the ways in which the physical environment affects performance has been applied to the formulation of design standards for some aspects of the ambient environment in buildings, such as illumination standards, and the evaluation of the functional effectiveness of innovative design plans, such as open-space classrooms and offices.

6 Coping with Environmental Stress

In 1976 the national news media carried the tragic story of an elderly couple who took their own lives in a Bronx apartment. A note left behind explained that they were unable to endure the stresses of city life any longer. For these two people, the city's cacophony of sight and sound, the polluted air and streets, and the ever present threat of violent crime (they had repeatedly been mugged and robbed) became too much. They were no longer able to marshal the energy and courage that are required to survive in today's urban environment.

While this tragic story is not typical, the central issue of urban challenge and urban stress is familiar to every city dweller. Here we encounter the physical environment not as a satisfying and enriching resource, but as a source of psychological stress. The human process involved is the challenge of coping effectively with extreme environmental demands. At risk are the psychological consequences of the failure to master the environmental stresses and eventually the hidden psychic costs of long-term adaptation to adverse environmental conditions.

The demands of urban life are particularly stressful for the elderly. Many everyday environments that are easily negotiated by the young—a dangerous traffic intersection, a steep flight of steps, poorly marked bus stops—can be a source of challenge and stress for the aged. And many older persons are faced with the stresses of institutional life in nursing homes, geriatric hospital wards, and sheltered-care settings.

The nature of environmental stress and the ways people cope with stressful environmental conditions represent an especially important and rapidly growing area of concern in environmental psychology. The social relevance of research on environmental stress is particularly apparent; while we are becoming more aware of the problems of environmental stress, our metropolitan areas continue to grow and the population of senior citizens becomes larger. Research on environmental stress offers essential aid in our efforts to humanize the urban environment and more effectively meet the needs of the elderly. Because the problems in this area are so severe and the long-term consequences so grave, the study of environmental stress is a uniquely important field of study for environmental psychologists.

THE NATURE OF ENVIRONMENTAL STRESS

In Chapter 5 we saw how the physical environment, including aspects of the ambient environment, can affect human performance. In this chapter we shall continue to focus a major part of our attention on features of the ambient environment (noise, extreme temperature, and air pollution), though here our emphasis will be on the ways these environmental variables can operate as *stressors* in people's lives. Aversive environmental conditions, such as intense noise and extreme heat or cold, can negatively affect people's health and emotional well-being, and disrupt their relationships with other persons.

Crowding can also be a source of stress. Because a vast body of research and theory has been addressed specifically to crowding, we shall examine it separately in Chapter 7. This treatment of ambient environmental stressors and crowding in separate chapters reflects only the need to organize two extensive bodies of literature efficiently, not a view that the two sources of stress are unrelated. In fact, we shall find that the negative effects of crowding on human functioning are similar to those exerted by ambient environmental stressors, and that some of the same theoretical perspectives have been called on to explain the effects of these two sources of environmental stress.

DEFINING ENVIRONMENTAL STRESS

The concept of stress is familiar in our popular vocabulary; we talk about stress at work, stress on the job, stress in interpersonal relationships. Yet our definition of stress is often vague, and the word may mean different things to different people. Psychologists who have studied environmental stress have found that an initial research task is to establish a precise definition of stress (see Appley and Trumbull, 1967; Cofer and Appley, 1964; McGrath, 1970). Two kinds of stress have come to be differentiated: systemic stress and psychological stress.

Systemic stress   The initial definition of stress, which evolved in medically oriented research, emphasized the systemic or physiological aspects of stress. The pioneering work in this area was carried out by Hans Selye (1956, 1973, 1976), who has conducted research in Canada on systemic stress for more than forty years. Selye defines *systemic stress* as the nonspecific response of the body to any environmental demands made on it. He calls the conditions in the environment that produce stress, such as toxic chemicals or extreme temperatures, *stressors.*

Central to Selye's definition of stress is the notion that stress involves a *nonspecific* response of the body. The body responds in an identical biochemical manner to a wide variety of environmental stressors, among them drugs, hormones, cold, and heat. It is immaterial whether the stressor is pleasant or unpleasant; a game of tennis or a passionate kiss will cause the same systemic stress reaction as a noxious chemical. The body's biochemical reactions to these different stressors are meant to cope with the increased demands on the body which they generate.

Selye identifies three distinct stages in the body's systemic response to stressful conditions, which he terms the *general adaptation syndrome,* or GAS. The first stage of the GAS is an *alarm reaction* of the autonomic nervous system, involving increased adrenaline secretion, heart rate, blood pressure, and skin conductance. This reaction represents a generalized "call to arms" of the body's defenses. The state of alarm is followed by a stage of *resistance* or adaptation, which includes a variety of physiological coping responses. These physiological changes are quite different from—and often opposite to—those of the alarm reaction. A third stage, *exhaustion,* will ensue if the stressor is intense and of sufficient duration, and if the resistive efforts fail to deal with it adequately.

Selye emphasizes the role of the GAS in coping with environmental stress. The GAS is the body's effort to continue to function in a steady state while under stress. The body's ability to maintain steady functioning despite changing external conditions is termed *homeostasis.* If the body is to remain healthy, none of its internal processes can be allowed to deviate far from their normal level of functioning. Homeostasis involves the many complex and coordinated biochemical mechanisms the body employs to ensure stable functioning despite changing environmental conditions.

The body uses two types of physiological reactions to maintain this homeostatic process. First, *syntoxic* reactions are called into play when an "aggressive" element attacks the body but does not present a serious threat to its functioning. These reactions operate to pacify the body's tissues, allowing them to function in a state of tolerance or "peaceful coexistence" with the aggressive agent. *Catatoxic* reactions are initiated when the body is attacked by an aggressive element that seriously threatens its functioning. The pathogen is attacked by chemical agents, such as destructive enzymes, that accelerate its metabolic breakdown. These adaptive bodily reactions presumably emerged in the course of evolution as the body learned to cope with a variety of aggressive elements. They rely on a complex pattern of stimuli and feedback involving chemical messengers (e.g., corticosteroids) and nervous stimuli.

Psychological stress   A more recent definition of stress has focused on its psychological aspects as well as its physiological ones. The most extensive work on psychological stress has been carried out by Richard Lazarus and his associates (Gal and Lazarus, 1975; Lazarus, 1966, 1968, 1971; Lazarus and Cohen, 1977; Lazarus, Cohen, Folkman, Kanner, and Schaefer, 1979; Lazarus and Launier, 1978). Lazarus emphasizes that stress involves an essential psychological component; the individual *appraises* the personal meaning and significance of the stressor.

Lazarus proposes the notion of cognitive appraisal as a mediating variable between the environmental stressor and the individual's adaptive reactions. Psychological stress occurs when the individual appraises the stressful environmental condition as threatening a level of harm that will strongly challenge or exceed his or her coping abilities. The individual's perception of the stressful situation is essential to the definition of stress; an objectively neutral situation that is *perceived* as threatening will cause psychological stress. Thus, according to Lazarus, cognitive appraisal is not a passive perception of the elements of a threatening situation, but an active psychological process in which the individual assimilates and judges the elements of the situation against an established pattern of ideas and expectations.

Psychological stress involves three types of cognitive appraisal. Lazarus defines *primary appraisal* as a mediating psychological process that serves to discriminate potentially threatening situations from beneficial or irrelevant ones. *Secondary appraisal* functions to assess the individual's resources for coping with the threatening situation. Finally, *reappraisal* is a change in the original perception of the situation due to either changing environmental conditions or changes within the person resulting from cognitive coping efforts. Reappraisal might, for instance, involve a changing perception that a situation that originally appeared benign is threatening after all.

Recently Lazarus and his colleagues (Folkman and Lazarus, 1980; Lazarus, 1980) have identified two major types of coping strategies individuals may employ in their attempts to deal with stressful situations. *Problem-focused coping* consists of behavioral or cognitive efforts to deal directly with the source of stress by altering the stressful environmental condition, changing one's own behavior in dealing with it, or both. *Emotion-focused coping* involves behavioral or cognitive efforts oriented toward reducing or better tolerating one's own emotional reactions to the stressful situation.

Imagine that a man buys a new home in New York City, only to discover after moving in that the house is located on an air corridor to La Guardia Airport. He may begin by reflecting on the potential seriousness of the problem (*primary appraisal*), deciding that the aircraft noise will be a major source of annoyance. Next, he considers how the noise problem might be coped with (*secondary appraisal*), and determines that it will be extremely difficult to deal effectively with the situation. This appraisal of the problem as both threatening and unmanageable triggers a psychological stress reaction; the man becomes physically upset, moody, and less interested in socializing.

Still, being a resourceful person, the man tries to deal with the noise problem as effectively as possible; he has double-paned windows installed to buffer the noise (*problem-focused coping*) and tries to reduce his negative feelings by focusing on the many positive features of his new home (*emotion-focused coping*). He is surprised to find that these coping efforts are quite successful. He discovers, in addition, that the air corridor is used only infrequently, during bad weather. Reflecting on this new information (*reappraisal*), the man decides that the situation is considerably more benign than it originally appeared, and the psychological stress reaction is successfully resolved.

It should be emphasized that the notions of psychological and systemic stress are not mutually exclusive concepts; the psychological stress process may include aspects of the systemic stress reaction. Richard Lazarus and Judith Cohen (1977) describe environmental stress as including somatic reactions (e.g., the secretion of catecholamines and corticosteroids), as well as behavioral and emotional ones. Similarly, Selye (1973) recognized that a harmless situation, if *interpreted* as threatening, will trigger a systemic stress response. Although the differences between psychological and systemic stress have typically been emphasized, it is probably more useful to consider the two concepts as emphasizing different aspects of a single stress reaction that includes both psychological and somatic elements.

IDENTIFYING ENVIRONMENTAL STRESSORS

Urban stressors   After defining stress, psychologists who have investigated environmental stress have sought to identify those environmental conditions that operate as stressors. The preponderance of work in this area has been concerned with identifying stressful features of the urban environment. The features of the ambient physical environment of cities that have been most fully investigated as stressors are noise, air pollution, and extreme temperatures. As Daniel Stokols (1979) has pointed out, the variables that have traditionally been labeled urban stressors are not restricted to urban settings, although they tend to be prevalent in cities.

An extensive body of literature deals with noise as an urban stressor (Cohen, Glass, and Phillips, 1979; Glass and Singer, 1972b). Researchers have been interested in learning about the potential effects on people's health and behavior of long-term exposure to the noise of heavy traffic, building construction, wailing sirens, and street repair (see box, "The Long-Term Costs of Environmental Stress"). A particularly interesting question has concerned the psychological effects of noise over which people are able to exert some personal control in contrast to noise that is uncontrollable.

Another urban stressor that has received considerable research attention is air pollution (Waldbott, 1973). The chemical composition of the city's polluted air is complex, and research has identified a growing list of polluting substances. The initial belief that urban air pollution consisted essentially of soot, dust, and pollen was replaced by a conviction that the chief pollutants were carbon monoxide and sulfur dioxide. Later research has revealed the presence in urban air of additional contaminants, including nitrogen oxides and photo-oxidants from automobile engines, asbestos particles from brake linings and building insulation, and colloidal substances generated by the action of tires on pavement. Yet more than half of the contaminants in urban air are as yet unidentified (Dubos, 1970). Research is needed to identify these additional pollutants, and to ascertain the long-term effects of continued exposure to air pollution on human health and behavior.

Further research has examined the role of extreme temperature, especially heat, as an urban stressor (Baron and Bell, 1976; Griffitt, 1970). Although heat may operate as a stressor in a variety of environmental settings, such as tropical cli-

The Long-Term Costs
of Environmental Stress

René Dubos, at Rockefeller University in New York City, has been especially concerned about the long-term consequences to human beings of living in stressful environmental conditions. His books *Man Adapting* (1965) and *So Human an Animal* (1968) have had a profound effect on public awareness of the human costs of our environmental problems. Dubos has argued eloquently that simply adapting to bad environmental conditions is not enough; we need to assess the long-term costs of such adaptation:

> Man is not on his way to extinction. He can adapt to almost anything. I am sure that we can adapt to the dirt, pollution and noise of New York City or Chicago. That is the real tragedy—we can adapt to it. As we become adapted we accept worse and worse conditions without realizing that a child born and raised in this environment has no chance of developing his total physical and mental potential. It is essential that we commit ourselves to such problems as a society and as a nation, not because we are threatened with extinction, but because, if we do not understand what the environment is doing to us, something perhaps worse than extinction will take place—a progressive degradation of the quality of human life.

[From R. Dubos, "Stimulus/Response: We Can't Buy Our Way Out," *Psychology Today*, 1970, *3:10*, pp. 20, 22, 86–87. Copyright 1970 Ziff-Davis Publishing Co. Reprinted with permission.]

Scientists are concerned about the long-term consequences to human life of adaptation to such stressful environmental conditions as air pollution and noise.

© *Peter Menzel/Stock, Boston.*

mates and some industrial settings, it is in cities that most of us encounter the concentration of buildings, industry, and people that can lead to an excessive buildup of heat. Heat waves have been found to be correlated with increases in the mortality rate in several American cities (Schuman, 1972). At a more speculative level, heat has received some attention as a possible aggravating factor in urban unrest during the "long, hot summers" of the 1960s (see the Report of the National Advisory Commission on Civil Disorders, 1968).

Environmental stressors and the elderly An especially important, though often neglected, issue in the study of environmental stress concerns environmental stressors as they are experienced by elderly persons (see Schooler, 1975). The urban stressors we have discussed are considerably more stressful for the elderly than for younger persons (Douglas, 1980). James Birren (1970) referred to the urban elderly as a "silent majority" trying to cope with the stresses of urban life, unable to join younger families in the movement to suburbia. Urban redevelopment often makes things worse rather than better for the elderly, who are faced with an array of new apartment buildings, shopping areas, and restaurants they cannot afford.

The effects of aging on perceptual acuity, physical strength and endurance, and ease of mobility make many environmental features that younger persons take for granted sources of challenge and frustration for the elderly (Carp, 1976; Lawton, 1977). Birren (1970) points out that many elements of the contemporary city can be discouraging obstacles for older persons, including high curbs, wide streets that are heavily traveled, fast-changing lights, poorly labeled buildings, and wind-tunnel effects between some high-rise buildings (Figure 6-1). As Susan Saegert (1976) has indicated, environments that demand high levels of effort, energy, and attention are stressful.

Investigators have been interested in learning how features of the housing environment, such as building type and structure and nearness to transportation facilities and other resources, affect the physical and psychological well-being of the elderly (Carp, 1976; Lawton, 1971). Researchers have also studied how the quality of institutional environments, such as nursing homes, geriatric hospitals, and sheltered-care facilities, affects the lives of their elderly residents (Lawton, 1977; Moos, 1980).

Our purpose in examining environmental stressors in the lives of elderly persons is not to exaggerate or overdramatize the "plight of the elderly" in today's society, and we do not suggest that older persons in general are helpless or defenseless in the face of environmental stress. In fact, for many persons the later years are a socially and personally meaningful period, in which cultural, recreational, and continuing professional interests are energetically pursued. Our goal is rather to consider realistically some sources of environmental stress that place significant adaptive demands on many elderly persons, and to encourage planners to design settings that can foster and maintain their adaptive potentials.

MEASURING ENVIRONMENTAL STRESS

Lazarus and Cohen (1977) have suggested three types of measures for assessing the psychological effects of environmental stress: somatic measures, behavioral mea-

Figure 6-1
Elderly pedestrians con-
front environmental stress
at wide intersections that
are poorly controlled.

© *Michael Abramson/Black
Star.*

sures, and subjective measures, corresponding to the physiological, functional, and affective components of stress reactions. Similarly, Stokols (1979) includes in his definition of stress the notion that stress is manifested in many physiological, behavioral, and emotional reactions.

Physiological measures   As Lazarus and Cohen explain, physiological measures can be used to assess both short- and long-term reactions to stress. Short-term in-dices are of three types. First are measures of bodily reactions that are controlled by the autonomic nervous system, such as cardiovascular changes, alterations in breathing rate, decreased skin resistance, increased muscle-action potentials, and changes in stomach motility. Second are measures of the secretion of catechol-amines, such as adrenalin (or the bodily reactions they stimulate), which are se-creted by the medullary portion of the adrenal glands. Finally, stress may be mea-sured by the secretion of corticosteroids by the cortex of the adrenal glands when they are stimulated by the pituitary gland.

Lazarus and Cohen point out that investigators can measure the long-term somatic reactions to stress in terms of stress-related diseases—diseases that are caused by prolonged exposure to internal stress. The accumulated effects of short-term stress reactions may result in serious somatic damage. Selye (1973) includes as stress-related diseases high blood pressure, headaches, upset stomach, and gastric and duodenal ulcers.

Behavioral measures   Lazarus and Cohen explain that people under stress do not behave as they ordinarily do, and these behavioral changes can be used to measure stress. Stress can be inferred from the *coping behaviors* that people employ to deal with it. Such coping behaviors might include actions to alter the stressful situation (quitting a stressful job) or to alleviate the symptoms of stress (taking tranquilizers). Stress may also be inferred from the *disorganized functioning* it causes, such as inappropriate actions, rigidity in behavior, and an inability to attend to the task at hand. One limitation of measures of disorganized functioning, however, is that they are influenced by people's ability to cope effectively with stress, as well as by the stress itself; stress does not inevitably impair functioning, and some people's functioning may even be enhanced by effective coping with a stressful situation. Another index of stress is the *expressive behaviors* it induces, such as fidgeting, lip biting, pacing, and tremulousness. Finally, stress may be assessed in terms of decrements in the performance of ongoing or subsequent tasks, such as those discussed in Chapter 5 (see Cohen, 1980).

Subjective measures   Stress can also be measured by subjective or self-report indices of its emotional or affective components. Lazarus and Cohen note, in fact, that subjective measures have been the most widely used indices of stress. Subjective measures may consist of an assessment of the degree of emotional distress caused by stress along a single dimension, or by ratings along a number of different dimensions of some of the distinctive emotional states associated with stress, such as anger, anxiety, or depression.

PSYCHOLOGICAL EFFECTS OF ENVIRONMENTAL STRESS

THE EFFECTS OF NOISE

Physiological effects   Research conducted primarily in laboratory settings has demonstrated conclusively that noise causes the systemic physiological reactions typically associated with stress. Noise has been shown to increase electrodermal activity (Glass and Singer, 1972b, 1973; Glass, Snyder, and Singer, 1973), the secretion of adrenaline (Frankenhaeuser and Landberg, 1977), and blood pressure (Cohen, Evans, Krantz, and Stokols, 1980; Jonsson and Hansson, 1977). The level of these physiological reactions tends to increase when the noise is intense, when it is aperiodic, and when it is uncontrollable. While there is some evidence that these physiological reactions abate as people become habituated to continued noise

(Glass and Singer, 1972), other data suggest that not all of them do (Jansen, 1969; Cohen, Evans, Krantz, Stokols, and Kelly, 1981).

Health effects   It is well established that long-term exposure to noise of high intensity can cause significant hearing loss. Boilermakers, for example, who are exposed to loud noise on the job are likely to suffer hearing loss (especially in the high-frequency range) with prolonged exposure (Beranek, 1966; McLean and Tarnopolsky, 1977). Some recent evidence suggests that even the typical sound levels in parts of hospitals and computer laboratories may threaten hearing loss (Falk and Woods, 1973; Shapiro and Berland, 1972). Even young people have been found to suffer minor degrees of hearing loss after frequent exposure to the sound levels at discotheques and rock concerts (Fearn, 1972).

Investigators have asked whether long-term exposure to noise may be related to health problems in the nonauditory realm. A survey study of Detroit and Los Angeles families found a correlation between reported exposure to noise and reported acute and chronic physical illnesses (Cameron, Robertson, and Zaks, 1972). The correlational and self-report nature of this research, however, suggest that additional data must be sought before the relationship of noise to nonauditory illness can be clearly established. Other studies have suggested some relationship between noise and some aspects of physical and mental health, such as headaches, nervousness, and insomnia (Crook and Langdon, 1974; Fog and Jonsson, 1968; Jones and Cohen, 1968). More recent correlational evidence suggests a slight association between exposure to aircraft noise in residential areas and the rate of admission to mental hospitals; however, socioeconomic differences between the areas investigated (e.g., Meecham and Smith, 1977) and inconsistent findings among the hospitals studied (e.g., Hand, Tarnopolsky, Barker, and Jenkins, 1980) limit the usefulness of these data.

After reviewing research on the effects of noise on health, investigators (Cohen, Glass, and Phillips, 1979; McLean and Tarnopolsky, 1977) conclude that such problems as sample selection and the validation of health indices make it impossible to conclude with certainty that noise causes either mental illness or nonauditory physical illnesses. Those studies that have shown some relationship between noise and mental illness do not clearly indicate whether noise causes psychopathology or merely aggravates existing psychological problems. Noise is probably related to health problems in susceptible individuals, particularly when it is experienced along with other sources of stress.

Behavioral effects   A growing body of research has shown that aversive noise has deleterious effects on many aspects of social behavior. When Lawrence Ward and Peter Suedfeld (1973) played taped traffic noise over loudspeakers at selected sites on the campus of Rutgers University, they found that students' classroom participation and attention declined, and professors asked less often for students' opinions. The noise also resulted in more disagreement, irrelevant dramatization, and more time used and statements made in group discussions. The authors speculate that the increased time and speech in group discussions may have been due to the increased difficulty of communicating over the noise.

In their survey of residents of San Francisco, Donald Appleyard and Mark

Lintell (1972) discovered that there was little social participation among residents on a street noisy with traffic, and a marked tendency to withdraw from the street socially. On a relatively quiet street with light traffic, in contrast, residents had three times as many friends and twice as many acquaintances, and showed a strong inclination to engage in the social life of the street. The correlational nature of this study leaves open the possibility that some of the observed differences may have been due to differences in the types of residents who lived on the streets in question.

Additional research has shown that noise reduces people's willingness to engage in helping behavior. Karen Mathews and Lance Canon (1975) found that subjects exposed to loud noise were less willing than those who experienced low noise to help an individual pick up accidentally dropped materials in a laboratory setting. When they used a field experimental design, the investigators found that people were less inclined to help someone pick up accidentally dropped books when a loud lawnmower was running than under quiet conditions. Richard Page (1977) also found that under noisy conditions, students on the campus of Wright State University in Ohio were significantly less willing to help another student who had dropped packages or to offer change for a quarter to a student who wanted to make a phone call than were students under low noise conditions.

Two laboratory studies have shown that people tend to pay less attention to social cues under noisy conditions. Sheldon Cohen and Anne Lezak (1977) demonstrated that although noise did not affect subjects' memory for task-relevant nonsense syllables in a laboratory task, noise did reduce their memory for task-irrelevant social information involving slides of people appearing calm or distressed. Judith Siegel and Claude Steele (1979) asked students in either a quiet or noisy setting to assign rewards to two persons they observed on tape playing a competitive word game. At a key point, the tape showed one player (the "harmdoer") inadvertently brushing the other player's (the "victim's") word solutions off the table. For half of the subjects in each condition the harmdoer was shown wearing an arm cast (low personal intent), while for the other half of the subjects in each group the harmdoer did not wear a cast (high personal intent). Siegel and Steele found that the social discriminations of the subjects were impaired in the high-noise condition; these subjects failed to distinguish between the harmdoer and victim roles or between high and low personal intent when they assigned their rewards. It seems possible that such a reallocation of attention away from social cues under noisy conditions may help to explain people's tendency to be less sociable and less helpful in noisy circumstances. We shall return to this important issue of how people allocate their attention under stress later, when we discuss theories of environmental stress.

Laboratory research has shown that people behave more aggressively toward one another in noisy conditions than in quiet conditions. Students threw more rubber balls at a "passive resistor" (actually an experimental accomplice) under noisy conditions (Knipmeyer and Prestholdt, 1973). Russell Geen and Edgar O'Neal (1969) found that students delivered more intense electric shocks to an experimental confederate (the shock represented an evaluation of the confederate's performance on a task) in a noisy setting when they had also just viewed an aggressive boxing film. Similarly, students gave more electric shocks to an experimental confederate under high noise conditions when they had also been previously angered (Donnerstein and Wilson, 1976; Konechni, 1975). While these findings on noise and aggression appear straightforward (noise increases aggression), we shall see later, when we consider the effects of extreme temperature on aggression, that there is probably an optimal range in which stress leads to aggression. It seems reasonable that at some point noise will become so aversive as to discourage any type of social activity, including aggressive acts. Furthermore, our interpretation of these findings needs to be cautious since laboratory responses, such as throwing a rubber ball or delivering an electric shock, may have low external validity as measures of aggression in real-life situations.

Subjective effects   People's subjective reactions to noise have typically been assessed by community surveys oriented toward particular noise problems, such as noise from traffic or from aircraft. Community surveys have generally focused on subjects' reports of "annoyance" as a result of noise. Investigators have tended to define annoyance as a form of mild anger that is often associated with people's feelings about activities that are disrupted by noise (Weinstein, 1976). In interpreting research findings in this area, we should bear in mind that the data are subject to some methodological limitations in that exposure to noise is correlational and willingness to participate in a community survey may bias the samples.

Several surveys have focused on residents' annoyance with traffic noise. Surveys in Great Britain (McKennell and Hunt, 1966) and in the United States (Bolt, Beranek, and Newman, 1967) have indicated that automobile traffic noise is the most bothersome and most frequently mentioned source of noise in urban settings. In a before–after study in Great Britain, residents were significantly more annoyed by traffic noise after the opening of a nearby highway than they had been before (Lawson and Walters, 1974). In all of these studies, however, considerable individual variation in annoyance was observed. Respondents were generally inclined to mention noise as a source of annoyance only when they were specifically asked about it; they seldom mentioned it spontaneously. Other features of the community (e.g., lack of local shopping and public transportation) were often seen as more annoying than traffic noise.

Other surveys have been conducted in the general vicinity of major airports throughout the world, including Great Britain (OPCS, 1971; Stockbridge and Lee, 1973), Sweden (Rylander, Sörensen, and Kayland, 1972), Los Angeles (Burrows and Zamarin, 1972), West Philadelphia (Bragdon, 1970), and Chicago, Dallas, Denver, and Los Angeles (Hazard, 1971). E. K. McLean and A. Tarnopolsky (1977) conclude that annoyance is related to level of exposure to aircraft noise; 60 to 70 percent of residents report annoyance in areas of high aircraft noise and only 3 to 4 percent register annoyance in areas of low aircraft noise. The areas where a very high proportion of residents are annoyed, however, are typically small and located very near the flight paths of aircraft that are landing or taking off. And as in the case of traffic noise, individual variation in reported annoyance with aircraft noise is considerable. McLean and Tarnopolsky point out that even at the lowest levels of aircraft noise, some people report being annoyed, while at the highest noise levels, some people are not annoyed at all.

Sheldon Cohen and Neil Weinstein (1981) point out that findings in this area are misleading in that they suggest a simple relationship between noise level and annoyance. In fact, actual sound level accounts for only from 10 percent to 25 percent of the variance in measures of annoyance. They suggest that additional intervening variables need to be explored to help us to understand how noise leads to annoyance, such as people's attitudes toward the source of the noise or their fears about its source (e.g., fear of plane crashes). They also suggest that some of the variability in indices of noise annoyance may be attributable to individual differences in *sensitivity* to noise (see box, "How Students Differ in Noise Sensitivity"). To complicate our interpretation of the meaning of measures of noise annoyance, some studies have revealed that even in areas where annoyance with noise is high, few people are inclined to move because of it, few express a willingness to incur a financial cost to reduce the noise level, and very few complain to responsible authorities (Fiedler and Fiedler, 1975; Goodman and Clary, 1976). Cohen and Weinstein (1981) note that while some initial research has examined the problem-fo-

How Students Differ in Noise Sensitivity

Neil Weinstein (1978) carried out a field correlational study to learn how differences in college students' sensitivity to noise affected their reported annoyance with noise in their dormitory. He began by mailing a survey concerned with noise sensitivity to all freshmen assigned to a dormitory at Rutgers University during the summer before they arrived on campus. From this sample Weinstein selected two subgroups: students who scored within the top 30 percent of noise-sensitivity scores (noise-sensitive students) and those who fell within the bottom 30 percent of scores (noise-insensitive students).

Next, Weinstein administered a questionnaire to find out how annoyed students actually were by the noise in their dormitory early in the school year and then again seven months later. The questionnaire assessed how disrupted students felt their behavior was by noise in the dormitory, as well as the negative feelings they experienced about dormitory noise. Weinstein discovered that the noise-sensitive students were in fact significantly more bothered by noise in their dormitory than were the noise-insensitive students. He also discovered that while the annoyance scores of the noise-insensitive group did not change over the seven-month period, the noise-sensitive group became increasingly more annoyed with dormitory noise during the school year.

To learn more about noise sensitivity, Weinstein conducted a follow-up survey study with students at Rutgers University and the University of California at Berkeley, oriented toward identifying some of the personality characteristics that are associated with noise sensitivity. He found that in comparison with noise-insensitive students, noise-sensitive students generally had a greater desire for privacy, were less comfortable and effective in social situations, and tended to express annoyance about a wide variety of nuisances. The size of the correlations between noise sensitivity and these personal characteristics was only moderate, however. Weinstein concludes that the tendency to be sensitive to noise is complex and not reducible to any single personality trait.

cused coping strategies people use to deal with noise problems (e.g., Appleyard and Lintell, 1972), such as changing bedrooms or installing double-paned windows, very little is known about emotion-focused coping in regard to noise. The need for research on how people cope with noise is underscored by Cohen and Weinstein's finding that people do not seem to adapt to noise in residential settings; long-time residents report being as bothered by noise as new arrivals.

THE EFFECTS OF EXTREME TEMPERATURE AND AIR POLLUTION

Health effects   Several correlational studies have reported a significant relationship between hot spells in urban areas and mortality rate. The mortality rate in some American cities rose sharply during a midsummer hot spell in 1976; the increase in mortality was as high as 50 percent in some areas (Schuman, 1972). Similarly, marked increases in mortality have been reported during heat waves in Los Angeles (Oechsli and Buechley, 1970) and New York City (Buechley, Van Bruggen, and Truppi, 1972). Rudolf Moos (1976) cautions, however, that the relationships reported in these correlational studies are based on normative statistics involving large numbers of people, and not on a direct investigation of the reasons for the deaths of particular individuals during heat waves.

A considerable body of literature dealing primarily with correlational studies has also demonstrated a relationship between various aspects of air pollution and physical illness. One approach has been to study how illness rates are affected during periods of excessive air pollution caused by temperature inversions. Evidence indicates a very strong relationship between temperature inversions and the occurrence of respiratory illnesses (Goldsmith, 1968).

Other studies have examined correlations between illness rates and fluctuations in air quality during more moderate periods of pollution. Data pointing to an association between the incidence of respiratory diseases (asthma, upper respiratory infections, the common cold) and relative increases in air pollution have been reported for Nashville (Zeidberg, Prindle, and Landau, 1961, 1964), Los Angeles (Sterling, Phair, Pollack, Schumsky, and De Groot, 1966), and New York City (Lebowitz, Cassell, and McCarroll, 1972). There is some evidence that such illness reactions to air pollution are most typical of "sensitive individuals" within the general population, such as older persons or those with respiratory diseases (Lebowitz, Cassell, and De Groot, 1966). A smaller number of studies have explored the possible relationship between air pollution and cardiovascular disease. There is some evidence from experimental studies conducted in laboratory and field settings that high levels of carbon monoxide reduce the time before symptoms appear during periods of exercise in patients with coronary artery disease (Anderson, Andelman, Strauch, Fortuin, and Knelson, 1973; Aronow, Harris, Isbell, Rokaw, and Imparato, 1972).

Recent reviews (Evans and Jacobs, 1981; Moos, 1976) have pointed out that the correlational nature of most of the work in this area makes it impossible to know which particular pollutants, combination of pollutants, or interactions between pollution and other weather conditions, such as high temperatures, are causally linked to the illness conditions under study. Further, there may be long-term consequences to health from living in polluted conditions that are not evident during short-term episodes of high pollution.

Behavioral effects   An important body of research based on laboratory experiments has suggested that excessively high temperatures affect people's social behavior. Two studies have demonstrated that people in hot conditions (dry bulb temperatures over 100° F) liked an anonymous stranger they evaluated from a questionnaire less than did persons in comfortable conditions (dry bulb temperature in the 70s) (Griffith, 1970; Griffith and Veitch, 1971). Later work (Bell and Baron, 1974, 1976), however, failed to find that heat reduced attraction toward another person who was actually in the room; it may be that when the other person shares the stressful condition, the positive effects of "shared suffering" counteract the otherwise negative interpersonal effects of heat (see Latané, Eckman, and Joy, 1966). Similar results have been reported when noise is the stressor; noise decreases attraction toward an anonymous stranger evaluated from a questionnaire, but actually increases attraction toward another person who has been continuously present in the aversive situation (Kenrick and Johnson, 1979).

A series of experimental research studies conducted by Robert Baron and his colleagues has investigated the effects of heat on interpersonal aggression. Baron (1972) discovered that, contrary to expectations, high temperatures *reduced* the level of aggression (delivery of shocks) people showed toward another person (an experimental confederate) in the same setting. Again, shared suffering may explain the findings, though Baron also noted that subjects could escape the uncomfortably hot environment more quickly by reducing the duration of the shocks they delivered. Baron and Sandra Lawton (1972) later showed that exposure to an aggressive model resulted in a significant increase in the number of shocks delivered in the hot condition but not in the comfortable condition. They concluded that high temperatures can facilitate the aggression-eliciting influence of an aggressive model.

A later study by Baron and Paul Bell (1975) resulted in the initially surprising finding that heat facilitated aggression in nonangry subjects, but *reduced* aggression in subjects who had been provoked to anger toward the victim. Ensuing studies (Baron and Bell, 1976; Bell and Baron, 1977) have provided an explanation of these findings. The investigators suggest that negative affect leads to aggression when the negative affect is at an *intermediate* level, but that aggression will not occur if the negative affect is either too low or too high. Thus either heat alone or provocation alone (intermediate affect) will lead to aggressive behavior, but heat and provoked anger together (very high affect) will not result in aggressive behavior. In one of the few studies to deal with uncomfortably cold conditions (64° F) and aggression, Bell and Baron (1977) showed that the effects of cold conform to their model for predicting the effects of heat on aggression.

In a field correlational study based on archival data, Baron and Victoria Ransberger (1978) attempted to apply this model of heat and aggression to analyze incidents of collective violence in the United States between 1967 and 1971. They compared 102 incidents of recorded collective violence with the maximum ambient temperature level at the time of the violence. Their data seemed initially to suggest that collective violence and ambient temperature levels were indeed curvilinearly related (an inverted U-shaped curve). It appeared that violence was most frequent in the 80° F range, and tended to decrease as temperatures either fell below or rose above this level.

More recently, however, J. Merrill Carlsmith and Craig Anderson (1979) have suggested that Baron and Ransberger's earlier analysis is misleading in that it did not take account of the fact that there are more days at some temperature ranges than others. Taking account of the number of days in various temperature ranges, Carlsmith and Anderson have demonstrated that the probability of collective violence increases monotonically with temperature (Figure 6-2). The authors do point out, however, that their analysis is restricted to the normal temperature range, and that it is likely that at some point the temperature will become so hot that the probability of violence will begin to decline.

Very little is known about the effects of air pollution on people's social behavior, although a small number of data from controlled experimental studies suggest that some aspects of poor air quality can reduce feelings of interpersonal attraction (see Evans and Jacobs, 1981). One study (Rotten, Barry, Frey, and Soler, 1978) conducted with college students at an Ohio university examined how one malodorous component of air pollution (ammonium sulfide) affected attraction toward a stranger. Consistent with experimental findings in regard to other environ-

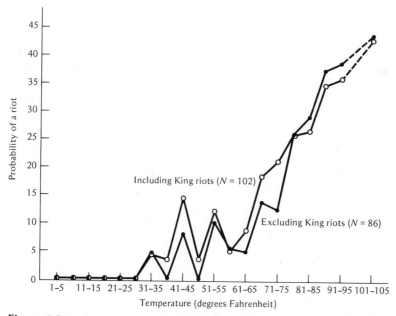

Figure 6-2
Likelihood of a riot as a function of ambient temperature.

From J. M. Carlsmith and C. A. Anderson, "Ambient Temperature and the Occurrence of Collective Violence: A New Analysis," Journal of Personality and Social Psychology, 1979, 37:337–44. Copyright 1979 by the American Psychological Association. Reprinted by permission of the publisher and author.

mental stressors (e.g., heat and noise), air pollution decreased feelings of attraction toward an anonymous stranger who did not share the noxious episode with the subject, while increasing attraction toward a stranger who appeared similar to the subject on a questionnaire and who seemed to share the same unpleasant experience as the subject.

A second experimental study (Bleda and Sandman, 1977) conducted with military personnel in the Washington, D.C., area revealed that secondhand cigarette smoke can reduce feelings of attraction on the part of nonsmokers. Nonsmokers were less attracted to a stranger they interacted with (actually an experimental accomplice) who smoked than to one who did not smoke. The authors attributed the reduced liking to the physical discomfort associated with secondhand cigarette smoke. In this situation, the environmental stressor reduced liking for another person actually in the setting with the subject, possibly because subjects did not assume that the smoker's experience was unpleasant and because the smoker's behavior was the direct cause of their own discomfort.

Subjective effects   Although measurement techniques are available for assessing the odorous effects of air pollution (see Berglund, Berglund, and Lindvall, 1976), assessment of the subjective effects of air pollution presents some measurement challenges. First, as we saw in Chapter 2, although people are able to perceive concentrations of particulate matter in polluted air, they are much less able to perceive such gaseous pollutants as sulfur dioxide and hydrocarbons. Mary Barker (1976) explains that ordinary citizens are likely to become aware of air pollution only when the pollution level affects visibility (e.g., smoke or haze), when clothing or personal property is soiled, or when the pollution carries offensive odors. Gaseous pollutants are likely to be perceived only at very high concentrations, when physical discomfort in the eyes and nose may be experienced. People's awareness of air pollution is further limited because air pollution is chronic and insidious rather than sudden and dramatic.

After summarizing available studies of the subjective effects of air pollution, Barker concludes that there are wide variations in the awareness and concern of individuals and of whole communities about air pollution. Fifty-three percent of citizens interviewed in Edinburgh identified an air-pollution problem, while a large majority of respondents in Sheffield denied the existence of an air-pollution problem; winter smoke concentrations in the two cities were, in fact, similar (I.G.U. Air Pollution Study Group for the U.K., 1972). Barker suggests that people's awareness of air pollution is only partially affected by direct sensory perception, and that indirect experience (e.g., news media or publicity campaigns) exert a major, possibly even dominant, influence. People with such respiratory problems as bronchitis and emphysema are more likely to recognize air pollution as a serious problem than people who have no such health problems.

It appears that many citizens employ emotion-focused coping strategies in dealing with air pollution as a potential area of concern in their lives. People may recognize a general air-pollution problem but deny that it affects them personally (Wall, 1973). Similarly, citizens may admit that air pollution is a national problem, but comfort themselves by pointing out that their community is less polluted than others. Or air pollution may be recognized as a local problem but evaluated as a less serious concern than other problems, such as unemployment and race rela-tions (Miller, 1972). Finally, Barker (1976) notes that because people often do not see well-defined channels for taking action against air pollution, they may cope with their feelings of frustration by denying the seriousness of the problem or by insisting that someone else, such as the government, is solving it (see Rankin, 1969).

Research on the subjective effects of temperature has primarily emphasized the human factors and performance aspects of subjective comfort, which are dis-cussed in Chapter 5.

EFFECTS OF ENVIRONMENTAL STRESSORS ON THE ELDERLY

Health effects   The health effects of high temperatures and air pollution are especially pronounced among the elderly. Researchers who have found an associa-tion between urban heat waves and rising mortality rates have reported that mor-tality among the elderly is especially high (Oechsli and Buechley, 1970; Schuman, 1972) (see box, "Victims of Heat: The Poor and the Old"). The relationship be-tween variations in air quality and illness rates is also especially pronounced for older urbanites (Zeidberg, Prindle, and Landau, 1961), and extremely poor air quality during temperature inversions can be fatal for some elderly persons (Gold-smith, 1968).

Mortality among the institutionalized elderly can increase dramatically after they are involuntarily relocated (Kasl, 1972; Lawton and Nahemow, 1973). Nor-man Bourestom and Leon Pastalan (1975) report an increased mortality rate of up to 100 percent among elderly persons forcibly transferred from one institution to

Victims of Heat: The Poor and the Old

Under a scorching sun, temperatures hung in triple digits this summer for as long as a month in some places, prompting a wave of press stories about 1,000-plus deaths resulting from the heat. The headlines may actually have been too cool. The Atlanta-based Center for Disease Control has now verified 148 such deaths during July in Kansas City alone. About 450 people received emergency care in hospitals, and untold numbers consulted private physicians.

The heat's chief victims were the poor and the old. The median age of those who died in Kansas City was 73 (72% were 65 or over). The death rate in low-in-come areas was 9.6 per 10,000, vs. .09 in wealthy districts. Apparent reason: the well-off elderly were not only healthier to begin with, but better supplied with fans, air conditioners and other aids to keep body temperatures below the 105° F (40.6° C) or so that helps trigger heat stroke.

(From "Victims of Heat: The Poor and the Old (65 Plus)," *Time*, September 1, 1980, p. 55. Reprinted by permission from *Time*, The Weekly Newsmagazine; copyright Time Inc. 1980.)

another. Mortality was related to both the amount of environmental change involved in the move and the degree of the individual's physical and mental decline.

Behavioral effects   Pastalan (1970) notes that an individual's world tends to shrink as he or she grows older and is less willing or able to cope with the demands of the environment. As one's energy and physical abilities decline, the range of one's activities contracts. M. Powell Lawton (1977) explains that the use of resources by older persons is a function of (1) physical distance to the resources, (2) barriers to the elderly persons' mobility, and (3) the salience of the resources. Table 6-1 shows that when resources are highly salient, such as medical facilities or the homes of children and relatives, they will be visited by older persons even when the physical distance entailed is considerable. In contrast, other resources that are nearby though less salient, such as beauty shops, barbershops, and restaurants, are underused because of the economic barriers they present to many elderly people. Similarly, entertainment is underused by the elderly because it is less salient, is costly, and is often located at some distance. Finally, resources that are both close by and highly salient, such as the homes of friends, churches, and grocery stores, are well visited by the elderly.

Frances Carp (1980) conducted a field correlational study that is especially helpful in suggesting some of the reasons *why* the mobility patterns of the elderly are constricted. She surveyed elderly retired people in San Francisco and San Antonio concerning their *actual* mobility patterns in the urban environment, as well as their *desired* use of the environment. She was particularly interested in learning how characteristics of the urban environment affect the mobility of the elderly. San Francisco is a spatially concentrated city, with many amenities available in local neighborhoods, and with excellent coverage of the urban area by the city transit system. San Antonio, in contrast, is spread out over a large area, has many shopping malls and other amenities located outside of the local neighborhood, and has limited bus service.

Carp discovered that the mobility patterns of elderly people in the two cities were remarkably different. In San Francisco, nearly two-thirds of older persons went to see friends at least once a week, while only 14 percent reported never visiting friends. In San Antonio, in contrast, just over one-third of the elderly visited friends at least weekly, while 42 percent said they never went to see friends. Similarly, while 21 percent of elderly San Franciscans went grocery shopping every day, only 5 percent of older persons in San Antonio shopped daily. Trips for entertainment were particularly noteworthy: 20 percent of the San Francisco sample and 81 percent of the San Antonio sample never made entertainment trips.

While these findings show that in both cities the elderly are not very mobile outside of their own neighborhoods, they also point to marked differences in mobility patterns between the two cities. Elderly persons were considerably more mobile in San Francisco than in San Antonio. These differences become especially significant when one learns that elderly people in San Antonio consistently indicated that they would make more trips to visit friends and relatives and for entertainment if they had transportation. Based on these correlational findings, Carp speculated that the more spread-out urban area and more distant amenities in San Antonio, coupled with limited city bus service, were the chief reasons why the el-

Table 6-1. Modal percentage of elderly persons using various resources, frequency of use, and travel time and distance required.

Resource	Median percent using	Modal frequency of use (users)	Modal frequency of use (all)	Modal travel time (users)	Modal use distance, NYC [a]	Modal nearest distance, PH [b]
Grocery	87%	2/week	1 or 2/week	7 min.	1 to 3 blocks	1 to 3 blocks
Physician	86	"several"/year	several/year	15 min.	>20 blocks	4 to 10 blocks
Visit one or more children	98	1/week	1/week to never	20 min.	<10? blocks	
Other shopping	70	1 or 2/month	never		4 to 6 blocks	
Church	67	1/week	1/week	12 min.	4 to 6 blocks	4 to 10 blocks
Bank	64	1/month			4 to 6? blocks	
Visit friends	61	2 or 3/week	never	7 min.		
Visit other relatives	57	several/year	never	35 min.		
Beauty/barber shop	40				>20 blocks	1 to 3 blocks
Restaurant	31	several/year	several/year		1 to 3 blocks	1 to 3 blocks
Park	30					
Clubs, meetings	29	1/month	never	15 min.	>20 blocks	
Entertainment	19	1/month	never	20 min.	>20 blocks	>11 blocks
Library	18		never			>11 blocks

[a] Cantor (1975), New York City poverty-area residents.
[b] Newcomer (1973), public housing tenants.

Source: M. P. Lawton, "The Impact of the Environment on Aging and Behavior," in J. E. Birren and K. W. Schaie (eds.), Handbook of the Psychology of Aging, Litton Educational Publishing, Inc., 1977, p. 278. Reprinted by permission of Van Nostrand Reinhold Company.

derly in San Antonio were less mobile than those in San Francisco. These findings are of particular relevance to transportation planning, and we shall return to this issue later, when we discuss applications of research knowledge to transportation planning for the elderly.

Subjective effects   A considerable amount of self-report information has been collected concerning housing satisfaction among older people. Surveys of housing satisfaction have been conducted in age-segregated housing for the elderly (e.g., Rosow, 1967) and planned housing for older persons (e.g., Carp, 1966). Overall, such surveys indicate that elderly residents have positive feelings about age-segregated and planned housing (Lawton, 1977). Carp (1976) cautions, however, that self-reported satisfaction with housing is greatly affected by respondents' perceptions of alternative housing choices. Alternatives for the elderly are often limited, and housing with which elderly residents report they are satisfied has often been rated as poor by investigators.

THEORETICAL PERSPECTIVES ON ENVIRONMENTAL STRESS

As we have seen, environmental stressors can exert powerful effects on our well-being, affecting our physical health, our interactions with other people, and our sense of satisfaction and morale. Environmental psychologists have been especially interested in understanding the psychological dynamics that underlie these diverse effects of environmental stressors in our lives. Exciting theoretical developments have emerged from efforts to understand the *cognitive* and *psychological* factors that mediate the effects of environmental stressors on people's behavior and experience. Psychologists have studied how people's ability to *predict* and to *control* stressful events affects their psychological consequences. Researchers have investigated how an inability to control stressful conditions may cause people to develop feelings of *helplessness* and to lose the determination to improve their circumstances. The effects of environmental stressors on both ongoing and subsequent activities have been explained in terms of an *overload* of the individual's capacity to process the information necessary to carry out those activities effectively.

PREDICTABILITY AND ENVIRONMENTAL STRESS

After a series of experiments on the psychological consequences of noise, David Glass, Jerome Singer, and their associates (Glass, Reim, and Singer, 1971; Glass and Singer, 1972a, 1972b; Glass, Singer, and Friedman, 1969; Glass, Singer, and Pennebaker, 1977) concluded that the psychological costs of noise are greater when the noise is unpredictable than when it is predictable. Glass and Singer initiated their research with the assumption, based on work by Selye (1956), that people expend "psychic energy" in the process of adapting to stressful environmental conditions. They reasoned further that people who had to adapt to an unpredictable stressor would need to expend more energy and would incur greater psychological costs than would people who adapted to a predictable stressor.

To test this assumption, Glass and Singer exposed subjects to either predictable (periodic) or unpredictable (aperiodic) noise in a laboratory setting, and observed their functioning during both the noise period and the period after the noise was terminated. They found that subjects in both conditions adapted successfully while the noise was on, but that *after* exposure to *unpredictable* noise, subjects showed impaired performance on laboratory tasks and a reduced tolerance for frustration.

You may recall from Chapter 5 that task performance during the postnoise period is more adversely affected by aperiodic than by periodic noise. What is important here is Glass and Singer's interpretation of these performance decrements in terms of a *stress* model of the effects of noise. The physiological data they collected (phasic skin conductance, vasoconstriction of peripheral blood vessels, and muscle-action potentials) indicated that the initial exposure to noise was stressful, but that subjects adapted to the noise as the stressful conditions continued. They note that their findings concerning the negative effects of unpredictable noise have also been shown for stressors other than noise, such as electric shock.

PERSONAL CONTROL AND ENVIRONMENTAL STRESS

Several investigators have argued that the negative psychological effects of environmental stressors can be reduced when individuals achieve personal control over the stressors. James Averill (1973) explains that there are three types of personal control people can exert over threatening circumstances. He defines *behavioral* control as the availability of a response that can directly modify a threatening event. For example, a person who becomes uncomfortably hot at home may modify the situation by turning on the air conditioner. *Cognitive* control refers to the way an individual interprets a threatening situation. An urbanite might, for instance, decide that air pollution is a small price to pay for the diverse cultural advantages of life in a big city. Predictability may even be considered an example of cognitive control in that it provides a form of "informational control" over a stressor. *Decisional* control is defined in terms of the range of choices available to an individual. For example, an elderly person who requires some form of specialized care might be allowed to choose whether to live in a nursing home, a sheltered-care setting, or a planned community.

Although Averill cautions that personal control will not always succeed in reducing stress, laboratory studies have uniformly found that behavioral control reduces the negative psychological effects of noise. Glass and Singer (Glass, Reim, and Singer, 1971; Glass and Singer, 1972a, 1972b; Glass, Singer, and Friedman, 1968; Glass, Singer, and Pennebaker, 1977) demonstrated that when subjects are told they can terminate aversive noise by pressing a button attached to their chairs (behavioral control), the adverse psychological aftereffects of unpredictable noise are substantially reduced (see box, "An Unintended Consequence of Personal Control"). The beneficial effects of personal control occur as long as subjects believe they can terminate the noise, even if the button is never pressed. *Indirect* control (behavioral control exercised through an intermediary who has direct control) also mitigates the negative psychological aftereffects of unpredictable noise.

An Unintended Consequence
of Personal Control

Occasionally researchers discover the importance of a variable through the *unintended* effects it has on people's behavior. Gerald Gardner (1978) has described such an experience in a series of research studies he carried out dealing with the negative aftereffects of aversive noise. Employing the research paradigm utilized by David Glass and Jerome Singer (1972b), he studied accuracy on a proofreading task for student subjects exposed to either a noise or a no-noise control condition. To Gardner's surprise, while the first three studies he conducted replicated Glass and Singer's finding that noise produces negative aftereffects on the speed and accuracy of proofreading, the next two studies he carried out failed to show any aftereffects.

In attempting to learn what "went wrong" in his last two studies, Gardner discovered that the procedures for involving subjects in the experiments had been different in the first three studies from those in the last two studies. When the last two studies were conducted, stringent federal guidelines designed to protect human subjects from any harmful effects of research studies had been instituted. Subjects were informed that they were free to withdraw from the experiment without penalty, that they were free to choose an alternative way to meet course requirements besides being subjects in a research experiment, and that they might bring any complaints about the experiment to identified university administrators.

Gardner speculated that the new procedures may have afforded the subjects in his last two experiments perceived personal control over the noise, and that this personal control eliminated the expected negative aftereffects of noise. To find out whether the new procedures had affected the experimental outcome,

Drury Sherrod and his colleagues (Sherrod, Hage, Halpern, and Moore, 1977) demonstrated that the deleterious effects of aversive noise on task performance and persistence on a postnoise task were reduced linearly as subjects' behavioral control over the noise increased. Subjects who were able to stop the noise performed better than those who could only start it, while subjects who could turn the noise both on and off performed better than those who could only stop it. Two additional studies have shown that behavioral control can reduce the adverse interpersonal consequences of noise. Sherrod and Robin Downs (1974) found that subjects who believed they could have aversive noise turned off were 50 percent more helpful to another student (an experimental confederate) who needed help in working some problems after the noise period than were subjects exposed to noise without personal control. Edward Donnerstein and David Wilson (1976) report that postnoise aggression (shocks delivered to an experimental accomplice) was significantly higher in angered subjects exposed to noise without personal control than in either angered subjects not exposed to noise or angered subjects exposed to noise they were told they could have terminated (Figure 6-3).

Recently the role of personal control in the reduction of stress has been extended to research with the elderly in institutional environments. Richard Schulz

he designed another study in which the old and the new procedures were systematically varied. While noise produced negative aftereffects in terms of reduced accuracy and speed of proofreading under the old procedures, there was no significant difference in accuracy or speed between the noise and no-noise conditions when the new procedures were employed.

Accuracy and speed of proofreading as a function of exposure to noise and subject procedures.

Condition	% accuracy	No. of lines read
Old procedures		
Silence	44.1	147.0
Noise	41.4	114.4
New procedures		
Silence	44.8	132.5
Noise	44.2	145.6

Source: G. T. Gardner, "Effects of Federal Human Subjects Regulations on Data Obtained in Environmental Stressor Research," Journal of Personality and Social Psychology, 1978, 36: 628–34. Copyright 1978 by the American Psychological Association. Reprinted by permission of the publisher and author.

(1976) investigated the effects of increases in both personal control and predictability on the well-being of residents in a retirement home in North Carolina. In a field experiment, he had college students visit the residents under three conditions: (1) residents in a *no-control* condition were visited on a random schedule; (2) persons in a *predictability* condition were informed when a visit would occur and how long it would last; and (3) individuals in a *behavioral-control* condition were allowed to determine the frequency and duration of visits. Residents who were not visited served as a baseline *comparison* group. Both the behavioral-control and predictability conditions had positive effects on the well-being of the elderly residents. People in both groups were more healthy physically, more positive psychologically, and more active than those in the no-control and comparison conditions. Note that while Schulz found control enhanced the residents' well-being, he did not find, as Glass and Singer did, that control was more beneficial than predictability alone. More research is needed to ascertain the relative benefits of control and predictability.

In another field experiment, conducted in a nursing home in Connecticut, Ellen Langer and Judith Rodin (1976) compared a group of residents whose *decisional control* and *behavioral control* were enhanced with a group whose control in

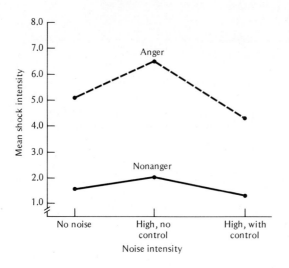

Figure 6-3

Mean shock intensity as a function of noise, control, and anger.

From E. Donnerstein and D. W. Wilson, "Effects of Noise and Perceived Control on Ongoing and Subsequent Behavior," Journal of Personality and Social Psychology, *1976, 34:774–81. Copyright 1976 by the American Psychological Association. Reprinted by permission of the publisher and author.*

these areas remained low. Residents in an *enhanced-control* condition were given a talk that emphasized that they were responsible for themselves, and were offered houseplants they could care for themselves. Persons in a *low-control* condition, in contrast, were told that the nursing-home staff was responsible for them, and then were given plants that were cared for by the staff. After three weeks behavioral measures and survey ratings revealed that residents in the enhanced-control condition were more alert, more active, and had a generally greater sense of well-being than people in the no-control condition (Table 6-2).

LEARNED HELPLESSNESS AND ENVIRONMENTAL STRESS

Some environmental psychologists have proposed that the negative psychological consequences of an inability to exert personal control over adverse environmental circumstances are mediated by the experience of helplessness. They have drawn on the theory of *learned helplessness*, originally developed in experimental psychology, in attempting to understand why people who are unable to control stressful conditions tend to demonstrate low persistence and a low level of tolerance for frustration after the stressful conditions have ended. The theory of learned helplessness has been most fully developed by Martin Seligman (1973, 1974, 1975). Seligman describes learned helplessness as a psychological condition in which an individual develops the expectation that his or her responses are independent of future (typically aversive) outcomes; that is, the person comes to believe that his or her actions will be unable to influence the outcome of unpleasant future events.

A study by Donald Hiroto (1975) provides an example of the standard research paradigm that has been employed in most learned helplessness studies with human subjects. In the first phase of the study, Hiroto assigned college student subjects to one of three experimental conditions. Subjects in a *controllable-noise* condition could terminate an aversive noise by pushing a button. Those in an *uncontrollable-noise* condition were faced with noise that was terminated independently of their responses. A third group served as a *no-noise* comparison. In the sec-

Table 6-2.   Mean scores on self-report, interviewer, and nurses' ratings for responsibility-induced and low-control comparison conditions among nursing-home residents.

Questionnaire responses	Responsibility induced (N = 24)			Comparison (N = 28)			Comparison of change scores (p <)
	Pre	Post	Change: Post–pre	Pre	Post	Change: Post–pre	
Self-report							
Happy	5.16	5.44	0.28	4.90	4.78	−0.12	0.05
Active	4.07	4.27	0.20	3.90	2.62	−1.28	0.01
Perceived control							
Have	3.26	3.42	0.16	3.62	4.03	0.41	—
Want	3.85	3.80	−0.05	4.40	4.57	0.17	—
Interviewer rating							
Alertness	5.02	5.31	0.29	5.75	5.38	−0.37	0.025
Nurses' ratings							
General improvement	41.67	45.64	3.97	42.69	40.32	−2.39	0.005
Time spent							
Visiting patients	13.03	19.81	6.78	7.94	4.65	−3.30	0.005
Visiting others	11.50	13.75	2.14	12.38	8.21	−4.16	0.05
Talking to staff	8.21	16.43	8.21	9.11	10.71	1.61	0.01
Watching staff	6.78	4.64	−2.14	6.96	11.60	4.64	0.05

Source: E. J. Langer and J. Rodin, "The Effects of Choice and Enhanced Personal Responsibility for the Aged: A Field Experiment in an Institutional Setting," Journal of Personality and Social Psychology, 1976, 34: 191–98. Copyright 1976 by the American Psychological Association. Reprinted by permission of the publisher and author.

ond stage of the experiment, all three groups were presented with aversive noise that they could terminate simply by moving a lever from one side of a shuttle box to the other. While subjects in the *controllable-noise* and *no-noise* conditions readily learned the escape response, those in the *uncontrollable-noise* group failed to escape, and instead listened passively.

Seligman explains that the theory of learned helplessness entails three sequential steps. First, the individual acquires information from the environment about the relationship between his or her responses and aversive environmental outcomes. In the second step, which is essential to learned helplessness, the individual processes the environmental information and transforms it into a cognitive representation of the contingency. Here the person becomes convinced that his or her efforts will not affect the outcome of future events. In the third step, this cognitive representation shapes the individual's behavior.

When individuals have determined that unpleasant future outcomes cannot be influenced by their personal actions, their *motivation* to respond is lowered, and they reduce their initiation of responses. Once people decide that they cannot influence future outcomes, it becomes more difficult for them to relearn at a *cognitive level* that their personal actions can in fact influence outcomes. Learned helplessness has *emotional* effects as well. The occurrence of a traumatic event generates fear, and the individual's awareness that he or she cannot control the event causes the fear to be replaced by depression.

An extensive body of research based on the learned helplessness paradigm has been carried out with dogs, cats, rats, and fish (see Maier and Seligman, 1976) and with human subjects (e.g., Burger and Arkin, 1980; Gatchel and Proctor, 1976; Hiroto and Seligman, 1975; Kranz, Glass, and Snyder, 1974; Miller and Seligman, 1975). The research with animals has involved escape responses to aversive stimuli, such as electric shock; human studies have entailed both escape responses to aversive stimuli (usually loud noise) and random reinforcement on difficult or unsolvable problems (such as anagram puzzles). The overwhelming evidence supports the theory of learned helplessness, although, as we shall see, research findings with human subjects have been more complex than those with animals.

In an interesting field experiment, Sheldon Cohen and his colleagues (Cohen, Evans, Krantz, and Stokols, 1980) applied the learned helplessness paradigm to assess the consequences of long-term exposure to aircraft noise on the cognitive functioning of schoolchildren. They studied children in the four noisiest schools in the air corridor of the Los Angeles Airport and children in three matched quiet schools. All of the children were tested in a noise-insulated trailer parked near the schools. The investigators employed a puzzle task in the first phase of the study (helplessness manipulation) and a similar puzzle in the second stage of the study (test for helplessness effects). They found that youngsters in the noisy schools were more likely to fail on the puzzles and to stop seeking solutions than were their counterparts in the quiet schools.

AN ATTRIBUTIONAL MODEL OF HELPLESSNESS

Some theorists have recently proposed a reformulation of the theory of learned helplessness to take account of the complexity of the learned helplessness phenomenon in human subjects. In the reformulated model, learned helplessness theory is merged with aspects of *attribution theory* from social psychology (see Weiner, Frieze, Kukla, Reed, Rest, and Rosenbaum, 1971). Lyn Abramson and her colleagues (Abramson, Seligman, and Teasdale, 1978) advance an attributional model of learned helplessness that expands and more fully describes the series of cognitive processes that mediate between objective lack of contingency between response and outcome and the psychological symptoms of learned helplessness. Figure 6-4 shows that the cognitive mediation process in learned helplessness involves three distinct components—perception, attribution, and expectancy. On the basis of the perception that outcome is not contingent on response, the individual attributes the noncontingency to some factor.

It is the attribution that determines the individual's expectations of future noncontingencies, and the resulting expectations determine the feelings of helplessness. People may attribute noncontingency to a variety of things. An attribution to some *stable* factor, such as personal ability, predicts that the expectation of future noncontingency will persist for some time. An attribution to an *unstable* factor, such as bad luck, indicates that the expectation may fade after a lapse of time. An attribution to a *global* factor, such as lack of intelligence, predicts that the ex-

Cognitive mediation process

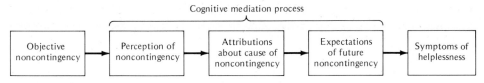

Figure 6-4

The learned helplessness model reformulated according to attribution theory.

Adapted from I. Y. Abramson, M. E. P. Seligman, and J. D. Teasdale, "Learned Helplessness in Humans: Critique and Reformulation," Journal of Abnormal Psychology, 1978, 87:49–74. Copyright 1978 by the American Psychological Association. Adapted by permission of the publisher and author.

pectation will recur even if the situation changes. An attribution to a *specific* factor, such as poor skills in math, indicates that the expectation may not recur when the situation changes (when verbal ability is tested, for example).

In a similar reformulation of the learned helplessness model, Ivan Miller and William Norman (1979) argue that the type of factor to which the individual attributes the lack of contingency between response and outcome determines the nature and generalization of the learned helplessness response. Specifically, the individual's expectation of future outcomes depends on attributions about the cause of the noncontingency (or failure experience) related to *locus of control* (internal vs. external), *stability* (stable vs. variable), *specificity* (specific vs. general), and *importance* (important vs. unimportant). In general, attributions to factors that are internal, stable, general, and important will maximize the severity and the generalization of the helplessness symptoms; attributions to factors that are external, variable, specific, and unimportant will minimize the helplessness effects.

Let us return to Figure 6-4 and see how this attributional model of learned helplessness can help us to understand the behavior and experience of an elderly woman in a stressful urban environment. Imagine first that the woman is living in a poorly ventilated apartment without air conditioning, and is exposed to considerable stress during a particularly hot summer. At first she tries to reduce the stress of heat by opening the windows and using the stove less often, but these measures help very little. Further coping efforts, such as sitting outside and using a small electric fan, prove equally unsatisfactory. There are some variations in daily temperature, of course, but they are unpredictable and out of her control; there is an *objective noncontingency* between the woman's behavior and the aversive outcome.

The *cognitive mediation* process begins when the elderly woman *perceives* this lack of contingency between her own coping efforts and the aversive summer heat. Next, the frustrated woman *attributes* this noncontingency to her own personal inability to deal effectively with life's demands (an internal, stable, general, and important attribution). On the basis of this attribution, she comes to *expect* that future unpleasant outcomes will also be beyond her control. She gives up trying to cope with the heat stress, becomes depressed, and later, when she is faced with other stress caused by increased traffic noise, gives up immediately without even trying to cope with it (*helplessness symptoms*).

A small number of experimental studies have directly examined the effects of variations in types of attributions on the psychological deficits associated with learned helplessness (e.g., Dweck, 1975; Tennen and Eller, 1977). These studies have demonstrated that attributions are central to the learned helplessness experience, and have generally been consistent with the particular theoretical predictions advanced by Abramson, Seligman, and Teasdale (1978) and Miller and Norman (1979).

Some indirect support for the role of attributions in affecting learned helplessness has been provided by two recent studies of the long-term effects of variations in personal control on the well-being of elderly institutional residents. Rodin and Langer (1977) assessed the long-term effects of their manipulation of personal control in a nursing home (Langer and Rodin, 1976) eighteen months after the initial study. The positive effects of control on residents' functioning found in the earlier study persisted over the eighteen-month period. In contrast, Richard Schulz and Barbara Hanusa (1978), after assessing the long-term effects (up to three and one-half years) of manipulations of control and predictability in a retirement home (Schulz, 1976), discovered that the initial beneficial effects of control and predictability on residents' well-being did not persist after the initial study was terminated.

Schulz and Hanusa used an attributional model of learned helplessness to explain the discrepant findings between these two studies. Although residents' attributions were not directly assessed in either study, the different strategies used to enhance residents' feelings of control probably caused quite different types of attributions. Schulz and Hanusa speculate that Langer and Rodin's emphasis on residents' *responsibility for themselves* probably encouraged residents to make internal, stable, and global attributions. In contrast, Schulz's use of a *particular visiting program* probably led residents to make external, unstable, and specific attributions. According to the attributional model of learned helplessness, we would expect that the beneficial effects of the Langer and Rodin study would endure after the initial study was terminated, while those from the Schulz study would not.

INFORMATION OVERLOAD AND ENVIRONMENTAL STRESS

Sheldon Cohen (1978, 1980) has advanced a model of environmental stress that explains the negative effects of stressors on ongoing and subsequent activities in terms of *information overload*. Exposure to an environmental stressor can result in an overload of the individual's capacity to process the information necessary to carry out task-related and social activities effectively. Cohen argues that unpredictable and uncontrollable environmental stressors substantially increase the demands placed on an individual's attention capacity because they are potentially threatening. In effect, such stressors demand a high level of sustained cognitive appraisal if the individual is to ascertain the nature of the stressful situation and determine appropriate coping responses.

When demands on the individual's attention capacity are prolonged, Cohen explains, the attention capacity will shrink, showing "cognitive fatigue." As cognitive fatigue reduces the capacity to process information, its effects are most evident on complex tasks that involve a considerable amount of information. Thus the negative effects of a stressor may be apparent in ongoing complex activities. With less complex tasks requiring less attention, the effects of cognitive fatigue may become visible only on a subsequent task that demands more attention and is less practiced than the previous activity. Here we see the aftereffects of environmental stress. Moreover, the effects of task complexity or "task load" and stress on subsequent tasks are cumulative; that is, the more demands a task places on an individual's attentional capacity and the more stressful the situation under which the task is completed, the greater the performance deficit will be on subsequent activities.

Some experimental laboratory studies have provided support for the information overload model of environmental stress. L. R. Hartley (1973) reports that subjects' performance on a laboratory task when reaction time was measured decreased as a cumulative function of both task load (length of time on the task) and prior exposure to noise. Similarly, people's ability to tolerate frustration has been found to decrease as a function of the cumulative effects of prior task load and prior noise exposure (Rotton, Olszewski, Charleton, and Soler, 1978). A recent study (Matthews, Scheier, Brunson, and Carducci, 1980) demonstrated that when the attentional demands of predictable noise are equated with those of unpredictable noise (subjects were asked to attend to the predictable noise to keep them from tuning it out), the performance decrements on an ongoing task are similar for the two types of noise. The investigators conclude that the frequent finding that performance is worse under unpredictable noise than under predictable noise is due to the fact that unpredictable noise demands more attention (thus reducing the available attention capacity for the ongoing task) than predictable noise.

At the present time research data are too limited to permit us to choose between the learned helplessness and information overload models of environmental stress. In fact, as Cohen (1980) suggests, it seems likely that both theories may accurately explain different aspects of the human consequences of environmental stress. It is possible that uncontrollable events that place relatively low demands on attention capacity may still exert negative aftereffects if the attributions they engender are stable and global. Alternatively, uncontrollable events that are attributed to unstable and specific factors may still cause subsequent performance decrements if they place high demands on attention capacity. In Chapter 7 we shall see that learned helplessness and especially information overload have also been applied to explain the psychological effects of crowding. It seems likely that both theories will play important roles in future research on the human consequences of environmental stress.

APPLICATIONS TO
ENVIRONMENTAL PLANNING

REDUCING URBAN NOISE

How might our knowledge about the stressful effects of noise be applied to environmental planning? Evidence that noise can have negative behavioral and experiential effects underscores the need to reduce the overall level of noise in the urban environment. Yet, as Susan Saegert (1976) has pointed out, environmental stressors often cannot be totally eliminated because the aspects of the environment that are stressful may serve some important uses, and may also contribute to the unique character of a place. For example, the combined noise of traffic, trains, and milling people in New York's Grand Central Station is intimately related to the station's role as a great transportation hub, and is also a part of the area's vitality and interest. The challenge of reducing noise in the urban environment is complex; we must learn how to manage and control the stressful aspects of noise without impeding the ongoing functions and services that are often associated with it. Moreover, the environmental psychologist who is interested in reducing urban noise must be realistic about the increased economic costs that may be incurred as a result of noise-conscious planning, such as the use of sound-absorbant materials or quieter machinery. Decision makers will need to be convinced that increased production costs will be balanced by human gains as a result of a less stressful urban environment.

Buffer zones   Michael Southworth (1969) encouraged planners to take account of the noise associated with activities in determining where to locate them. Noisy traffic interchanges or thoroughfares and airplane takeoff and landing routes should not be located near human environments they would seriously disrupt, such as hospitals, schools, and quiet parks. Leo Beranek (1966) encouraged the use of zoning legislation to place buffer zones between residential areas and noisy superhighways and airports. Airplane takeoff patterns can be planned to minimize their disruption to residential areas. For example, Beranek noted a plan to require planes leaving Washington National Airport to climb as steeply as possible to 1,500 feet while flying over the Potomac River, and then to coast to 3,000 feet on minimal power. He noted that such carefully planned flight patterns will be especially important as supersonic transport becomes more popular, and the noise of jet engines is compounded by the prospect of sonic booms (explosive reports caused by the pressure waves generated by supersonic flight).

Neil Weinstein (1976) argues for the systematic application of psychological measures of noise annoyance in communities that might be adversely affected by a new project, such as a highway extension. When an established community stands to be disrupted by a new noise source, he contends, the decision to build or not to build should be based on the needs of the community in question and not on fixed noise standards developed in other settings. This admonition is also consistent with a finding we discussed earlier, that community annoyance with noise is a function of a variety of attenuating factors, such as people's attitudes toward the noise source, and only partially of actual sound levels. Weinstein notes, for instance, that a residential area that has been sought out by its residents precisely because it is quiet should have its more stringent noise requirements respected by planners.

Noise-insulating designs   Beranek (1966) criticizes the features of contemporary design that tend to make modern buildings poor barriers against noise. The extensive reliance on glass and open space in contemporary buildings, while aesthetically attractive, leaves these settings especially susceptible to noise both from adjacent areas within the building and from the outside. He encourages design strategies that would more effectively insulate against noise, such as thicker or multilayered walls, "floating" floors, acoustically insulated ceilings, and better designed ventilation ducts that would reduce the transmission of noise. He notes that many European countries have developed excellent acoustical building codes that have been applied in the major rebuilding programs since World War II. Beranek adds, however, that similar acoustical standards will not be easy to achieve in the United States because each local community establishes its own building codes, and because builders in a competitive housing market are often more interested in economic shortcuts than in qualitatively better designs.

Southworth (1969) proposes the innovative design strategy of masking or distracting people's attention from moderate noise by overlapping it with interesting sounds. In fact, such an approach has been applied successfully in Paley Park and Greenacre Park in New York City (Figure 6-5). These small, "vest-pocket" parks in the heart of the city have creatively used the sound of attractive waterfalls to mask surrounding city noise. Southworth points out that the shapes of structures and the wall materials employed along the street can either reduce and absorb noise or create a resounding chamber of amplified noise if they are poorly designed. The use of new types of road surfaces can also serve to reduce traffic noise.

Designing quieter transportation equipment   Beranek (1966) and Southworth (1969) have urged improvements in the design of transportation equipment that would help to reduce noise. Motor vehicle noise might be reduced by the im-

Figure 6-5
This vest-pocket urban park uses an attractively designed waterfall to mask the noise of the surrounding city environment.

© *George W. Gardner.*

proved design of engines, mufflers, tires, and noise-dampening housing for engines. Such improvements would be especially beneficial in trucks and buses. Airplane noise might be reduced through the improved design of engines and wings, and by silencing devices fitted to the exhaust ports of jet engines. Beranek cautions, however, that such costly design changes will probably not be undertaken until the responsible government agencies establish noise standards for the various transportation modes.

Weinstein (1976) proposes that when a regulation would entail costly noise-control programs in established industries, it should be based on a fixed noise level and not on the unique needs of each locality affected. Otherwise, standards for an airline would vary from one airport to another. Even in such remedial programs, however, human evaluations of noise annoyance should be taken into account when appropriate standards are determined. Noise surveys might be carried out for a variety of noise sources at many locations. The measurement approach employed in perceived environmental quality indices (see Chapter 4) can be applied in such an endeavor. For instance, a traffic noise index has been developed that relates both the intensity and the variability of traffic noise to people's subjective reactions (Griffiths and Langdon, 1968).

DESIGNING FOR THE ELDERLY

Planned housing for the elderly   Our knowledge of environmental stressors in the lives of older persons may also be applied in environmental design for the elderly. A major source of stress for many elderly persons is the challenge of finding a decent and affordable place to live. Several research studies over the last fifteen years have evaluated the relative advantages of planned housing for the elderly—typically, high-rise, low-rent housing for the poor and retirement communities for middle-income people. The strong consensus of this body of work is that planning housing specifically for the elderly is better for their overall well-being than leaving them to fend for themselves in the open housing market (e.g., Carp, 1966; Lawton and Cohen, 1974; Messer, 1967; Rosow, 1967). When investigators surveyed elderly tenants and project managers in over 150 public housing sites, they found that, when a variety of personal and environmental factors were statistically controlled, age segregation was positively related to housing satisfaction, participation in on-site activities, neighborhood mobility, and overall morale (Teaff, Lawton, Nahemow, and Carlson, 1978). In fact, 83 percent of respondents expressed a preference for living in a setting limited to persons 62 and over.

Recent reviews of research on age-segregated housing for the elderly (Carp, 1976; Lawton, 1977; Michelson, 1976) have offered some explanations for this finding. Age segregation increases the number of age peers living in close proximity to an older person, and so provides the resident with a choice of friends. While easy access to friends is especially important for elderly persons whose mobility may be restricted, we shall see in Chapter 10 that proximity plays a role in the formation of friendships for all age groups. In addition, age segregation fosters an age-appropriate normative system that allows older people the option of participating in social activities and functions in accordance with their personal needs and wishes. In a setting with mixed age groups, high activity and participation norms are generally set by younger persons, and low activity levels are negatively evaluated, so that the self-esteem of older persons is lowered.

Facilitating access to resources   Because reduced mobility is often associated with aging, a particular planning need is to facilitate older people's access to resources. Carp (1976) emphasizes that in planning for the elderly, "housing environment" must be broadly defined to include the total life setting of older persons. The total living environment of the elderly should be envisioned as including supportive social and medical services, shopping resources, and social and recreational opportunities, as well as an acceptable housing unit.

Lawton (1977, 1979) points out that transportation is an especially vital resource for the elderly because it provides the means of facilitating access to other essential resources in the total living environment. A paramount planning goal is the development of safe and affordable public transportation systems that are accessible to older persons. Public officials should make a special effort to route buses through areas where there are high concentrations of the elderly, and to see that such routes include a shopping center, medical facilities, and a senior recreation center. Route signs at transportation stops and on vehicles need to be clearly legible and informative for older users. In view of the multiple sources of environmental stress that characterize much of today's public transportation—noise, heat, crowds, physical barriers—it is impossible to overemphasize the planning challenges in this area. These recommendations are consistent with Carp's (1980) finding that elderly people would like to make more trips to visit friends and relatives and for entertainment if transportation were available. Carp emphasizes that the preferred type of transportation for older persons is buses. Similarly, Victor Regnier and Karen Rausch (1980), who studied neighborhood mobility among low-income elderly persons in Los Angeles, conclude that, after walking, buses provided the most important means of getting about for the elderly.

We have recently witnessed a growing social awareness of the physical needs of older and physically handicapped persons. National design standards have been established to make buildings and other facilities accessible to people with physical disabilities. These standards have attempted to take account of the "manifestations of the aging processes that significantly reduce mobility, flexibility, coordination, and perceptiveness" (American National Standards Institute, 1961:6). They have been applied to ground grading, public walks, parking facilities, ramps, building entrances, floor surfaces, toilet facilities, public telephones, and building and room identification in buildings and facilities used by the public.

More research is needed to identify the features of the everyday environment that present barriers to the normal functioning of older and handicapped individuals (see Bednar, 1977). Gundez Ast (1977), who studied a group of older persons with a variety of physical disabilities in Moline, Illinois, concludes that many features of the city environment (such as exterior stairways to buildings, the location of bus stops, and nonautomatic doors) are discouraging and even dangerous to such people. Several studies have reported that many elderly people experience considerable physical difficulties in carrying out a variety of everyday household tasks (Lawton, 1977) (Table 6-3). Lawton emphasizes the need for human factors research to improve the design of such everyday furnishings as sinks, bathtubs,

Table 6-3.   Percentages of elderly residents reporting difficulty in carrying out a variety of household tasks, by age groups in three samples.

Task	National sample[a]				National sample[b]			Low-income communities[c]	
	65–69	70–74	75–79	80+	65–69	70–74	75+	Poor	Near-poor
Doing things around the house					13	28	37		
Bathing self	7	8	13	19	8	11	17	9	8
Household cleaning								28	23 (females)
Doing laundry								32	28
Preparing meals								16	13 (females)
Dressing	8	8	7	13	10	12	12		
Climbing stairs	26	29	33	42	9	18	20		
Going outdoors					7	13	20		

[a] Shanas, Townsend, Wedderburn, Friis, Milhøj, and Stehouwer, 1968.
[b] Schooler, cited in *Indicators of the Status of the Elderly in the United States* (1972).
[c] National Council on the Aging (1971).

Source: M. P. Lawton, "The Impact of the Environment on Aging and Behavior," in J. E. Birren and K. W. Schaie (eds.), Handbook of the Psychology of Aging, 1977, p. 282. Reprinted by permission of Van Nostrand Reinhold Company.

toilets, storage spaces, chairs, and appliances, so that they might be more easily and safely used by elderly persons.

Enhancing control in institutional environments   On the basis of findings that personal control can reduce the adverse psychological effects of institutional environments on elderly residents, we might consider some ways in which personal control in such settings can be enhanced. M. Powell Lawton (1971, 1974, 1979) has addressed considerable attention to ways in which the physical design of institutional environments for the elderly might be improved to meet the social and psychological needs of residents. We shall concentrate here on his recommendations that pertain to enhancement of residents' sense of personal control. Lawton explains that residents' control can be enhanced by designs that generate a range of social options within the setting. For instance, a geriatric setting might be designed with a large communal space for active group participation, a smaller lounge area where more personal conversations might be pursued, and private or semiprivate bedrooms where residents might think, meditate, or read by themselves.

Lawton suggests that a variety of spaces might be planned in institutional environments that would allow residents to maintain and practice personal skills and competencies, and to pursue personal interests or hobbies. For instance, a setting might include facilities for cooking and planning meals; washing, drying, and ironing clothes; and personal activities, crafts, and hobbies. More competent residents might be allowed to engage in these activities freely, while physically or mentally impaired individuals might be provided appropriate supervision and guidance.

Lawton encourages design features that facilitate recognition and orientation in the environment, permitting residents to function more independently as they move about the setting. He notes that at the Philadelphia Geriatric Center's Weiss Institute, which was designed for older people with serious behavioral deficits, residents' rooms are color-coded to facilitate recognition; the elevator area on each floor has a distinctive color; large three-dimensional room numbers are located outside each room; names of occupants are affixed to room doors in large letters; and large signs and maps indicate the location of important areas on each floor.

Alan Lipman and Robert Slater (1979) have provided architectural plans for nursing homes that are intended to minimize residents' dependence on staff and to encourage independent functioning. They favor small-group living units that are physically separated from staff accommodations (Figure 6-6). Staff services are minimized, and the plans encourage residents to care for their own daily needs. The authors prefer to sacrifice large communal areas, such as a lounge for twenty persons or a dining area for thirty persons, in favor of smaller, dispersed accommodations. The group living unit for seven persons shown in Figure 6-6 provides appropriate spaces for residents to engage in a variety of normal day-to-day activities, such as preparing their food, eating or socializing in small groups, and bathing themselves in privacy. In conventional nursing homes, built at an institutional scale, residents need to ask for staff assistance even for such a simple activity as preparing a cup of tea. The small-group unit, separated from staff accommodations, encourages residents to assist one another when help is required instead of relying exclusively on staff support.

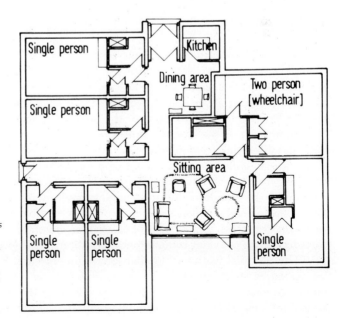

Figure 6-6

This plan for a unit in a nursing home, separated from staff accommodations, enables residents to maintain considerable independence and personal control in carrying out day-to-day activities.

From A. Lipman and R. Slater, "Homes for Old People: Towards a Positive Environment," in D. Canter and S. Canter (eds.), Designing for Therapeutic Environments: A Review of Research, 1979, p. 286. Reprinted by permission of John Wiley & Sons Ltd., Chichester, England.

Rudolf Moos and his colleagues (Moos, 1980; Moos and Lemke, 1979) have developed a comprehensive assessment instrument called the Multiphasic Environmental Assessment Procedure (MEAP), which can be used to evaluate the overall environmental context of institutions for the elderly. The MEAP includes measures of an institution's architectural features, policy orientation, social environment, and resident and staff characteristics. It is appropriate for use in planning for and evaluating the types and levels of resident control that may characterize a setting.

The MEAP's *resident control scale* affords an assessment of the degree of behavioral control available to residents in a facility. This scale measures the extent to which the institution allows residents formal and informal input in running the facility (e.g., residents help to plan menus and play a role in making the rules about attendance at activities). An index of *policy choice* assesses the level of decisional control residents have by measuring the degree to which the institution provides options from which residents can select individual patterns for daily activities (e.g., residents can choose the times they get up, have meals, and go to bed).

Moos (1980) used the resident control and policy choice scales of the MEAP to examine the relationships between personal control and resident well-being in a field correlational study of ninety-three residential care facilities for the elderly. He found that settings in which residents had more personal control were characterized by more cohesive and conflict-free social atmospheres than those where residents were afforded less personal control. Residents in such settings also had more social and functional resources than those in facilities that permitted residents less personal control.

SUMMARY

Systemic stress is the nonspecific response of the body to any environmental demands made on it; the conditions in the environment that produce stress are called *stressors*. The *general adaptation syndrome,* or GAS, consists of three stages: an alarm reaction, resistance, and eventual exhaustion. *Psychological stress* involves an essential psychological component; the individual *appraises* the personal meaning and significance of the stressor. *Primary appraisal* is a mediating psychological process that discriminates potentially harmful situations from beneficial or irrelevant ones. *Secondary appraisal* functions to assess the individual's resources for coping with the threatening situation. *Reappraisal* involves a change in the original perception of the situation due to changing environmental or personal conditions. *Problem-focused coping* consists of behavioral or cognitive efforts to deal directly with the source of stress, while *emotion-focused coping* involves behavioral or cognitive efforts to reduce or better tolerate one's emotional reaction to stress.

The physical features of urban environments that have been most fully investigated as stressors are noise, air pollution, and extreme temperatures. Such environmental stressors are particularly salient in the experience of elderly persons. Problems of mobility make many environmental features that younger persons take for granted sources of challenge and frustration for the elderly. Two aspects of the physical environment that have received considerable research attention as stressors for the elderly are the housing environment and institutional settings. The psychological effects of environmental stress may be assessed through somatic, behavioral, and subjective measures.

Research conducted primarily in laboratory settings has demonstrated conclusively that noise causes the systemic physiological reactions typically associated with stress, such as increased electrodermal activity, the secretion of adrenaline, and elevated blood pressure. Long-term exposure to high-intensity noise can cause significant hearing loss. The possible relationship of noise to nonauditory physical or mental illness, however, is still not known. A growing body of research evidence has shown that noise does have deleterious effects on social behavior. Noise has been found to lower social interaction, to reduce helping behavior, and under some conditions to increase interpersonal aggression. Community surveys have found that people are annoyed by noise from traffic and aircraft, but that extensive annoyance is found only in the most noisy areas. Individual variation in annoyance is considerable, reflecting attitudes and fears toward the source of the noise as well as individual differences in sensitivity to noise.

Researchers have reported a significant relationship between hot spells in urban areas and a rising mortality rate. A considerable body of research has demonstrated a relationship between various aspects of air pollution and physical illness, primarily respiratory diseases. Much evidence gleaned from laboratory experiments suggests as well that excessively high temperatures affect people's social behavior. Heat has been shown to affect interpersonal aggression, although the effects are complicated, and additional factors (such as the presence of an aggressive model or provocation to anger) interact with the effects of heat in producing aggression. It appears that when a combination of stressful conditions produces an *intermediate* level of negative affect, aggression will occur, but that negative affect that is either below or above that level will not lead to aggression. Little is known about the effects of air pollution on social behavior, although a small number of experimental data suggest that some aspects of poor air quality can reduce interpersonal attraction. Awareness of and concern about air pollution vary widely, and many citizens employ emotion-focused coping strategies to deal with it, such as denying the problem or shifting responsibility to other people.

The health effects of high temperatures and air pollution are especially pronounced among the elderly. The behavioral effects of stress in the lives of elderly persons may be seen in a narrowing of the normal range of activities, such as reduced mobility or the underutilization of resources. The use of resources by older persons is a function of (1) physical distance, (2) barriers to elderly persons' mobility, and (3) the salience of the resources. Surveys of housing satisfaction have indicated that elderly residents have positive feelings toward age-segregated and planned housing for the elderly, but residents' reports of satisfaction may be affected by their perceptions of alternative housing choices.

Theoretical developments concerning environmental stress have grown out of efforts to understand the *cognitive* and *psychological* factors that mediate the effects of environmental stressors on people's behavior and experience. Experiments on the psychological consequences of noise have demonstrated that negative psychological effects are greater for *unpredictable* noise than for noise that is predictable. For example, after exposure to unpredictable noise (but not after predictable noise), subjects showed impaired performance on laboratory tasks and a reduced tolerance for frustration. Investigators have also argued that the negative psychological effects of environmental stressors can be reduced when people achieve personal *control* over the stressors. Personal control may involve direct *behavioral* control over a threatening event, *cognitive* control over the interpretation of a threatening situation, or *decisional* control over the choices available. Laboratory studies of behavioral control (even perceived behavioral control) over aversive noise have uniformly found that control reduces the negative psychological effects of noise. Field studies of behavioral and decisional control in institutions for the elderly have similarly demonstrated that control has positive effects on the well-being of residents.

Some environmental psychologists have proposed that the negative psychological consequences of an inability to exert personal control over adverse environmental circumstances are mediated by the experience of *learned helplessness*. Learned helplessness is a psychological condition in which an individual who has been unable to control stressful conditions develops the expectation that his or her responses are independent of future outcomes. Learned helplessness involves a lowered *motivation* to respond, difficulty in relearning at a *cognitive* level that personal actions can influence outcomes, and feelings of depression at an *emotional* level. A recent reformulation of the learned helplessness theory emphasizes that the nature of the factors to which the individual *attributes* the lack of contingency between response and outcome determines the nature and generalization of the learned helplessness response.

An alternative model for explaining the negative effects of environmental stress is *information overload*. This model proposes that because unpredictable and uncontrollable environmental stressors are potentially threatening, they substantially increase the demands on an individual's attention capacity. "Cognitive fatigue" (a shrinking of attention capacity) results from the cumulative effects of both task load and stress, and can adversely affect ongoing complex tasks as well as subsequent tasks.

Research knowledge about environmental stress has been applied to the reduction of urban noise and the planning of housing, transportation, barrier-free designs, and institutional settings for the elderly.

7 Coping with Crowding

In recent years we have grown increasingly aware of a uniquely serious environmental problem—the potential human consequences of the spiraling growth of the world's population. Although the concomitant patterns of industrialization and urbanization that have led to a concentration of the human population in urban centers are not new, sensitivity to the social and psychological implications of these events is a recent phenomenon. Today we are confronted on all sides by newspaper and magazine articles, best-selling books, distinguished lecture series, and television specials on the "population explosion." The warning: at the present rate of growth, the population of our already crowded world will more than double by the end of this century.

When we think about the problem of overpopulation, we tend to limit our concern to the poorer, less developed countries of the world; we imagine hunger in Asia, poverty in Africa, unemployment and illiteracy in Latin America. In fact, however, the social and psychological consequences of crowding are also very real in the industrially advanced nations of the world. Reflect for a moment on those situations in which you commonly cope with crowding in your own daily life. The hassles and high costs involved in finding a place to live while attending college are directly related to the large number of people in the housing market. The persistent slowdowns on urban and suburban highways are the consequences of the vast number of people in need of transportation. The lines we have grown accustomed to at the theaters, the long waits at our favorite restaurants, and the crowds of shoppers at department stores and shopping malls all reflect the same problem—crowding (see box, "The Ubiquitous Queue").

A dramatic example of the psychological effects of crowding is apparent in a newspaper account of traffic congestion on America's highways (Robins, 1978). In areas where massive traffic congestion has grown rapidly, frustrated drivers have sometimes vented their anger in outbursts of violence. State police in Texas and California have reported fist fights between enraged drivers stalled in bumper-to-

The Ubiquitous Queue

In an intriguing article entitled "Learning to Live with Lines," Leon Mann offers a rich description of the ever present queue in the life of a major city:

Beyond a shadow of doubt, New York is the uncrowned queueing capital of the U.S. Over the years there have been some memorable lines on Broadway, outside the Met Opera and the Metropolitan Museum, at Madison Square Garden, and outside Yankee Stadium.

New Yorkers turn out in force when there is an astronaut to be greeted or a hero to be farewelled. During the twenty-four hours when the body of Senator Robert Kennedy lay in St. Patrick's Cathedral (June, 1968), thousands of New Yorkers in lines over one mile long waited patiently to file past the casket. The average wait in line was seven hours, but many who could stay only the one hour of their lunch break, and knew they had no chance of viewing the casket, stood in line as a mark of respect for the slain leader.

On February 7, 1963, despite rain, slush, and bone-chilling cold, a crowd of 23,872 New Yorkers queued up at the Metropolitan Museum of Art to get a glimpse of the *Mona Lisa*. The lines that day stretched over three city blocks. During its three and one-half week sojourn at the Metropolitan, half a million people passed in front of the *Mona Lisa*. When *I am Curious (Yellow)* opened at the Rendezvous theater on West 57th Street on March 10, 1969, it quickly drew large crowds and long queues; people were queuing six deep for one and a half blocks for the evening sessions and lines for the 10:00 AM session were forming at 7:00 o'clock in the morning. Waiting lines for Broadway hit shows are often themselves smash hits. *South Pacific, My Fair Lady,* and *Hello, Dolly* drew remarkably long lines during their record runs. When the box office for the musical *Coco* opened on November 3, 1969, the line stretched along 51st Street, turned the corner onto Broadway, headed uptown, and finally came to an end half way up 52nd Street.

[From Mann, L., "Learning to Live with Lines," in J. Helmer and N. A. Eddington (eds.), *Urbanman: The Psychology of Urban Survival*, The Free Press, 1973, p. 45. Reprinted by permission of the author.]

bumper traffic. One highway incident in Dallas resulted in a fatal stabbing, while a confrontation on a Houston boulevard ended in a shooting. A California Highway Patrol officer commented that the violent driver is often a "meek little office worker . . . the guy who lives next door to you, until somebody cuts him off, and then this mild mannered man turns into a revenge-seeking fool" (p. 20). A Utah study reported that 15 percent of Salt Lake City motorists admitted that at times they have been so angry on the highway that they "could gladly kill another driver" (p. 1).

The psychological effects of crowding and the ways people cope with it have become the topics of an important and especially interesting body of research in environmental psychology. Environmental psychologists have examined how crowding affects people's physical health, as well as social processes such as cooperation, help giving, withdrawal, and aggression. Other investigators have focused on the ways people cope with crowding at both a group and a personal level. The psychological study of crowding can make an important contribution to the design of multifamily housing and public institutions, planning for urban and suburban transportation, and planning to meet people's recreational needs.

THE NATURE OF CROWDING

What image first comes to mind when you think about crowding? You may find yourself thinking about a large number of people, packed together like sardines, in a limited amount of space (Figure 7-1). Yet with a little more reflection it will become apparent that the meaning of crowding is more complex than it seems. Does crowding refer to a large number of people or to a restricted amount of space? Are 25,000 fans at a baseball game crowded in the same way that ten people on an elevator are crowded? And does it matter whether the people involved feel crowded or not? People crowded into a subway car are not likely to feel the same about their circumstances as people crowded together at a party.

Environmental psychologists, too, have discovered that the meaning of crowding, which at first seems straightforward, is actually quite complex (Sadalla, Burroughs, and Staplin, 1978). In fact, a serious problem with much of the earlier research on crowding was investigators' failure to define crowding clearly in each study. Thus they sometimes reported conflicting results simply because their definitions of crowding differed. Environmental psychologists now agree that a precise and clear definition of crowding must be the first step in the investigation of this complex environmental concern.

DEFINING CROWDING

Crowding vs. density   Some environmental psychologists believe that in order to understand the effects of crowding on people, we need to know how the people being studied experience the crowded situation. Daniel Stokols (1972a, 1972b, 1978) has proposed that researchers adopt a psychological definition of crowding. He notes that researchers have often defined crowding exclusively in terms of

Figure 7-1
Crowding is a familiar part of students' lives on most large college campuses.

© *Lawrence Frank 1976.*

spatial restrictions—that is, density—ignoring the personal experiences that may mediate between the spatial aspects of crowding and the resultant effects on human behavior.

Stokols advocates that environmental psychologists distinguish between density and crowding. He proposes that the term *density* be restricted to the strictly physical or spatial aspects of a setting (that is, the number of persons per spatial area), while the term *crowding* should be used to refer to the psychological or subjective factors in a situation (that is, individuals' perceptions of spatial restrictions). According to this definition, density (spatial constraint) is a necessary but not sufficient condition for crowding (the subjective response to tight space). For example, spectators at a musical concert might be brought together in a dense spatial arrangement but fail to experience any sense of being uncomfortably crowded.

Stokols points out that in order for a high-density situation to be experienced as crowded, additional personal and social factors need to be present. People are more likely to report feeling crowded when the spatial constraints in an area cause social interference, such as competition, between individuals in the setting. Similarly, people will feel crowded when the restriction of movement in the setting interferes with their ability to carry out coordinated social activities.

Other investigators have agreed with Stokols' distinction between density and crowding (see Altman, 1975; Sundstrom, 1978b). Amos Rapoport (1975) has gone a step further in arguing for a subjective emphasis in crowding research; he has proposed that density, too, be viewed in subjective terms. He notes, for example, that density is a perceptual experience: the individual perceives the number of people in a setting, the amount of space available in the area, and the way in which the available space is organized. Density is the direct perception of available space, while crowding is a subjective evaluation that the perceived amount of space is inadequate.

Although most researchers now agree that it is advisable to distinguish between objective density and subjective crowding, a few investigators have pointed to some potential problems with this approach. While researchers can readily define the objective parameters of density in a research experiment, it is very difficult to define the complex subjective elements that lead to the perception of crowding (Insel and Lindgren, 1978). Jonathan Freedman (1975) has argued even more forcefully against the density-crowding distinction. He believes that crowding should not be restricted to a subjective perception, but rather should refer simply to the amount of objective space available per person.

Freedman's position is based on his view that crowding is not necessarily a negative experience, but may be either positive or negative depending on the particular circumstances in which it occurs. If the term "crowding" is used to refer to the amount of objective space per person, it is clear that people's experience of crowding may be either pleasant or unpleasant. If, however, "crowding" is restricted to mean the subjective perception of the situation, the experience can only be psychologically negative. "Three's a crowd," Freedman points out, is not meant to be inviting.

In this chapter we shall generally follow the definitions of density and crowding suggested by Stokols and advocated by most environmental psychologists. "Density" will be used to refer to the physical or spatial aspects of a situation; "crowding" will refer to individuals' subjective perceptions of situations involving density. At the same time, however, Freedman's caution about assuming that high density is necessarily negative is important. While our discussion of crowding (the subjective perceptions) will typically concern people's negative social and psychological reactions, our consideration of density (the spatial aspects) may involve either positive or negative human reactions.

Social density vs. spatial density   Even after we have agreed that the definition of density should be restricted to the number of people per spatial area, we find again that our definition is in fact more complex than it at first appears. If we begin with the assumption that high density exists when there are too many people for an available spatial area, it is apparent that the density level may be reduced in two ways—by a decrease in the number of people or by an increase in the amount of space. In fact, with this issue in mind, some environmental psychologists have proposed that a distinction be made between social density and spatial density (Loo, 1973a, 1973b; McGrew, 1970; Saegert, 1973, 1974).

These investigators have defined *social density* in terms of the number of people in a given area, *spatial density* in terms of the available space in a particular setting. With this distinction in mind, researchers involved in studying the psychological effects of various levels of density have engaged in two quite different experimental strategies. To study the effects of social density, investigators have varied the size of a social group in a spatial setting of constant size. For example, a group of five persons might be compared with a group of ten persons in a moderate-sized room. Spatial density, in contrast, has been investigated by studying a social group of constant size in spatial settings of various sizes. Here a group of seven persons might be compared in a small and a large room.

The distinction between social and spatial density is important because these two types of density can lead to different psychological reactions. The factors to which individuals attribute their discomfort in dense settings may differ depending on whether the situation is one of social density or spatial density (see Stokols, 1976). If, for instance, the discomfort is seen as arising from the presence of too many other people (social density), the individual may blame those other people for interfering with his or her good feelings, and react to them with antagonism. In contrast, if the discomfort is viewed as a product of too little space (spatial density), the individual may blame the environmental arrangement for his or her discomfort, and be somewhat less inclined to react with negative feelings toward the other persons in the area. Because individuals' reactions to social and spatial density may differ, I shall attempt to indicate the particular type of density the research in question has examined.

Inside density vs. outside density　Environmental psychologists have pointed to another important distinction in defining social density. This distinction becomes apparent when we consider the difference between high social density in animal and in human populations. For the ethologist who studies the effects of high density on a herd of sika deer, density may be readily defined as the number of deer per spatial area. For the environmental psychologist who investigates density in a human population, however, the definition is more complex because humans have built shelters to separate themselves from one another. Thus we need two different measures of density—density in the broader geographic area, which includes a large number of dwelling units, and density within the dwelling units themselves.

With this concern in mind, a number of investigators have suggested a distinction between inside density and outside density (Carey, 1972; Carson, 1969; Galle, Gove, and McPherson, 1972; Jacobs, 1961; Schmitt, 1957, 1966; Zlutnick and Altman, 1972). *Inside density* has been defined as the number of people per spatial area within a dwelling unit; for example, the number of people per room or the number of people per residence. *Outside density* is the number of people (or residences) within a broader geographic area; for example, the number of people (or residences) per acre or per census tract.

When we reflect on inside and outside density, it becomes evident that the distinction between the two is quite important. Steven Zlutnick and Irwin Altman (1972) present a two-dimensional model (Figure 7-2) to show some of the types of residential environments that can be identified when both inside and outside density are varied from low to high levels. At one extreme we find the urban ghetto, which is characterized by high levels of both inside and outside density. At the other extreme is the suburban environment, which has low levels of both inside and outside density. Between these two extremes are poor rural areas, where inside density is typically high while outside density is low, and urban luxury environments, where inside density is low and outside density is high. The distinc-

Figure 7-2

Inside and outside density in four residential environments.

From S. Zlutnick and I. Altman, "Crowding and Human Behavior," in J. F. Wohlwill and D. H. Carson (eds.), Environment and the Social Sciences: Perspectives and Applications, p. 51. Copyright 1972 by the American Psychological Association. Reprinted by permission of the publisher and author.

		"Inside" density (within residential, etc., units)	
		Low	High
"Outside" density (neighborhood and community)	Low	I. Suburbia	II. Rural area
	High	III. Urban luxury area	IV. Urban ghetto

tion between inside and outside density is clearly important; we would not expect people living in an urban ghetto, a poor rural setting, and urban luxury housing (all of whom are confronted with some aspect of high density) to have the same social and psychological reactions to their environmental circumstances. For this reason I shall attempt to specify which particular type of density is being examined in each study under discussion.

CROWDING AS A STRESSOR

The stress models introduced in Chapter 6 to explain the behavioral effects of noise and extreme temperatures can also help us to understand the psychological processes through which crowding affects human behavior. For example, many environmental psychologists believe that the relationship between crowding and outcome behaviors, such as negative forms of social activity, is mediated by a psychological stress reaction. Yakov Epstein (1981) reviews studies of crowding stress, exploring the role of personal control in reducing the adverse psychological consequences of crowding. We shall see later in this chapter that two important theories of crowding, information overload and behavioral constraint, are based on an underlying assumption that crowding induces psychological stress. In addition, some investigators (Evans and Eichelman, 1976; Worchel and Teddlie, 1976) have suggested that crowding stress is mediated by heightened *arousal* resulting from the invasion of personal space that often occurs with high density. We shall explore these models more fully in Chapter 9, in discussing the stress model of personal space.

Stress   Let us briefly review the components of a stress reaction. According to Hans Selye (1956), stress involves a complex series of bodily reactions, termed the *general adaptation syndrome*, or GAS. The GAS is initiated by an automatic alarm reaction that involves increased adrenaline secretion, heart rate, blood pressure, and skin conductance. This general state of alarm is followed by a stage of resistance in which coping responses occur at the physiological, behavioral, and cognitive levels. If these resistive efforts fail to deal with the stressor adequately, the individual will proceed to a state of exhaustion.

Psychological stress in humans involves an important cognitive component; the individual *appraises* the personal meaning or significance of the stressor (Lazarus, 1966). During primary appraisal potentially harmful situations are discriminated from beneficial or irrelevant ones; during secondary appraisal the individual's resources for coping with the threatening situation are assessed. Reappraisal involves a change in the original perception of the situation due to changing conditions. Problem-focused coping is an attempt to deal directly with the sources of stress, while emotion-focused coping is oriented toward reducing or better tolerating the emotional reaction to stress.

Daniel Stokols (1972a, 1976) has envisioned crowding as a form of psychological stress in which an individual's perceived need for space exceeds the space available. The feeling of being crowded involves a fear that inability to acquire more space will result in personally unpleasant consequences. If the foreseen consequences represent rather a slight inconvenience, such as the limitations on movement experienced at a crowded concert, the stress reaction will be minor. Stokols points out, however, that if failure to obtain more space threatens one's sense of physical or psychological security, as when one is close to a violence-prone person, the stress reaction can be intense.

Chalsa Loo (1977) has presented a model of crowding stress that is especially congruent with our earlier discussion of definitions of crowding. First, her model is applicable to stress associated with both social density and spatial density. Second, consistent with the notion that high density is not necessarily negative, her model shows that a state of harmony between the individual and the environment will result from an *optimal* level of density. The level of density that will be optimal for a particular person at a particular time is a function of the individual's need for social and spatial resources at the time, and may vary considerably.

Loo's social-spatial model of crowding stress is shown in Figure 7-3. Loo explains that people have both social needs and spatial needs. Social needs concern the number of people that an individual desires to be with and the amount of personal space needed. Spatial needs relate to the amount of physical space and the type of boundaries a person wishes. Crowding stress occurs when the individual's needs in these two areas are out of step with environmental realities. Thus, according to Loo's model, psychological stress can be induced by settings that are undercrowded as well as by those that are overcrowded. A state of harmony is achieved when the environment is both socially and spatially uncrowded.

Thus, while high density *can* operate as a psychological stressor, it will not inevitably result in psychological stress. Whether or not high density leads to psychological stress will depend on the individual's social and spatial needs in a particular situation, as well as the particular characteristics of the situation, such as whether it is viewed as generally pleasant or generally unpleasant (see Freedman, 1975).

Coping   Stokols (1972b) has pointed out that crowding may be viewed in two quite different ways. First, crowding may be seen as a *stressor situation*; that is, environmental factors, such as the amount of space available to a group, may be viewed as independent variables that cause people to feel crowded. Alternatively, crowding may be envisioned as a *stress syndrome*; that is, the psychological experience of being crowded may be seen as a dependent variable that results from particular environmental factors. An especially important aspect of crowding as a

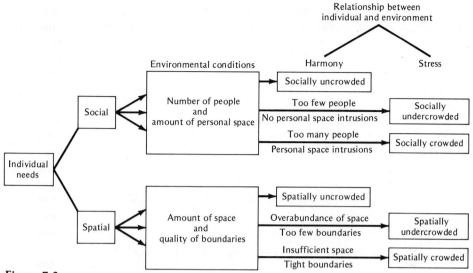

Figure 7-3
A social-spatial model of crowding stresses.

From C. Loo, "Beyond the Effects of Crowding: Situational and Individual Differences," in D. Stokols (ed.), Perspectives on Environment and Behavior: Theory, Research, and Applications, 1977, p. 163. Reprinted by permission of Plenum Publishing Corp.

stress syndrome consists of the ways in which people attempt to *cope* with crowding.

Stokols (1972a) explains that when people feel crowded, they are motivated to alleviate the perceived constraints of the crowded situation. The specific ways in which they try to alter the situation depend on which factors in the situation can be modified. When spatial factors can be altered, the individual may attempt to increase his or her supply of available space. If spatial constraints cannot be directly altered and the perceived limitations are extreme, the individual may leave the crowded situation and seek a less crowded area. When one is unable either to alter the environment directly or to move to another area, one may cope with the crowded situation in a perceptual or cognitive manner. One may, for example, readjust one's standards of how much space is necessary or find ways to make the task one is engaged in more attractive. Stokols concludes that if people are unable to alter the perceived environmental restraints either behaviorally or perceptually, they will manifest behaviors that are symptomatic of general stress, such as discomfort or aggression.

Altman (1975) has speculated that even successful coping with the stress of crowding may, over an extended period of time, accumulate physiological and psychological costs. Referring to earlier work by René Dubos (1965, 1968) and John Cassel (1970, 1971), he explains that interpersonal stresses in dense environments disturb the natural ecology of the organism. In the process of coping with the stress of crowding, a long-term chain of events may occur: social stimulation → initial coping responses → psychological stress → further coping efforts →

eventual costs. Altman's position is consistent with the finding that the psychological effects of crowding, like those of ambient environmental stressors, sometimes show up as reduced performance on tasks that *follow* the crowding experience (see Evans, 1979; Sherrod, 1974).

We shall see later that the idea of crowding as a coping process is central to much of the empirical research and theoretical development in this area. In our consideration of the psychological effects of crowding, for example, we shall find that people sometimes respond to crowding with aggression and sometimes with social withdrawal—both coping mechanisms. In our discussion of psychological theories of crowding, we shall examine theoretical views that are based on an underlying conception of crowding as a coping process. Theorists have proposed that crowding behavior reflects the individual's attempt to cope with the "stimulus overload" induced by crowded settings or with the "behavioral constraints" imposed by crowding.

RESEARCH STRATEGIES FOR STUDYING CROWDING

The psychological effects of density can be assessed by the same kinds of measures that are used to study the human effects of ambient environmental stressors—somatic measures (physiological reactions and health-related indices), behavioral measures (social behaviors and task performance), and subjective measures (indices of affect, mood, and comfort) (see "Measuring Environmental Stress," in Chapter 6). Of particular interest has been the progressive increase in the sophistication of the research *strategies* that have been employed in research on crowding. The weaknesses of earlier research techniques have typically been corrected in the newer research strategies that have replaced them. This fascinating development of revised and improved research strategies continues to be evident even in the research conducted during the last decade.

Correlational studies based on census data   The first research strategy devised by investigators to study the human consequences of crowding was to analyze the relationship between density and social pathology as reflected in census data or other archival records, such as crime statistics, rates of hospitalization, and indices of mental illness. This research strategy, introduced in the 1920s, was correlational, and involved comparisons of very large geographic areas. For example, investigators might compare rates of mortality, mental illness, and serious crime in the densely populated central portion of a large city and in its more sparsely populated suburbs. Such large-scale correlational studies generally revealed that social pathology of various types was greater in the heavily populated central city than in low-density suburbs (see Michelson, 1976).

Altman (1975) has pointed out that the correlational nature of these studies made it impossible to conclude that density *caused* social pathology. In fact, several other potential influences on people's functioning tended to vary systematically between central city and suburb, including economic status, educational level, and the quality of public services. In addition, early correlational studies often failed to define density clearly, and were generally limited to indices of outside density, with no mention of corresponding levels of inside density. Finally, the correlational studies tended to focus on the social outcomes associated with high density and to ignore the social processes, such as the quality of group interaction, that might mediate the effects of crowding on social pathology.

Altman notes that more recent long-term correlational studies, conducted since the 1960s, have begun to overcome the limitations of earlier studies. Some recent studies have attempted to control statistically the effects of outside influences on people's social functioning, such as socioeconomic status (e.g., Galle, Gove, and McPherson, 1972). Recent studies have also systematically defined density, including indices of both outside and inside density. Finally, recent work has attempted to consider some of the ways social processes in the home might mediate the effects of high levels of density on human functioning.

Experimental studies in laboratory settings   A more recent research strategy in the study of crowding, dating from the 1960s, has involved experimental studies in laboratory settings. Some early studies were conducted among animal populations, such as rats (e.g., Calhoun, 1962b, 1967), in an effort to learn how high density might affect social and physical pathology among animals in controlled settings. Later laboratory studies have investigated the effects of crowding in human populations on a wide range of psychological outcomes, such as interpersonal attraction, aggression, withdrawal, mood, and task performance. The goal of such studies has been to establish a causal link between crowding and behavior by systematically controlling the potential influences of extraneous factors. Altman (1975) has noted that recent laboratory studies have been more successful than earlier ones in identifying the effects of crowding on ongoing social processes, such as group cooperation, and in separating the effects of social density and spatial density.

Studies of crowding in small-scale field settings   The most recent strategy in the study of crowding has involved research in small-scale field settings, such as dormitories and department stores (Saegert, Mackintosh, and West, 1975). These field studies have employed a diversity of psychological measures, including behavioral observations, interview surveys, performance on social and cognitive tasks, and indices of physiological responses. This strategy has allowed researchers to study the effects of long-term exposure to crowding in complex real-life settings. In addition, studies of this type have permitted a precise definition of crowding and density, along with the use of more sophisticated psychological measures.

Diverse research strategies   Each of the research strategies used to study crowding has its strengths and its weaknesses. As Altman (1975) has suggested, the various strategies tend to complement one another, and a full understanding of the complex phenomenon of crowding will require the advantages of diverse research methods. While correlational studies have afforded a picture of some important correlates of crowding, such as mental and physical health, in real-world settings, they have not allowed researchers to draw a clear causal picture. Laboratory studies, in contrast, have afforded a clear causal analysis of the effects of crowding, but have not permitted easy generalization from the artificial purity of the laboratory

to the complex realities of the real world. A chief strength of the studies of crowding in small-scale field settings has been their ability to incorporate in a single research design aspects of both the naturalistic richness of correlational studies and
the precise measurement and analysis of laboratory studies, balancing the requirements for both external and internal validity.

PSYCHOLOGICAL EFFECTS OF DENSITY

ILLNESS AND SOCIAL PATHOLOGY IN ANIMALS

Although our primary interest here is the effects of density on humans, it is important to briefly review research dealing with the effects of density on animal
populations because the animal studies have greatly influenced later research with
humans. While animal studies of density have been conducted with a wide variety
of animal species in both laboratory and naturalistic contexts, the findings that

Some Dramatic Effects of Density on Laboratory Rats

Research on the psychological effects of crowding has been greatly influenced by a fascinating series of experiments on high social density in animal
populations conducted by John Calhoun (1962a, 1962b, 1967, 1971). He placed
groups of Norway rats in an experimental enclosure where they were free to
breed and multiply. When the number of rats in the enclosure reached eighty,
he held the population size constant by removing the new offspring. Calhoun
observed a remarkable breakdown in the caged animals' use of the available
space; the rats tended to congregate in limited parts of the enclosure, so that
the level of density in those areas was excessively high. Calhoun termed this behavioral irregularity a "behavioral sink."

The manifestations of the behavioral sink in Calhoun's rats were serious and
dramatic. Abnormalities were observed in the female rats' ability to bear and
care for their offspring. Many females were unable to carry a pregnancy to full
term, while others died in the process of delivering their young. In addition,
many females were unable to carry out the normal maternal behaviors essential
to raising their offspring. Their normal activities in building nests for their young
became highly disrupted. Some female rats, instead of constructing the usual
cup-shaped nests, simply piled the nest materials in a careless heap. Others lost
interest in building a nest before the job was finished, and some females failed
to construct nests altogether, choosing instead to bear their offspring on the
sawdust floor of the cage. Calhoun reported that 96 percent of infants born in
the behavioral sink area died before weaning.

Male rats also showed dramatic manifestations of the behavioral sink. Some
males engaged in unusual patterns of sexual behavior, which Calhoun described as "pansexual." They failed to discriminate between appropriate and
inappropriate sex partners, making sexual advances to females who were not in
heat, to juvenile rats, and to other males. Another group of male rats, which

have most influenced human research on density have typically come from controlled laboratory studies with small animals, such as mice, rats, and rabbits (see box, "Some Dramatic Effects of Density on Laboratory Rats"). Findings from a large number of animal studies have indicated that high social density can cause abnormalities in physiological functioning, typically involving the adrenal glands, brain, and reproductive processes (see Christian, 1975; Davis, 1971). Animal research has also demonstrated that high social density can have negative social and behavioral effects, including social disorganization, intraspecies aggression, learning decrements, and reduced exploratory behavior (e.g., Goeckner, Greenough, and Mead, 1973; Southwick, 1967).

At the same time, however, Freedman (1979b) has cautioned that many of the findings of such studies are quite complex and should not be interpreted as indicating that social density always has negative effects on the physiological and social functioning of animals. In some animal studies, the size of a group has been more important than the actual amount of space available to it, and the effects of density have interacted with other variables, such as species type, the sex of the animals, the situation, and the social structure of the group.

Calhoun designated "probers," became frenetically hyperactive, engaging at times in cannabalistic behavior. Still other males became pathologically withdrawn, and emerged to eat and drink only when the other rats in the population were asleep. Calhoun concluded that the abnormalities associated with the behavioral sink were so severe that, over time, the breakdowns in normal reproductive functions of the animals would have resulted in the extinction of the rat colony.

Because Calhoun's research has been extraordinarily influential in later studies of the physical and social effects of high density, it is important to be especially clear about what his findings actually show. Most important, his data do not demonstrate that high social density among Norway rats will inevitably result in a behavioral sink. In fact, Calhoun designed his experimental enclosure to *encourage* an unequal distribution of rats over the various parts of the enclosure by "biasing" design features, such as electrified partitions separating some areas, different numbers of ramps leading into various areas, and contrasting elevations of the burrows in various sections. He deliberately made eating a difficult process in areas where the rats were likely to meet other rats by covering the food hoppers with wire grating. These unusual design features of the experimental apparatus deliberately fostered the high concentration of rats in limited sections of the enclosure.

Thus, although the controlled experimental nature of Calhoun's research allowed a high degree of internal validity, the artificial nature of the experimental apparatus limits the study's external validity. This research is important because it demonstrates that unusually high density can lead to extreme forms of social pathology among Norway rats. It does not suggest, however, that the types of high density likely to occur among rats in their natural habitats will cause the forms of social deterioration observed in the laboratory.

ILLNESS AND SOCIAL PATHOLOGY IN HUMANS

Physiological reactions If density operates as a psychological stressor for humans, it should be possible to measure the types of physiological reactions that are typically associated with stress. In fact, investigators have demonstrated that density can in some situations induce stress-related physiological responses in humans. Gary Evans (1979a) has found that high spatial density causes heart rate and blood pressure to increase. Other studies have shown that high density induces physiological arousal as measured by increased skin conductance (Aiello, DeRisi, Epstein, and Karlin, 1977). Finally, high social density in a shopping environment led to increased adrenocortical activity in males (Heshka and Pylypuk, 1975).

A study of Swedish commuters demonstrated a positive relationship between adrenaline level and social density aboard rush-hour trains (Lundberg, 1976). In a similar vein, investigators found that the distance and speed of industrial workers' travel to work (variables presumably associated with the stress of traffic congestion) in Southern California were positively correlated with commuters' blood pressure (Novaco, Stokols, Campbell, and Stokols, 1979; Stokols, Navaco, Stokols, and Campbell, 1978). Studies conducted in prisons have indicated that inmates in high-occupancy dormitory settings have more complaints of illness (McCain, Cox, and Paulus, 1976) and higher blood pressure levels (D'Atri, 1975) than prisoners in single or double cells, and higher levels of blood pressure as a function of increasing spatial density (Paulus, McCain, and Cox, 1978).

Physical illness On the basis of the evidence that social density has sometimes been related to physical symptomatology in animal populations, researchers have tried to learn whether high density is also associated with physical illness for humans. Two correlational studies, conducted in different parts of the world, have pointed to a relationship between inside density and physical illness. A study among Filipino men demonstrated an association between inside density and psychosomatic forms of illness (Marsella, Escudero, and Gordon, 1970). Another study conducted in Chicago (Galle, Gove, and McPherson, 1972) found a very small relationship between inside density and mortality rate, which reflects predominantly death due to disease.

Some studies conducted in small-scale field settings have also supported the finding that crowding bears a relationship to physical illness. A student health center was visited more often by students who lived in dormitories of high rather than low social density (Baron, Mandel, Adams, and Griffin, 1976), and by students who perceived the level of residential crowding to be high rather than low (Stokols, Ohlig, and Resnick, 1978). Similarly, illness among enlisted navy men has been shown to be positively correlated with both social and spatial density aboard ship (Dean, Pugh, and Gunderson, 1975, 1978).

Can we conclude on the basis of these correlational and field studies that crowding represents a serious health hazard? At the present time, environmental psychologists believe that crowding offers a partial explanation of physical illness in some settings, but that the relationship between crowding and health is influenced by a wide range of other environmental and personal factors. In fact, studies have shown that when the influences of other factors in the situation or the individual are sufficiently strong, crowding is not significantly related to illness. Correlational studies that failed to find a relationship between crowding and serious physical illness have been conducted in various parts of the world, including Hong Kong (Michelson, 1976; Mitchell, 1971; Schmitt, 1963), Southwest Africa (Draper, 1973), Hawaii (Schmitt, 1966), the Netherlands (Levy and Herzog, 1974), and Chicago (Winsborough, 1965). Investigators have concluded that the relationship between crowding and illness in human populations is strongly influenced by the way people have learned to cope with crowding both individually and by means of social organization.

Social pathology   Several correlational studies have explored the potential association between crowding and mental illness (Figure 7-4). Robert Faris and Warren Dunham (1965) report a relationship between outside density in Chicago and the level of mental illness. Omer Galle and his associates (Galle, Gove, and McPherson, 1972) point to an association between inside density (rooms per housing unit) and admissions to mental hospitals. Other investigators have found correlations between outside density and the rate of suicide (Langer and Michael, 1963; Sainsbury, 1956; Schmid, 1955).

Further correlational studies have reported an association between crowding and other forms of social pathology, such as crime and juvenile delinquency. Researchers found higher crime rates in central cities than in less densely populated suburbs (Schmid, 1969, 1970; Schmitt, 1957, 1963, 1966; Shaw and McKay, 1942). Galle, Gove, and McPherson (1972) have also reported a relationship between juvenile delinquency and inside density (persons per room).

Does the evidence from these correlational studies permit us to conclude that

Figure 7-4
Researchers have investigated whether crowding in cities may be related to physical and mental illness.

© *George W. Gardner.*

crowding leads to serious social pathology? The most current assessments by environmental psychologists have determined that such an assumption is unwarranted (Fischer, 1976; Freedman, 1975). First, those studies that have reported a relationship between crowding and social pathology have generally found that the association is not very strong (Altman, 1975). More important, when alternative influences, such as socioeconomic status, have been controlled for, the assumed relationship between crowding and social pathology has often disappeared (Freedman, 1975; Kirmeyer, 1978). Studies in Chicago (Winsborough, 1965) and New York (Freedman, Heshka, and Levy, 1975) found that when economic level and ethnicity were equated, density was not significantly correlated with social pathology.

SOCIAL BEHAVIOR

Aggression   Many of the everyday situations in which people are confronted with high density, such as congested streets or a busy department store, are associated with feelings of frustration and anger. Environmental psychologists have sought to learn whether a high level of crowding will in fact lead to aggressive behavior between individuals in a crowded setting.

An important series of experiments has investigated the effects of crowding on aggressive behavior among children in play situations. One study of crowding in an experimental play setting (Hutt and Vaizey, 1966) found that as social density increased, the level of aggression between normal children and children who were brain-damaged also increased, although density was unrelated to aggression in children who were autistically withdrawn. Later investigations of children in play situations, however, have shown a more complex picture. One study, involving behavioral observations of children in playgrounds of differing sizes, indicated that the frequency of fighting increased as spatial density rose (Ginsberg, Pollman, Wanson, and Hope, 1977). Two studies of children in controlled play situations, however, failed to find more aggression as density was increased (Loo, 1972; Price, 1971).

Investigators now believe that the relationship between density and aggression is influenced by both situational factors and individual differences. William Rohe and Arthur Patterson (1974) found that high density led to aggression among children when few toys were available for them to play with, but that when the supply of toys was ample, density alone did not cause aggression. Chalsa Loo (1978) reports that the level of density needs to be taken into account; her research indicates that while moderate levels of density will lead to aggression, very high levels of density will not. This finding is consistent with our earlier discussion of temperature and aggression, where we found that moderately uncomfortable temperatures lead to aggression, though extremely uncomfortable ones do not.

Several studies have found that some people are more likely than others to respond to density with aggression. Studies conducted in both laboratory and field settings indicate that while high density causes men to feel more aggressive and to act more competitively, high density is not associated with increased aggression or competition in women and may even lead women to respond to one another in more positive and friendly ways (Freedman, 1971; Freedman, Levy, Buchanan, and

Price, 1972; Saegert, Mackintosh, and West, 1975; Stokols, Rall, Pinner, and Schopler, 1973). These gender-related differences in response to high density are probably associated with normative differences in sex-role behavior in our society.

Yakov Epstein and Robert Karlin (1975) suggest that while high density is uncomfortable for both men and women, women come to feel more positive toward one another under high density because social norms allow them to share their feelings of discomfort with one another; men feel more negative toward one another because social norms discourage them from sharing their distress. In a somewhat similar vein, Jonathan Freedman (1975) speculates that high density may accentuate gender-related patterns of social behavior in our society, leading men to respond to other men as rivals while encouraging women to respond to other women in a nonhostile and friendly manner. Studies with children have indicated that children with behavioral problems, such as those who are brain-damaged or anxious and impulsive, show relatively more aggression under high density than do normal children (Hutt and Vaizey, 1966; Loo, 1978).

Withdrawal Research studies conducted in both laboratory and field settings have consistently found that as density increases, people are inclined to withdraw socially from one another. Studies have shown that as density in an experimental play setting increases, children tend to withdraw socially and to interact less with one another (Hutt and Vaizey, 1966; Loo, 1972; McGrew, 1970). Eric Sundstrom (1975), who studied male college students in an experimental interview situation, found that subjects' willingness to discuss intimate topics was inversely related to the level of density. Fascinatingly, even the expectation of being exposed to high social density causes people to exhibit social withdrawal. Laboratory studies have shown that people who believed that they would be joined by a relatively large group of other people in a research study tended to reduce their eye contact and to face away from the other people with whom they were waiting (Baum and Greenberg, 1975; Baum and Koman, 1976).

Field studies in diverse settings have provided further support for the finding that high density leads to social withdrawal. Researchers who studied college students in university dormitories found that students who lived in more socially dense environments were less talkative, less sociable, and less group-oriented than students who lived in less dense settings (Baum, Harpin, and Valins, 1975; Baum and Valins, 1977; Valins and Baum, 1973). Research in psychiatric hospitals demonstrated that adult patients engaged in less social behavior as the number of persons who shared a hospital bedroom increased (Ittelson, Proshansky, and Rivlin, 1970), and that child patients used their bedrooms less frequently as room occupancy increased (Wolfe, 1975). Finally, comparisons of the amount of eye contact between people in urban and suburban or rural areas reveals that eye contact decreases as a setting's density increases (McCauley, Coleman, and DeFusco, 1978; Newman and McCauley, 1977).

Reduced helping If high density is associated with a tendency to withdraw socially, is it also related to an unwillingness to offer a helping hand to another person who is in need of assistance? The social significance of this question was made clear to psychologists by a shocking event that occurred in New York City in the 1960s. A young woman, Kitty Genovese, was murdered in a socially dense residential neighborhood while returning home from work. Frighteningly, despite the fact that her cries were heard by thirty-eight local residents over an agonizingly long period of time, not one of those people responded to her pleas for help (Rosenthal, 1964).

Social scientists wondered whether the socially dense living conditions of urban life might provide a partial explanation for this horrifying event. An initial series of laboratory studies demonstrated that as social density rises, people become increasingly unwilling to take personal action to help in a social emergency. In a classic social-psychological laboratory experiment, John Darley and Bibb Latané (1968) observed how individuals responded to hearing what sounded like another person experiencing a severe epileptic-like seizure in an adjacent room. In one experimental condition, subjects thought they were the only person who was aware of the emergency; in another condition, they were led to believe that four other people had also heard the sounds made by the "victim." The investigators discovered that subjects who believed they were alone with the victim were much more likely to help and responded much more quickly to the emergency than did those who thought that other persons were aware of the situation.

A series of field experiments involving incidents such as helping a person locate an address or aiding someone who had dialed a wrong telephone number have been conducted in densely populated urban environments and in less populated rural settings. In general, earlier studies tended to support the notion that urbanites are less willing than residents of rural settings to lend a helping hand (Gelfand, Hartman, Walder, and Page, 1973; Korte and Kerr, 1975; Milgram, 1970). Later findings, however, failed to support this belief (Forbes and Gromoll, 1971; Korte, Ypma, and Toppen, 1975; Weiner, 1976). Two recent studies (Holahan, 1977; House and Wolf, 1978) report that helping behavior in the city is influenced more by a situationally based fear for personal safety than by a personal norm of unresponse on the part of city residents.

Some related field experiments in small-scale field settings, such as university dormitories and shopping centers, have also examined the association between high social density and altruistic behavior. Researchers found that students who lived in socially dense university housing were less willing than residents of less dense settings to mail a stamped letter that had apparently been lost and less willing to join in cooperative group projects (Bickman, Teger, Gabriele, McLaughlin, Berger, and Sunaday, 1973; Jorgenson and Dukes, 1976). The level of social density in a large shopping center was inversely related to people's willingness to help and to the amount of time they helped when they encountered another person (an experimental accomplice) who claimed to have lost a contact lens (Cohen and Spacapan, 1978).

In summary, a considerable body of research, involving laboratory studies as well as field experiments in both urban and small-scale field settings, has indicated that increasing social density can reduce people's willingness to help someone in need. At the same time, however, research findings on altruism and social density have been complex, especially those of studies that have compared helping behavior in urban and rural contexts; such research findings have been inconsistent and even contradictory. Later, when we discuss theoretical perspectives on crowding, we shall consider some of the complex social psychological processes that mediate the effects of density on altruistic behavior.

Attraction   Environmental psychologists have asked whether the reported effects of density on negative social behavior might reflect an underlying relationship between density and interpersonal attraction. Studies in both laboratory and field contexts have tended to support this hypothesis. Stephen Worchel and Charles Teddlie (1976) found that as spatial density decreased, men tended to perceive other members of a short-term group as more friendly. Robert M. Baron and his associates (Baron, Mandel, Adams, and Griffen, 1976) reported that students who were housed three to a room in rooms originally intended for two occupants expressed more dissatisfaction with their roommates than did students housed in double rooms.

As we saw earlier, Epstein and Karlin (1975) found that spatial density was negatively related to interpersonal attraction in male groups, but that increased density led to greater liking in female groups. Speculating that this sex difference might have occurred because it was easier for women to share their discomfort with other group members, they conducted a second experiment (Karlin, McFarland, Aiello, and Epstein, 1976) in which the female group members were unable to interact with one another. Under this condition, the feelings of attraction among the group members were reduced.

TASK PERFORMANCE AND MOOD

Task performance   Have you ever felt that high density in the classroom may affect your performance on examinations? Environmental psychologists have tried to find out whether high density negatively affects performance on a wide range of tasks. Earlier studies, which involved performance on relatively simple tasks in laboratory contexts, failed to demonstrate that density significantly affected performance in group discussions, creative problem solving, and psychomotor tests (Freedman, Klevansky, and Ehrlich, 1971; Freedman, Levy, Buchanan, and Price, 1972; Rawls, Trego, McGaffey, and Rawls, 1972; Stokols, Rall, Pinner, and Schopler, 1973). More recent laboratory studies, however, have indicated that under some circumstances, high density can adversely affect task performance. It has been shown that when a task is complex, such as the solving of a very difficult maze or processing information at a high rate, social and spatial density can cause performance decrements (Bray, Kerr, and Atkin, 1978; Evans, 1979; Paulus and Matthews, 1980; Paulus, Annis, Seta, Schkade, and Matthews, 1976). Density has also been shown to reduce performance when people must interact with one another to carry out a task (Heller, Groff, and Solomon, 1977). Moreover, high spatial density has been found to have delayed *aftereffects* on tasks designed to assess tolerance for frustration (such as attempts to solve insoluble puzzles) similar to the aftereffects associated with ambient environmental stressors (Evans, 1979a; Sherrod, 1974). Fascinatingly, however, as in the case of ambient environmental stressors, the negative aftereffects from high density are attenuated when people are permitted to feel that they have personal control over the stressor (Sherrod, 1974). We shall return to the important issue of personal control when we consider theoretical perspectives on crowding.

Evidence from studies in small-scale field settings also offers support for the contention that in some circumstances, high social density can lead to poorer task performance. Susan Saegert and her colleagues (Saegert, Mackintosh, and West, 1975), in a field experiment, tested subjects' performance on several practical and cognitive tasks under socially dense conditions in a railway terminal and in a department store. They found that density interfered with people's ability to carry out tasks that involved knowledge about and manipulation of the environment. Density also reduced the accuracy of subjects' maps of the environment—an indication that high density may have reduced the clarity of people's environmental images.

Two field correlational studies have reported that social density in university dormitories can adversely affect students' college performance. One study (Glassman, Burkhart, Grant, and Vallery, 1978) examined the effects of a housing shortage at Auburn University that required some ordinarily double rooms to be used to accommodate three students each. First-quarter grades for students in the three-person rooms were significantly lower than those of students in two-person rooms. Similar results have been reported from a study (Karlin, Rosen, and Epstein, 1979) at Rutgers University. There the grades of students in triple rooms were significantly depressed during their first year in school, but after students were reassigned to less crowded accommodations their grades improved significantly, and could no longer be distinguished from those of students who had been housed in double rooms from the beginning of the year.

In summary, there is evidence that both social and spatial density can negatively affect task performance under some circumstances. The specific nature of the relationship between density and task performance is quite complex, however. The effects of density on performance appear to be influenced by other factors, including the complexity of the task, the timing of the performance measures, the likelihood of the task to induce stress, and individual differences in respondents. More research is needed before we can understand how each of these variables interacts with density to influence task performance.

Mood   Most of us can point to personal experiences in which the need to cope with high density has dampened our mood. We may recall occasions when sitting in bumper-to-bumper traffic or being bumped and jostled in a packed supermarket have caused us to feel tense and uncomfortable. Studies in controlled laboratory settings have demonstrated that negative feelings are associated with both spatial density (Smith and Lawrence, 1978; Stokols, Rall, Pinner, and Schopler, 1973) and social density (Evans, 1975; Sundstrom, 1975). High density associated with a laboratory task also led people to project feelings of anxiety (Baxter and Deanovitch, 1970).

Susan Saegert and her associates (Saegert, Mackintosh, and West, 1975) found that social density in a railway terminal and in a department store caused subjects to feel anxious. S. Smith and William Haythorne (1972) observed two- and three-person groups of navy men in twenty-one-day periods of isolation under conditions of high and low spatial density, and found some effects of high density on perceived stress in the three-person groups. Some studies in controlled settings have indicated that the negative effect of high density on mood is more characteristic of men than of women (Freedman, Levy, Buchanan, and Price, 1972; Ross, Layton, Erickson, and Schopler, 1973). This finding is consistent with the view that gender-related social norms permit women to share feelings of discomfort associated with high density more readily than men (see Epstein and Karlin, 1975). In addition, Lou McClelland and Nathan Auslander (1978), using slides of public settings, found that the unpleasantness of high density varied according to context, with high density perceived as most unpleasant in shopping and work settings.

THEORETICAL PERSPECTIVES ON CROWDING

CROWDING AND INFORMATION OVERLOAD

Have you ever found it difficult to study for an exam because of too many intrusive activities competing with your effort to study? For example, while you studied in your dorm or apartment room, a group may have been carrying on a heated conversation or playing a stereo loudly nearby. You may have found that in your attempt to tune out these unwanted distractions, you inadvertently failed to notice something important that occurred, such as a friend's attempt to get your attention. Environmental psychologists have proposed that a similar process occurs when people try to cope with crowding. People in a crowded setting may try to deal with the excess information that is impinging on them by screening out much of the activity around them. To explain this process, environmental psychologists have applied the theory of *information overload*, which we encountered in our discussion of ambient environmental stressors in Chapter 6. As we shall see, information overload can help us to understand why density can sometimes lead to a breakdown in positive forms of social behavior or interfere with task performance.

Overload in the city   In an important paper, "The Experience of Living in Cities" (1970), Stanley Milgram discusses information overload as a theoretical model for explaining how city residents cope with urban crowding. In presenting his argument, Milgram draws on and extends the sociological theories of urban life of Louis Wirth (1938) and Georg Simmel (1950). Milgram proposes that urbanites are exposed to information overload from three sources: (1) a large number of people, (2) high population density, and (3) a greatly heterogeneous population. These three features of urban life provide a flood of informational inputs whose rate and quantity exceed the individual's capacity to process them.

Adapting to overload   Milgram argues that the urbanite's effort to adapt to information overload leads to the socially aloof and interpersonally unresponsive stance that often characterize social life in the city. Specifically, Milgram proposes that in order to cope with information overload, city people develop adaptive social mechanisms. They learn to minimize the time devoted to some social inputs (e.g., a curt hello between neighbors leaving for work) and to disregard inputs they have learned to assign a low priority, such as a drunk passed out on the sidewalk. They develop techniques for blocking off or discouraging social inputs, as by assuming an unfriendly attitude. Finally, they learn to filter social inputs in such a way that only superficial forms of social involvement are permitted (e.g., the cold, ritualized exchanges between salespersons and customers in city department stores).

Milgram contends that these adaptive responses to information overload are the basis of the urbanite's stereotypic cold and unresponsive social attitude. Long-term adaptation to information overload leads to the development of generalized

A Field Study of Social Anonymity

Philip Zimbardo (1969) conducted a fascinating field experiment in order to assess the consequences of differences in social anonymity between an urban and a rural setting. He proposed that in settings where the social conditions of life destroy the sense of individual identity and make people feel anonymous, negative forms of social behavior will occur. He suggested, for example, that such feelings of social anonymity may explain differences in the amount of vandalism between urban and rural contexts.

To test this hypothesis, Zimbardo abandoned similar automobiles in contrasting urban and rural settings. One auto was left in New York City, across from the Bronx campus of New York University. The second car was deserted in Palo Alto, California, near Stanford University. Zimbardo removed the license plates from both cars and opened the hoods to provide what he termed "releaser signals" for potential vandals. He then observed how the two abandoned automobiles fared in the contrasting environmental contexts.

Zimbardo was amazed by what he found. In the Bronx, the first looters arrived within ten minutes—a father, mother, and eight-year-old son. While the father and son searched the car and removed the battery and radiator, the mother served as a lookout. In the first twenty-six hours the car was totally stripped of all usable parts, down to a set of jumper cables and a can of car wax. Most astonishing, Zimbardo observed that many of the city vandals were well-dressed, clean-cut adults, and that it was not uncommon for passers-by to stop and chat with the looters. Over three days the Bronx car suffered twenty-three separate incidents of vandalism, and was eventually battered into a useless pile of metal.

How did the Palo Alto car fare? It was untouched, except that when it began to rain, a passer-by closed the hood to keep the motor from getting wet.

Of course, Zimbardo's field study provides only a very gross comparison of urban-rural differences in vandalism, and probably reflects many other factors (such as socioeconomic differences between the two communities) in addition to differences in social density. The study is important because it provides a vivid picture of one way in which urban and rural life can differ. At the same time, it would be a mistake to overgeneralize these findings into a statement of "the urban condition," or to attribute the results to density alone.

social norms that curtail and minimize the breadth and intensity of urban social contacts. For Milgram, these adaptive mechanisms explain why urbanites may fail to lend a hand in a crisis and are sometimes unwilling to offer helpful advice or assistance to strangers. Overload explains why urban life sometimes lacks ordinary civility and social etiquette, as when one person fails to apologize for bumping into another. It also helps to explain why urban life has come to be characterized by highly stylized and superficial role relationships, such as those between a professional person and clients or between a merchant and customers (see Sadalla, 1978). Finally, information overload can help us to understand the social anonymity that is typical of urban life (see box, "A Field Study of Social Anonymity").

Overload may exert direct and immediate effects on the behavior of city dwellers as well as the long-term adaptational effects emphasized by Milgram. A field experiment carried out in the Netherlands (Korte, Ypma, and Toppen, 1975) suggests that altruistic behavior is not predicated on the urban-rural distinction in itself, but rather on the actual level of environmental input or stimulation in a particular setting at a specific time. While cities generally provide more environmental input than rural areas, altruism may also be expected to vary *within* both urban and rural settings as a function of ongoing levels of environmental input. In a similar vein, Holahan (1977b), using a paper-and-pencil index of college students' willingness to respond to another person's request for assistance, found that students from urban backgrounds did not exhibit a generalized social norm of nonresponse. Rather, all subjects (whether from urban or rural backgrounds) based their willingness to offer assistance on the particular characteristics of the situation, such as whether the person asking for help was a stranger, whether personal danger was involved, and the requester's overall need.

Diffusion of responsibility   Of course, explanatory factors other than information overload must also be taken into account if we are to develop a full understanding of the phenomenon of the unresponsive urban bystander. Darley and Latané (1968) suggest that their finding that social density reduces an individual's willingness to take personal action in a social emergency is based on a *diffusion of responsibility*. An individual who encounters a social emergency when no one else is about will recognize sole responsibility for coping with the situation. But if other people seem to be aware of the emergency as well, the individual may feel less personal responsibility for taking action and be less likely to offer assistance. Diffusion of responsibility is also based on the individual's perception of the relative costs and benefits involved in taking personal action (Morgan, 1978). If fewer benefits (e.g., reciprocation or enhanced feelings of self-esteem) than potential costs (e.g., lost time or personal danger) seem likely to result from an offer of assistance, the individual will avoid taking personal action when others are present.

Crowding as information overload   Environmental psychologists have advanced models of crowding that are based on the notion of information overload. J. A. Desor (1972) views crowding as excessive stimulation from social factors. Similarly, Aristide Esser (1972) suggests that crowding results from information overload from unfamiliar or inappropriate social sources. Amos Rapoport (1975) also envisions crowding in terms of information processing, proposing that crowding is caused by excessively high social or sensory stimulation. Using an overload model, Stuart Miller and Kathleen Nardini (1977) compared people who scored high in arousal seeking with people who scored low. They found that the high scorers placed more simulated human figures in a model room before they perceived the room as crowded.

Susan Saegert and her associates (Saegert, 1978; Saegert, Mackintosh, and West, 1975) have used the notion of information overload to explain the psychological consequences of social and spatial density. They suggest that as the number of people in a setting (social density) increases, cognitive complexity is increased by the added social information each newcomer represents; but when available space decreases while the total number of people in the setting remains the same, cognitive complexity will not necessarily increase. Saegert and her colleagues point out that the psychological effects of information overload are most pronounced when people are engaged in activities that require a precise knowledge of the total social and physical context. Their own research in a Manhattan department store demonstrates that when social density is high, customers are able to recall fewer details about the store's merchandise and physical layout than when density is low.

CROWDING AND BEHAVIORAL CONSTRAINT

Some environmental psychologists have proposed that the negative consequences of crowding are caused by the limitations that high social and spatial density impose on people's behavioral freedom. According to this view, the amount of distress you would experience and the extent to which your studying would be disturbed by a group engaged in a noisy conversation nearby would depend on the degree of choice you felt in the situation. If you felt confined to the noisy setting because you had no alternative place to study, your psychological distress might be considerable. But if you knew you could leave the setting and study somewhere else, even if you did not actually exercise that option, you would feel better and your studying would be more effective. To explain this psychological process, environmental psychologists have proposed the theory of behavioral constraint. This model can help us to understand how crowding affects people's moods as well as their performance at various tasks.

Freedom of choice   Harold Proshansky and his colleagues (Proshansky, Ittelson, and Rivlin, 1976) have suggested that underlying the psychological effects of crowding is the freedom of choice people experience in crowded settings. They argue that crowding as a psychological phenomenon is only indirectly related to numbers of people. What is essential to the experience of crowding is the feeling that the presence of other people is frustrating one's efforts to achieve a particular objective. Thus crowding occurs when the number of people in a setting restricts an individual's freedom of choice, as when a noisy conversation makes it impossible to study or an overcrowded department store makes it impossible to shop efficiently. A survey among almost 700 Southern California residents provides some empirical support for the view that freedom of choice is significantly related to people's perception of being crowded (Schmidt, Goldman, and Feimer, 1979).

Similarly, Altman (1975) has proposed a theoretical model of crowding that is based on an underlying notion of behavioral constraint and on the coping strategies that constraint initiates. As we shall see in Chapter 8, Altman believes that the individual's effort to achieve a desired level of privacy is central to the related social processes of territoriality, personal space, and crowding. According to this view, we feel crowded when privacy-regulation mechanisms fail to work effectively, causing us to experience more social contact than we desire. High density increases the possibility that our territory will be intruded upon, that our ongoing activities will be interfered with, and that our access to desired resources will be blocked. To cope with these threats, we engage in complex boundary-regulation mechanisms that involve verbal, paraverbal, nonverbal, personal space, and territorial behaviors.

Behavioral interference   John Schopler and J. E. Stockdale (1977) propose that earlier models of crowding, such as the notion of freedom of choice, might be subsumed under the concept of *behavior interference*. The behavior interference model views high density as a necessary but not sufficient condition for crowding stress. When high density interferes with our goal-directed behavior, we experience crowding stress. A large number of people (social density) or limited space (spatial density) in a setting may interfere with an individual's access to resources in an environment, such as materials necessary to complete a task or easy verbal access to other people. Behavioral interference from high density increases the costs (e.g., time, effort, energy) incurred in completing the goal-directed behavior.

A later study (McCallum, Rusbult, Hong, Walden, and Schopler, 1979) examines the effects on crowding stress of varying the importance of the goal-directed behavior that is interfered with by high density. The behavioral interference model suggests that our crowding stress will increase in proportion to the importance we assign to the goal we are trying to reach. In effect, the more important our goal-directed behavior, the more time, energy, and effort we are willing to exert (that is, the greater costs we are willing to incur) in coping with the high-density condition and in completing our goal-directed behavior.

To test this hypothesis, the investigators devised a laboratory experiment that involved a simulated clerical task (using a card filing system) in a mock office environment. By varying social density (three or six female college students using the "office"), the researchers made it relatively easy (low behavioral interference) or relatively difficult (high behavioral interference) for the subjects to use the filing system. To vary the importance of the goal-directed behavior, the investigators paid subjects in one experimental condition 15 cents for each name they located in the files (high importance of behavior), while other subjects worked without pay (low importance of behavior). As the behavioral interference model predicted, the highest level of crowding stress was reported by subjects in the condition that involved both high importance of behavior and high behavioral interference.

Psychological reactance   When people's freedom of choice is constrained by high social or spatial density, they will attempt to cope with the situation by increasing their behavioral options. Daniel Stokols (1976) has applied Jack Brehm's (1966) theory of *psychological reactance* to this aspect of crowding. According to the theory of reactance, a perceived restriction on behavioral freedom leads to psychological reactance, a motivational state involving feelings of infringement and efforts to reestablish the threatened behavioral freedom. An individual whose behavioral freedom is restricted by crowding will engage in behavioral, cognitive, and perceptual attempts to cope with the situation. For example, a behavioral attempt to cope with crowding might involve an effort to restructure the physical setting or to move to an alternative setting. A cognitive or perceptual strategy might involve an effort to alter the way the crowded situation is viewed or understood.

Stokols applied this reactance-based model of crowding in developing a typology of human crowding experiences (Figure 7-5). The typology shows that crowding involves two sorts of "thwartings" (personal and neutral) that may occur in two types of environments (primary and secondary). Thwarting consists of the arousal of frustration as a result of interference with a person's activities or motives. A *personal* thwarting is interference from another person which is seen as deliberately directed against oneself; a *neutral* thwarting is interference from an environmental setting rather than another person, and is not viewed as deliberately

ENVIRONMENT

		Primary		Secondary
THWARTING	**Personal**	Antecedents:	violation of spatial and social expectations in the context of continuous, personalized interaction	Antecedents: violation of spatial and social expectations in the context of transitory, anonymous interaction
		Experience:	rejection, hostility, alienation; high intensity, persistence, and generalizability	Experience: annoyance, reactance, fear; moderate intensity, low persistence, and low generalizability; tendency toward "neutralization"
		Behavior:	behavioral withdrawal, aggression, passive isolation	Behavior: self-defense, leave situation
		Example situation:	antagonistic suitemates occupying mutual living space	Example situation: approach by threatening strangers on a crowded street
	Neutral	Antecedents:	violation of spatial expectations in the context of continuous, personalized interaction	Antecedents: violation of spatial expectations in the context of transitory, anonymous interaction
		Experience:	annoyance, infringement, reactance, moderate intensity, persistence, and low generalizability; tendency toward "personalization"	Experience: annoyance, reactance, low intensity, persistence, and generalizability
		Behavior:	behavioral withdrawal, improve coordination with others, augmentation of psychological space	Behavior: improve coordination with others, augmentation of psychological space
		Example situation:	family confined to a small apartment	Example situation: attendance at a crowded concert; laboratory experiment

Figure 7-5
A typology of crowding experiences.

From D. Stokols, "The Experience of Crowding in Primary and Secondary Environments," Environment and Behavior, *8:49–86, © Sage Publications, Beverly Hills, with permission of the Publisher.*

directed at oneself. In general, personal thwartings are more likely to occur in situations involving high social density, while neutral thwartings are more likely in high spatial density.

Stokols defines *primary* environments as those in which we spend much of our time, relate to other people on a personal level, and pursue a wide range of personally important activities. Residential settings, classrooms, and workplaces are primary environments. *Secondary* environments are places where our encounters with other people are relatively transitory, anonymous, and inconsequential—commercial transportation and recreational settings, for example. Figure 7-5 includes a description of the antecedent conditions for each type of crowding, along with a description of the way each type of crowding is experienced and responded to. The experience of crowding and reactions to it are more negative in primary than in secondary environments, and when they are associated with personal thwartings rather than neutral thwartings.

ECOLOGICAL MODELS OF CROWDING

The theories of crowding we have considered so far—overload and behavioral constraint—have been based on psychological explanations of the consequences of crowding. Other theorists have proposed conceptualizations of crowding that are based primarily on the notion of social process. These models have been broadly referred to as "ecological" models of crowding, and they bear some important similarities to ecological principles in the fields of botany and zoology. Michael Micklin (1973) identifies the general properties shared by ecological models of human behavior. First, ecological theories of behavior focus on the adaptive *interrelationships* between people and their environments. Second, the unit of analysis in ecological models is the social *aggregate* rather than the individual, and the notion of social organization plays an important role in such models. Third, ecological conceptions of behavior emphasize the distribution and use of material and social *resources*. Ecological models of crowding can help us to understand how crowding affects the organization of social groups and the effects of crowding on social processes within large groups.

Some ecological models of crowding have been stated in broad and somewhat speculative terms. John Calhoun (1971) proposes that we are able to tolerate high density in the spatial environment because our form of social organization has evolved ways of putting "conceptual" space between people. Responsible choice among ideas has replaced the search for spatial resources, and the commitment to abstract values has replaced the aggressive defense of physical resources. In a similar vein, Nathan Keyfitz (1974) has suggested that human forms of social organization help us to cope with high density by developing "specifications" in occupation and personality that allow us to share settings without aggressively competing for available physical resources.

Claude Fischer (1976), after reviewing various theories of crowding, suggests that models that emphasize the effects of crowding on the distribution of resources seem to provide the best explanation of available data. As long as social organization is adequate and resources are sufficiently distributed, density alone will not result in negative psychological consequences. This view is consistent with Rohe and Patterson's (1974) finding that when ample resources (toys) were available, crowding did not lead to aggression among children.

Overmanning   An empirically based ecological model of crowding has been advanced by Allan Wicker, of the Claremont Graduate School (Wicker, 1973, 1979; Wicker and Kirmeyer, 1976; Wicker, McGrath, and Armstrong, 1972). He has termed his model the theory of "manning," and has based his conception on an extension of Roger Barker's (1968) work in ecological psychology. Wicker emphasizes the adaptive aspects of the theory of manning, pointing out that the theory represents an elaboration of the idea that behavior settings function to counteract threats to their ongoing programs.

Wicker explains that the level of manning in a setting is determined by three factors: (1) *applicants*, or eligible people who seek to participate in a setting; (2) *maintenance minimum*, or the fewest people necessary to carry out the setting's programs; and (3) *capacity*, or the largest number of people the setting can accommodate while maintaining its programs. We can determine the level of manning in a setting by contrasting the number of applicants with the setting's maintenance minimum and capacity. Wicker's model (Figure 7-6) shows how the level of manning in a setting is related to the setting's maintenance minimum and capacity. When the number of applicants is equal to or exceeds the maintenance minimum without exceeding the setting's capacity (points *b, c,* and *d* in the figure), the setting is adequately manned. If the number of applicants exceeds the setting's capacity (point *e* in the figure), the setting is overmanned. Note that the setting can also be undermanned (point *a* in the figure) if the number of people who want to participate in the setting falls below its maintenance minimum.

The consequences of overmanning   Wicker and his associates conducted a series of studies in various field settings to examine the consequences of overmanning. One series of studies was carried out in large and small high schools (Wicker, 1968, 1969b). Wicker discovered that the level of participation in school activities, as well as the quality of school experiences, varied in the two types of environments. Students in small schools entered a wider range of behavior settings

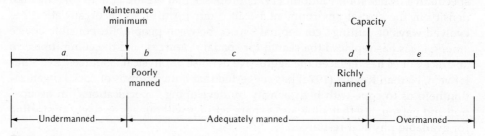

Figure 7-6
A continuum of degrees of manning.

From A. W. Wicker, "Undermanning Theory and Research: Implications for the Study of Psychological and Behavioral Effects of Excess Human Populations," Representative Research in Social Psychology, *1973, 4:185–206. Reprinted by permission.*

and assumed more positions of responsibility in their schools than did students in large schools. Students in small schools also tended to feel more needed, challenged, and self-confident than students in large schools.

Another series of field studies focused on the effects of differing levels of manning in church settings (Wicker, 1969c; Wicker, McGrath, and Armstrong, 1972; Wicker and Mehler, 1971). Members of small churches entered more church behavior settings and assumed more responsibilities than did members of large churches. In addition, new church members reported feeling more easily assimilated in small churches than in large churches. Wicker also found that members of small churches attended church more regularly and donated more money to the church than did members of large churches.

Wicker conducted a third series of studies in Yosemite National Park, in California's Sierra Nevada Mountains (Wicker, 1979). During July and August so many people visit the park that many of its facilities become overpopulated. Wicker found that by the end of the busy summer months, park rangers showed signs of reduced involvement in their work. He speculates that a "burnout" or negative attitude toward their jobs may have begun to develop as a response to the overpopulated conditions.

CROWDING AND PERSONAL CONTROL

We saw in Chapter 6 that the negative psychological effects of ambient environmental stressors are reduced when people are able to maintain a degree of personal control over the environmental stressor. Psychological stress is mitigated, for example, when an individual can modify a threatening event directly (*behavioral control*) or interpret the situation in such a way that it no longer appears threatening (*cognitive control*). Similarly, environmental psychologists have learned that the negative psychological consequences of high density can be mitigated when people experience some degree of personal control over it (Cohen, Glass, and Phillips, 1979; Moos, 1981; Sherrod, 1974). In fact, the notion of personal control is essential to some of the theories of crowding that we have considered, especially to the information overload and behavioral constraint theories of crowding.

Overload The notion that loss of personal control is an essential aspect of the psychological experience of crowding is implicit in the information overload model; the experience of crowding is related to the individual's inability to control the level of social and spatial information in high-density situations (Schmidt and Keating, 1979). An excellent example of the role of personal control in the information overload model is afforded by a series of laboratory and field studies conducted by Stuart Valins and Andrew Baum (1973) with residents of dormitories at the State University of New York at Stony Brook.

Valins and Baum systematically compared the psychological effects of two contrasting student housing designs that differed markedly in the levels of social density they generated. In one plan, a *corridor design*, thirty-four students were housed in seventeen double-occupancy rooms. Each student shared a common bathroom and lounge area with thirty-three other students. In the alternative design plan, a *suite design*, four to six students were housed in two- or three-bedroom suites, each with its own lounge and bathroom area. While the amount of space per person was comparable in both designs and all students had double-occupancy bedrooms, the students in the suites experienced a much lower level of social density, since each student shared common facilities with only three to five other students.

Valins and Baum discovered that the corridor and suite plans had very different effects on students. Students in the corridor arrangement reported feeling more crowded and were more sensitive to crowding pressures than students in suites, and were less sociable and less group-oriented. Essential to the investigators' interpretation of their findings was the notion that the corridor-design dormitory was an *overloaded* environment, where students were unable to exert effective *behavioral control* over the amount of social contact they had with other residents. The corridor-design dormitory required students to meet other residents in the hallways whether they wanted interaction or not; a student on the way to the bathroom or study lounge was likely to encounter another resident with whom he or she had not yet established a comfortable mode of relating. The corridor-design dormitory led to a general loss of personal control over social contacts in the hallways.

In a particularly interesting field experiment, Ellen Langer and Susan Saegert (1977) demonstrated that the adverse psychological effects of crowding can be reduced through a strategy of gaining a sense of *cognitive control* in the crowding situation. Drawing on available research concerning stress and health, they suggest that people can be helped to cope with the stress of crowding if they are provided with accurate information about the likely physiological and psychological reactions to crowding. In a field study conducted in New York City supermarkets, Langer and Saegert recruited eighty female shoppers and asked them to select the most economical product for each item on a prearranged grocery list. The shoppers completed the assigned task under conditions of high and low social density. In addition, the investigators provided half of the subjects with increased cognitive control by informing them beforehand that the crowded situation might cause them to feel aroused or anxious.

Langer and Saegert found that the aversiveness of the high-density situation was reduced through increased cognitive control. The shoppers who were provided with increased cognitive control felt more comfortable and were able to complete more of the assigned tasks successfully than were subjects who had not been provided with prior information. Langer and Saegert conclude that prior knowledge about the physiological and psychological effects of crowding can help people to develop anticipatory adjustments to the crowded situation.

Constraint   The notion that personal control plays a central role in the crowding experience is also implicit in the behavioral constraint model; the experience of crowding results when one is unable effectively to manage social and spatial interference or thwartings of one's goal-directed behavior in high-density situations (Schmidt and Keating, 1979). Objectively defined conditions of high density are translated into the psychological experience of crowding when the behavioral constraints associated with high density restrict the individual's sense of personal control—even tacit feelings of personal control that may occur below the level of conscious awareness.

Judith Rodin and her associates (Rodin, Solomon, and Metcalf, 1978) provide empirical evidence that perceived crowding in both laboratory and field settings is related to people's level of perceived control. In a field experiment Rodin and her colleagues investigated how *behavioral control* affected students' experience of crowding on an elevator in the Yale University Library. Four experimental accomplices systematically jockeyed subjects into either a position of high control (a place near the elevator control panel) or low control (a place away from the control panel). After subjects completed a postexperiment questionnaire, the investigators found that, although the elevator was actually equally dense in both experimental conditions, subjects in the high-control position reported feeling significantly less crowded than those in the low-control position.

In the laboratory context, Rodin and her associates examined how *behavioral control* affected college students' experience of crowding in a simulated discussion group. They varied control by allowing some students to start and stop the group discussion (high control) and preventing others from doing so (low-control). Again the investigators found that subjects who had been in the high-control condition felt that the setting was less crowded than those in the low-control condition.

Helplessness   In Chapter 6 we saw that the inability to achieve a sense of personal control over ambient environmental stressors can result in the psychological experience of learned helplessness (see Seligman, 1975). Learned helplessness involves a reduction in motivation to persist at a task, a resistance to relearning at a cognitive level that personal actions can successfully manage environmental outcomes, and depression. Environmental psychologists have extended the learned helplessness model of environmental stress to explain some of the psychological consequences of the loss of personal control in high-density situations.

Andrew Baum and Stuart Valins (1977) found that students who were unable to exert *behavioral control* over the nature, frequency, and duration of social contacts in their dormitories, in contrast to residents of dormitories where social contacts could be more easily controlled, demonstrated the symptoms of learned helplessness. The low-control residents were less likely than high-control residents to exercise a choice over the experimental condition in which they would participate, and were more likely to show withdrawal responses in a laboratory social game.

A follow-up study (Baum, Aiello, and Calesnick, 1978) demonstrated that the learned helplessness responses of students in the low-control environment became more pronounced as length of residence in the dormitory increased. Although low-control residents attempted initially to establish some degree of personal control over their social environment, by the end of the third week of residence they recognized that they were unable to control their social experiences in the dormitory. After seven weeks their expectations of control were quite low and their behavior demonstrated withdrawal and helplessness.

Judith Rodin's (1976) findings were similar when she studied black 6- to 9-year-old boys who lived in a low-income housing project. She reasoned that increases in inside social density would be related to reduced personal control over

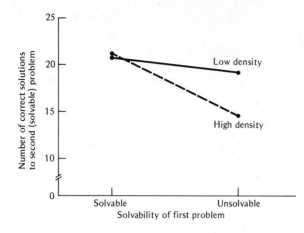

Figure 7-7

Number of correct solutions to a solvable problem achieved by children from low- and high-density residential environments after first being given either a solvable or an unsolvable problem.

From J. Rodin, "Density, Perceived Choice and Response to Controllable and Uncontrollable Outcomes," Journal of Experimental Social Psychology, *1976, 12:564–78. Reprinted by permission.*

such things as achieving desired quiet and going to sleep at a chosen hour. She speculated further that long-term residence in a low-control housing environment would generate symptoms of learned helplessness. Rodin found in an experimental game that the low-control youngsters exerted *behavioral control* (using a "choice" key associated with candy or marble rewards on a prearranged schedule) less frequently than children from high-control environments. Moreover, the low-control children were more inclined to use the key to earn experimenter-administered candy rather than self-selected candy.

Rodin conducted a follow-up study with a sample of junior high school students, some of whom lived in high-density residential environments while others did not. First she asked the youngsters to solve a problem that was either solvable (the control condition) or unsolvable (a condition that can induce learned helplessness). Then she asked both groups to solve a second problem that was solvable. Correct solutions on the second problem were similar for youngsters from high- and low-density residential environments when the first problem had been solvable. When the first problem had been unsolvable, however, the youngsters from high-density environments performed more poorly on the second problem than youngsters from low-density environments (Figure 7-7). Rodin concludes that the youngsters from high-density environments had come into her laboratory with learning histories that already included extensive helplessness conditioning. In effect, the high-density subjects began the task with a well-learned expectation that their ability to control events would be low and that they would perform poorly.

APPLICATIONS TO ENVIRONMENTAL PLANNING

How might research findings concerning the psychological effects of high density be applied to environmental planning and design? First, in those settings where high density has been found to have some negative psychological effects, environmental planners might be encouraged to design environments that avoid high social and spatial density. For example, designers of houses, apartment buildings, and dormitories might seek to avoid the creation of "overloaded" social environments in which residents are unable to manage their contacts with other persons effectively. In those settings where high density cannot be avoided or is a planning objective, design strategies and programming decisions might be directed toward the enhancement of people's sense of personal control and the avoidance of potential sources of crowding stress.

Designing Uncrowded Living Environments

Student housing   Research evidence from a series of studies conducted by Baum and Valins and their associates (Baum, Aiello, and Calesnick, 1978; Baum, Harpin, and Valins, 1975; Baum and Valins, 1977; Valins and Baum, 1973) indicates that high social density in university dormitories has negative social and psychological effects on residents' functioning when they are unable to control their contacts with other residents. Additional research comparing university housing characterized by high and low social density has also produced evidence of the negative social-psychological consequences of high social density in housing that features long double-loaded corridors and common facilities shared by large numbers of residents (Holahan and Wilcox, 1978; Wilcox and Holahan, 1976). These findings argue against the prevailing tendency to house college students in high-rise, socially dense settings that are characterized by commonly shared spaces where social contacts among large numbers of residents are virtually unmanageable. An alternative design philosophy might emphasize lower residential densities and more personal control over social contacts.

Of course, many universities already have large, socially dense residential environments that are constructed according to the traditional double-loaded plan. What can be done to improve the functioning of these existing facilities? A recent field study by Andrew Baum and Glenn Davis (1980) offers some encouraging planning suggestions. Baum and Davis were interested in exploring how a socially dense long-corridor dormitory for women at a small liberal arts college might be improved through relatively simple design modifications. Initially the dormitory consisted of rooms arranged along a long corridor shared by more than forty residents (Figure 7-8a). The design modification, carried out on one floor of the dormitory, involved the conversion of three central bedrooms to lounge space, creating two separate social groups of about twenty students each (Figure 7-8c).

In order to evaluate the effects of this design modification, Baum and Davis systematically assessed residents' behavior and experiences during a three-month period on the modified floor, on another long corridor in the same dormitory that was not changed, and on a short corridor in a comparable dormitory (Figure 7-8b). The investigators found that the design modification of the long corridor significantly improved the social-psychological functioning of residents on the remodeled floor. The modified corridor design was characterized by positive patterns of social interaction that were more like those on the short corridor than on the unchanged long corridor. Similarly, students on the remodeled floor and on the short corridor were more able to regulate their social contacts with other residents effectively and experienced less crowding stress than residents on the unmodified long corridor.

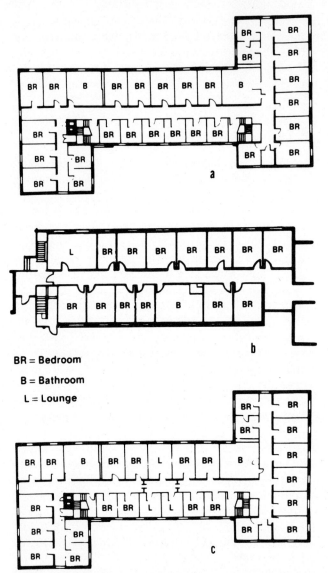

Figure 7-8

These contrasting dormitory floor plans show a traditional long-corridor plan (*top*), a short-corridor plan (*middle*), and a modified long-corridor plan (*bottom*).

From A. Baum and G. E. Davis, "Reducing the Stress of High-Density Living: An Architectural Intervention," Journal of Personality and Social Psychology, 1980, 38:471–81. Copyright 1980 by the American Psychological Association. Reprinted by permission of the publisher and author.

BR = Bedroom

B = Bathroom

L = Lounge

In addition to physical design features that can enhance residents' sense of personal control, the *process* of arriving at residential design decisions might be conducted so as to enhance residents' personal input, involvement, and influence. Drury Sherrod and Sheldon Cohen (1979) encourage residential planners to allow environmental users direct involvement in the planning process. After the building is occupied, planners should consult residents in order to assess how well the residential environment actually meets their needs. They emphasize that participation in the design process can enhance residents' sense of personal control in the residential environment, and that this enhanced sense of control can help to reduce the potential negative consequences of residential crowding (see box "Low-Density Housing Built by Students").

Urban housing   Jonathan Freedman (1979a) has suggested that the potential negative effects of high-rise urban housing might also be alleviated if conventional
long corridors were replaced by short corridors. Short corridors would avoid the
sense of uncontrollable anonymous social contacts that are typical in long-corridor
buildings, and would facilitate a sense of neighborliness and community among
the five or six families that shared a corridor. The short-corridor design would not
require any fewer floors in the building or fewer apartments on each floor; all that
would be needed would be a few more elevators. Freedman is convinced that the
cost of additional elevators would be worth the gain in the quality of the building's
social life and in its enhanced security.

Amos Rapoport (1975) has suggested several physical properties of settings
pertaining to both inside and outside density that affect the likelihood that they
will be perceived as crowded. His proposals were based on a belief that certain
types of information tend to be interpreted as indications that an environment is
densely populated. Rapoport's suggestions are especially relevant to the design of
urban housing. He notes, for example, that tall buildings tend to indicate a higher
density than low buildings even when other information indicates that the two
types of settings are equal in density. Rapoport also suggests that the availability
of adjacent nonresidential spaces, such as parks, pubs, and shops, tends to make a
residential area seem less dense. Density will also appear to be lower in residential
areas where "defenses" are available to control social interaction, such as fences,
compounds, and courtyards. Finally, settings with natural greenery will be judged
as less dense than areas without natural greenery.

DESIGN STRATEGIES IN DENSE ENVIRONMENTS

Although planners can reduce the levels of density to which people are exposed in
a variety of ways, it is apparent that the distribution of population and resources
will make high density in some areas a continuing aspect of contemporary life.
How might our knowledge of the environmental psychology of crowding be applied to optimize the physical design of settings that must serve a large number of
people? One solution is to design small-scale architectural features within dense
environments so as to reduce the level of perceived crowding in those areas.

Allen Schiffenbauer and his associates (Schiffenbauer, 1979; Schiffenbauer,
Brown, Perry, Shulack, and Zanzola, 1977) attempted to determine design factors
other than room size that modify the experience of feeling crowded in student
housing. They found that dormitory rooms that received a great deal of sunlight
were judged to be less crowded than rooms that received less sunlight. This finding is similar to Baum and Davis' (1976) observation that light-colored dormitory
rooms were seen as less crowded than rooms painted in dark colors. Schiffenbauer
and his colleagues also found that dormitory rooms with more usable floor space
and those located on higher floors were perceived to be larger than rooms with less
usable floor space and those located on lower floors. Finally, they encourage the
use of design elements to differentiate the space in front of residents' doors in
double-loaded corridors from the surrounding corridor space. Students complained that fellow residents typically intruded into their closed rooms without

——Low-Density Housing Built by Students——

Judith Corbett (1973) describes a remarkable university housing program at the University of California, Davis, that simultaneously achieved a low-density design and significant involvement of residents in the design process. The experimental program involved the construction of fourteen polyurethane-foam domes coated with a fireproofing layer of fiber glass. Each dome is occupied by two students and provides ample living space, with both a ground level and a second-story loft. Most residents used the loft as bedroom space and the ground floor for other day-to-day functions. Most important, in terms of the enhancement of residents' sense of personal control in the innovative housing environment, the student residents played a major role in designing and building the domes. Each pair of residents chose their own dome, selected the exterior color, and with some guidance from the contractor designed and constructed the interior furnishings. The university paid the students $2.47 per hour for their labor.

After the domes had been occupied for six months Corbett administered a self-report questionnaire to their residents and to a sample of residents of conventional dormitories on and near the Davis campus. She discovered that the new domes were seen as more satisfactory than the conventional dormitories in several important respects. The dome residents reported that the domes were quieter and more private than the conventional dormitories, and several commented that the extra ceiling height of the domes gave them a very spacious feeling. Moreover, the dome residents were more likely than dormitory residents to know all of the other students in their housing environment, because the total number of residents was fewer than in the conventional dormitories. Finally, the students who lived in the domes showed more group spirit and a greater feeling of community than did residents of the conventional housing.

---

hesitating. Design features that distinguished the areas near room entrances (such as a change in the color or texture of flooring immediately in front of each room or even the use of doormats) would act as a "warning" or cue to potential intruders that the room space was a private area.

J. A. Desor (1972) used a scaled-down environmental model in a laboratory setting to investigate the effects of several design features on perceived crowding. It was found that partitioning an enclosed space reduced the level of perceived crowding in the model, and that the effects of the partitioning were equally beneficial whether the partition consisted of a waist-high barrier, a glass wall, or a solid wall. Crowding was also reduced when the number of doors in a room was decreased and when a room's linear dimensions were unequal—that is, a rectangle rather than a square. Desor interprets the beneficial effects of these design features in terms of an overload model of crowding, suggesting that these design alterations were able to reduce the perceived levels of social stimulation in the crowded situations.

Allan Wicker (1979) conducted a field experiment at an overmanned shuttle-

The student residents of these innovative domes played a major role in their design and construction.

Photo courtesy of Judith Corbett.

bus stop in Yosemite National Park (Figure 7-9) designed to make boarding the bus safer, more efficient, and more equitable. The experiment involved the construction of a queuing device formed by a series of chain-linked iron posts that would serve to channel passengers onto the bus. Through behavioral observations at the bus stop and a survey of bus drivers, Wicker determined that the experimental queuing device succeeded in decreasing pushing and running at the bus stop, while increasing the efficiency and orderliness of passengers boarding the bus.

Susan Saegert and her associates (Saegert, Mackintosh, and West, 1975) propose that clear orientation in design elements is essential in overloaded high-density environments. The number of choice points in high-density settings should be low, and paths should be especially clear. The number of signs and messages should not be excessively high in environments that are already overloaded. Daniel Stokols (1976) adds that the potential negative effects of crowding can be reduced through architectural features that allow people optimum flexibility in coping with changing densities, such as movable walls and ceilings.

Figure 7-9
Visitors attempting to board an overcrowded shuttle bus in Yosemite National Park.

From A. W. Wicker, An Introduction to Ecological Psychology, p. 179. Copyright © 1979 by Wadsworth, Inc. Reprinted by permission of the publisher, Brooks/Cole Publishing Company, Monterey, California.

PROGRAMMING FOR HIGH-DENSITY SETTINGS

When planners are faced with high density in settings where the number of design modifications that can be made is limited, an alternative strategy to reduce the adverse effects of crowding is to program the use of the environment (Saegert, Mackintosh, and West, 1975). Activities that require concentration and that might suffer from irrelevant stimulation, such as reading, studying, or intimate social interaction, should not be planned for spatially dense settings. The difficulties caused by social interference reported by teachers in open-space school settings are relevant here.

Another strategy for programming in high-density settings is to increase the social or group cohesiveness in such settings. When Andrew Baum and his colleagues (Baum, Harpin, and Valins, 1975) examined the relationship between perceived crowding and group cohesiveness in a dormitory characterized by high social density, they discovered that group cohesiveness played an important role in moderating the psychological consequences of high density. Among students who perceived their dormitory floor as socially cohesive, only 19 percent reported that they felt the floor was crowded. In contrast, among residents who saw their floor as low in group cohesiveness, 76 percent reported that the floor was crowded. Baum and his colleagues conclude that efforts to facilitate the development of cohesive social groups in high-density settings can help to make those settings seem less crowded.

Wicker (1979) proposes that school planners attempt to develop extracurricular activity programs in large schools that might provide students with the types of involving and responsibility-building experiences that characterize small schools. Such activities need to be planned at the level of a unit smaller than the whole school in order to achieve the desired psychological benefits. Wicker recommends that large schools develop innovative programs that are responsive to students' expressed interests rather than traditional academic programs. School facilities might be made available to students to run an "underground" newspaper, to work on their own automobiles, or to display students' artwork.

SUMMARY

An essential first step in the study of crowding is to develop clear and precise definitions of crowding-related phenomena. Investigators have argued, for example, that crowding should be distinguished from density. They have proposed that *density* be restricted to references to the strictly physical or spatial aspects of a setting; that is, the number of persons per spatial area. *Crowding* should refer only to the psychological or subjective factors in a situation; that is, individuals' perceptions of spatial restrictions.

Researchers have investigated *social density*, defined as the number of people in a given area, by varying the size of a social group in a spatial setting of constant size. Other researchers have studied *spatial density*, defined as the available space in a particular setting, by varying the size of a spatial setting in which the size of a social group remains constant. *Inside density* has been defined as the number of people per spatial area within a dwelling unit; for example, the number of people per room or the number of people per residence. *Outside density* refers to the number of people (or residences) within a broader geographic area; for example, the number of people (or residences) per acre or census tract.

Many environmental psychologists believe that the relationship between high density and resulting behaviors, such as negative forms of social activity, is mediated by a psychological stress reaction. Crowding has been envisioned as a form of psychological stress in which an individual's perceived need for space exceeds the space that is available. While high density *can* operate as a psychological stressor, it will not inevitably result in psychological stress. Whether or not high density leads to psychological stress depends on an individual's social and spatial needs in a particular situation and the particular characteristics of the situation.

Crowding represents both a *stressor situation* (the environmental factors that cause people to feel crowded) and a *stress syndrome* (the psychological experience of being crowded). An important aspect of the psychological process of crowding is concerned with the individual's efforts to cope with crowding. In this sense, crowding may be viewed as a motivational state that is intended to alleviate the perceived spatial constraints of the crowded setting.

The first research strategy used by investigators to study the human consequences of crowding involved an analysis of the correlation between density and social pathology, as indicated by census data and other archival records. A limitation of correlational studies was that the covariation of other factors, such as socioeconomic status, made it impossible to conclude that density caused social pathology. More recent researchers have attempted to control statistically the effects of outside influences in correlational studies of the relationship between crowding and social pathology.

One research strategy in the study of crowding is the use of experimentation in laboratory settings. The goal of laboratory studies has been to establish a clear causal link between crowding and behavior by systematic control of the potential influences of extraneous factors. The most current strategy in the study of crowding has involved research in small-scale field settings, such as college dormitories, department stores, and hospitals. These field studies have been able to incorporate aspects of both the naturalistic richness of correlational studies and the precise measurement and analysis of laboratory studies.

Findings from a large number of animal studies have indicated that high social density can lead to physiological, social, and behavioral abnormalities. In some animal studies, however, the size of the group has been more important than the amount of space available to it, and in many studies in which density has been a significant factor, the effects of density have interacted with other variables, such as species type, the sex of the animals, the situation, and the social structure of the group.

Research studies of density with humans have revealed a complex picture. Correlational and field studies have indicated that crowding is a partial explanation of physical illness in some settings, but that the relationship between crowding and health is influenced by a wide range of other environmental and personal factors. An especially important factor is the manner in which humans have learned to cope with crowding both individually and by means of social organization. Although some early correlational studies showed a relationship between crowding and social pathology, most researchers now believe that this reported relationship is a function of the covariation of other factors, such as socioeconomic status.

Laboratory research has shown a relationship between crowding and aggression, though investigators believe the crowding-aggression relationship will occur only under particular conditions. In addition, some laboratory and field studies have indicated that the relationship between crowding and aggression is characteristic of men but not of women. A consistent finding in laboratory and field studies has been an association between crowding and social withdrawal. Field studies in both urban and university environments have indicated that under some circumstances, crowding can lead to reduced helping behavior. Laboratory and field studies have also shown that crowding can lead to a reduction in interpersonal attraction.

Earlier laboratory studies involving relatively simple tasks failed to show negative effects of crowding on the performance of tasks. Later studies that used complex tasks in both laboratory and field settings, however, have demonstrated that crowding leads to decrements in task performance under some conditions. Laboratory and field studies have shown a relationship between crowding and negative mood, though some studies have suggested that this relationship is stronger in men than in women.

Several environmental psychologists have advanced theoretical models of crowding that are based on the notion of *information overload*. According to this model, crowding presents the individual with so many informational inputs so rapidly that the individual is unable to process them. Central to the overload model is the belief that the individual's efforts to adapt to information overload cause the negative social behaviors that are associated with crowding.

An alternative theory of crowding, *behavioral constraint*, proposes that the negative psychological consequences of crowding are caused by the limitations that high social and spatial density impose on people's behavioral freedom. According to this model, high density is a necessary but not sufficient condition for crowding stress. When high density interferes with goal-directed behavior, the individual will experience crowding stress. When people's freedom of choice is constrained by high social or spatial density, they will attempt to cope with the situation by increasing their behavioral options.

An additional theoretical approach to crowding has produced *ecological models* that focus on the adaptive interrelationships between people and their environments as they attempt to obtain material and social resources. One ecological model is the theory of manning, which views crowding as overmanning. Overmanning has been defined as the condition in which the number of eligible participants in a setting exceeds the setting's capacity to accommodate them while maintaining its programs.

Environmental psychologists have learned that the negative psychological consequences of high density can be mitigated when people feel some degree of personal control over it. The notion of personal control is implicit in the information overload model of crowding; the experience of crowding is related to inability to control the level of social and spatial information with which the individual is confronted in high-density situations. The idea of personal control is also implicit in the behavioral constraint model of crowding; the experience of crowding results when one cannot effectively manage social and spatial interference or thwartings of one's goal-directed behavior in high-density situations. Several environmental psychologists have extended the learned helplessness model of environmental stress to explain some of the psychological consequences of the loss of personal control in high-density situations.

Findings from the psychological study of crowding have been applied to the design of university and urban housing, to the design of small-scale architectural features for dense environments, and to programming for high-density settings.

8 Privacy and Territoriality

PRIVACY

Attaining personal privacy in today's society can be a challenge; our privacy is threatened from many quarters in subtle yet serious ways. Electronic advances in "snooping" hardware, from devices for telephone wiretaps to concealed microphones, make access to people's private lives an easy matter. Complex data banks from which information can be transferred many thousands of miles instantaneously through modern computer technology give people the capability of monitoring other people's private transactions.

Although we generally tend not to think about our personal privacy, it is an essential aspect of our lives. For example, in studying for an important exam, you may expend considerable effort to find a place where you can study undisturbed. Consider also your pleasure at the discovery of a private corner where you can read a good book, reflect, or simply daydream, unassailed by the pressures of daily life.

236

We will find in this chapter that privacy is more complex than it may at first appear. Privacy sometimes means solitude, but at other times it calls for a place where two or more people can engage in personal conversation or share special feelings, secure from outside intrusions. The search for privacy sometimes leads us to seek a quiet corner inside a designed setting, as when we reserve a study carrel at the library or take the telephone into another room. At other times we seek privacy outdoors, in the natural environment—on a deserted seashore or at a placid mountain lake. Although privacy represents a relatively new area of interest for environmental psychologists, we shall see in this chapter that the investigation of privacy has grown rapidly into an exciting research area.

Privacy lacks a simple and universally accepted definition. Environmental psychologists who have studied privacy have found that its meaning varies according to context. The "privacy" we speak of in everyday usage is not the "privacy" of the lawyer, the politician, or the behavioral scientist (Margulis, 1977).

EVERYDAY MEANINGS OF PRIVACY

Some investigators have attempted to identify and categorize the popular meanings of privacy in our everyday language. To investigate the meanings of privacy among students at a junior college near San Francisco and their parents, Nancy Marshall (1970, 1972) developed the Privacy Preference Scale, which consists of a series of statements about privacy in a variety of situations. She discovered that the students and their parents had six distinct orientations toward privacy, which may be grouped under two major headings—withdrawal and control of information (Figure 8-1). As we shall see, withdrawal and control of information may also be used to categorize the major ways in which privacy is viewed in the law and in the behavioral sciences. Three of the orientations identified by Marshall emphasize withdrawal: *solitude, seclusion,* and *intimacy.* Marshall defines "solitude" as the desire to be alone. "Seclusion" refers to people's wish to live out of sight and sound of neighbors and traffic. "Intimacy " involves getting away from other people with one's family or a special person.

The three remaining orientations identified by Marshall emphasize the idea of control of information: *anonymity, reserve,* and *not-neighboring.* "Anonymity" involves keeping others from "knowing everything about you." She defines "reserve" as the desire not to disclose much about oneself, especially to people one does not know well. "Not-neighboring" is a preference for noninvolvement with neighbors, and a dislike of having people drop in without warning.

In another interesting survey study, Maxine Wolfe and her colleagues (Wolfe and Laufer, 1974; Laufer, Proshansky, and Wolfe, 1976) investigated the meanings of privacy held by schoolchildren from age 4 to late adolescence. They were especially interested in learning how the meaning of privacy develops during childhood and adolescence. They discovered that by age 5 children already have some sense of the meaning of privacy. Interestingly, the two broad meanings of privacy we have discussed—withdrawal and control of information—are associated with particular developmental periods. The idea of withdrawal and solitude is common in the earliest definitions of privacy and is never lost. The notion of privacy as the control of information develops slightly later, and is the most frequent definition

Figure 8-1
Sometimes our need for privacy requires us to withdraw from other people in order to think and reflect alone.

© *Sylvia Johnson 1981/Woodfin Camp & Assoc.*

of privacy by age 7—the age at which most children first become aware that they are able to exercise control. While very young children tend to view privacy in terms of the possession of personal objects, such as their toys, older youngsters define privacy according to their ability to control the use of their possessions by others.

Ross Parke and Douglas Sawin (1979) conducted a survey study to learn how the privacy behaviors that children engage in at home develop from childhood into adolescence. They discovered that as children develop, they make progressively more use of physical privacy markers, such as closing the door of a bedroom or bathroom they are occupying. They also found that the use of privacy rules, such as knocking on a closed door, becomes more common as a child grows older. Children's regulation of their privacy by restricting the access of parents and siblings to the bathroom while they are occupying it also increases with development. The largest increase in restricted access to the bathroom occurs during early adolescence with the development of secondary sex characteristics.

Children's privacy behaviors in the home are influenced by environmental variables, such as the number of bedrooms and bathrooms in the home (Table 8-1). The accompanying table shows that the proportion of children who reported

Table 8-1.   Mean percentage of children who closed bedroom and bathroom doors, by age and number of bedrooms and bathrooms in child's home.

Occupant's age	Number of bedrooms		Number of bathrooms		
	2–3	4–7	1	2	3 or more
2–9	0.17	0.50	0.58	0.75	0.92
10–17	0.33	0.58	1.00	1.00	0.92

Source: R. D. Parke, and D. B. Sawin, "Children's Privacy in the Home: Developmental, Ecological, and Childrearing Determinants," Environment and Behavior, 11: 87–104, © 1979 Sage Publications, Beverly Hills, with permission of the Publisher.

keeping their bedroom door closed was positively associated with the number of bedrooms in the home. A similar relationship was found in the case of bathroom doors, though only for younger children. The proportion of children aged 2 to 9 who reported closing the bathroom door was positively related to the number of bathrooms in the house. Older youngsters (aged 10 to 17), however, generally closed the bathroom door regardless of the number of bathrooms in the home.

PRIVACY AND THE LAW

While legal definitions of privacy are complex, some important aspects of the legal view of privacy may be subsumed under the two headings we have encountered in everyday definitions of privacy, withdrawal and control of information (Margulis, 1977). In common law, for example, individuals are protected from unwarranted intrusions on their seclusion and personal affairs and from the loss of personal control over the public disclosure of private information (see Kalven, 1966; Miller, 1972).

Alan Westin (1967) has provided an analysis of privacy from the perspective of the political scientist, which emphasizes the relationship between privacy and individual freedom. In discussing the various aspects of privacy, Westin includes both withdrawal and control of information. The individual has a right to *solitude*—to be alone and secure from observation by other persons—and a right to a state of *reserve*, in which one cannot be forced to reveal highly personal aspects of oneself.

Carol Warren and Barbara Laslett (1977) have drawn a distinction between privacy and secrecy that is founded on moral concerns. They point out that privacy is consensually agreed on by members of society, and is a right of the individual. Secrecy, in contrast, involves the concealment of something that is negatively valued by society; there is no corresponding individual right to secrecy. For example, while a family has a right to privacy in maintaining the personal relationships of family life, individuals plotting a crime do not have a right to secrecy in formulating their unlawful plans.

BEHAVIORAL SCIENCE VIEWS OF PRIVACY

Irwin Altman (1975) notes that the notions of withdrawal and control of information may also be used to categorize behavioral-science definitions of privacy. Those behavioral-science definitions of privacy that emphasize a withdrawal from other persons tend to view privacy in terms of seclusion and an avoidance of interpersonal interaction. Altman points out, for instance, that Sidney Jourard (1966b) defined privacy in terms of an individual's desire to withhold from others knowledge about personal experiences, actions, and intentions. Similarly, Alexander Kira (1966) and Leo Kuper (1953) viewed privacy as restricting visual and auditory interaction.

In the second broad category of privacy, control over personal information, privacy is viewed in terms of the ability to open and close access to oneself, depending on personal feelings, one's relationship with other persons, and characteristics of the situation. Altman points out that this second category is broader than the notion of withdrawal, since control over access to the self may be exercised by either withdrawal or nonwithdrawal, depending on one's intentions.

Altman notes that in their definition of privacy, William Ittelson and his colleagues (Ittelson, Proshansky, and Rivlin, 1976) emphasize the freedom to control the information about oneself that is communicated to other persons. Similarly, Westin (1976) defines privacy in terms of the individual's right to control the type of information about him- or herself that is communicated to others. Conversely, Georg Simmel (1950) proposed that privacy includes the ability to control stimulus input from other persons.

A BASIC DEFINITION OF PRIVACY

Because privacy is so complex and has so many meanings, we need a basic definition to guide our discussion. Irwin Altman has offered such a definition, one that is broad enough to encompass most of the meanings of privacy we have encountered in everyday, legal, and scientific usage. Altman defines privacy as the *"selective control of access to the self or to one's group"* (Altman, 1974:24; 1975:18).

Notice that Altman's definition of privacy emphasizes the notion of control of information. It may, however, also encompass withdrawal, since one way a person or group can restrict access to itself is by withdrawal. Yet Altman's definition goes beyond the simple notion of withdrawal because it emphasizes *selective* control. The self can be made more or less receptive to other persons by the manner in which the individual or group systematically regulates its openness, depending on the particular circumstances.

Notice also that Altman's definition allows us to consider privacy in relation to a variety of social units—individuals, groups, or individuals and groups in interaction. Finally, Altman's definition permits us to view privacy in terms of both the control of outputs from the self to others and the control of inputs from others to the self. We shall return to Altman's definition when we discuss theoretical views of privacy.

Privacy and territoriality   Julian Edney and Michael Buda (1976) point out that the concepts of privacy and territoriality appear intuitively to be related. They conducted two questionnaire studies with college students (one involving a laboratory manipulation of privacy and territoriality) to explore differences between the two concepts. They found that students distinguished between privacy and territory. For example, for some activities students desired territory without privacy, such as when they were relaxing or eating.

In their laboratory manipulation of privacy and territoriality, Edney and Buda found that privacy and territoriality produced distinct psychological effects on college students. Privacy proved more important than territoriality in making an area feel "stimulating" and "free," and was a greater stimulus to creativity. Territoriality, however, played a stronger role than privacy in causing people to attribute their behavior to "their own personality" rather than to "the influence of other people."

Yet privacy and territoriality are not completely independent, in that territoriality may sometimes be employed to enhance privacy. For example, people who want to preserve their privacy may retreat to a territory of their own, such as their own bedroom or office. In fact, as Edney and Buda note, other investigators have considered that territoriality functions in the service of privacy (see Altman, 1975; Proshansky, Ittelson, and Rivlin, 1976). Later in this chapter we shall see that Altman (1975) views territoriality as one mechanism that may be used to achieve a desired level of privacy. (We shall see in Chapter 9 that personal space offers another such mechanism.)

We should keep in mind, however, that although territoriality may sometimes be used to achieve privacy, the two concepts are not identical. While people do often employ territory to achieve privacy, they can also attain privacy without establishing their own territory (e.g., by using radio earphones). And territory can serve psychological functions other than privacy (such as maintaining a position of dominance in a social pecking order).

RESEARCH METHODS FOR STUDYING PRIVACY

The principal research method for studying privacy has involved self-report *survey* and *questionnaire* measures concerned with people's experience of privacy in a variety of real-world settings. In their study of the developmental aspects of privacy, Maxine Wolfe and her associates (Wolfe and Laufer, 1974; Laufer, Proshansky, and Wolfe, 1976) conducted personal interviews with schoolchildren in New York City and Milwaukee. They asked the children to tell all the things the word "privacy" meant to them; to describe a time when they felt very private; to "describe what each of the following is like for you: a private place, a private thought, a private feeling, a private thing, and a private talk." Similarly, in her investigation of popular meanings of privacy among college students and their parents, Nancy Marshall (1970, 1972) employed a paper-and-pencil questionnaire consisting of eighty-six items that assessed respondents' preferences for various aspects of privacy in a variety of settings.

Self-report measures have predominated in studies of privacy because the experience of privacy is colored by an individual's subjective impressions and needs. Environmental psychologists have found that the most effective way to learn about privacy in particular settings is to ask people about their own experiences of pri vacy in those contexts. Because the experience of privacy is a function of each individual's desire for a particular level of privacy, it will necessarily vary according to individual preferences and needs. Further, many effects of privacy are felt predominantly at a cognitive level, and are translated into overt behavior only in subtle ways.

A small number of investigators have employed *naturalistic observation* and even *unobtrusive measures* (see "Research Methods for Studying Performance," Chapter 5) to study privacy. Eric Sundstrom and his associates (Sundstrom, Burt, and Kamp, 1980) used both naturalistic observation and unobtrusive measures, along with questionnaire indices, to investigate privacy in a variety of work environments in Tennessee. They used naturalistic observation to provide an objective index of privacy in an office by systematically observing the locations where employees conducted their work tasks during the day, the degree to which workers' activities were open to surveillance by their supervisors, and the relationship of work spaces to commonly used pathways. The researchers also unobtrusively measured architectural characteristics of the office that could affect privacy by making a worker susceptible to noise, distracting activities, or outside surveillance (Table 8-2).

Table 8-2. The variables on the left and the accompanying physical characteristics of work spaces on the right provide unobtrusive measures of privacy in an office environment.

Variables	Characteristics of work spaces and their definitions
Enclosed sides[a]	Number of sides of work space enclosed by wall or partition (0, 1, 2, 3, or 4)
Distance from neighbors[b]	Floor space (in square feet)
	Distance[a] to the nearest adjacent work space
Number of neighbors[b]	Number of work spaces in the room
	Number of work spaces visible during work
	Number of work spaces within 25 ft., walking distance[a]
	Number of work spaces within 50 ft., walking distance
	Number of work spaces accessible from the same common entrance
Distance from pathways[b]	Distance[a] to nearest common entrance
	Distance to nearest common pathway
Visible to supervisor	Work space visible from supervisor's work space (yes or no)

[a] Distances refer to chair-centers when chairs are placed as during work, measured to the nearest 0.5 ft. Distances between work spaces were measured only if they were not separated by walls or partitions.
[b] Composite variables.

Source: E. Sundstrom, R. E. Burt, and D. Kamp, "Privacy at Work: Architectural Correlates of Job Satisfaction and Job Performance," Academy of Management Journal, 1980, 23: 101–17. Reprinted by permission.

PSYCHOLOGICAL FUNCTIONS OF PRIVACY

Privacy serves two important psychological functions. First, privacy plays a role in managing the complex social interactions that take place between people. Second, it helps us to establish a sense of personal identity.

MANAGING SOCIAL INTERACTION

Interaction management   Irwin Altman (1975) explains that an important psychological function of privacy is to regulate interaction between a person or group and the social world. Maxine Wolfe and her colleagues (Wolfe and Laufer, 1974; Laufer, Proshansky, and Wolfe, 1976; Laufer and Wolfe, 1977) note that the management of interpersonal relationships is central to the experience of privacy in our everyday lives. For example, the children these authors interviewed discussed privacy in terms of controlling access to spaces and avoiding being "bothered" by other people. When the children talked about the experience of having their privacy invaded, they spoke of episodes in which they had been unable to manage the boundaries between themselves and other persons. In fact, invasions of privacy were a very real part of these children's lives, and occurred despite closed doors, locks, and signs imploring, "Keep out." The children reported that such invasions of their privacy made them feel "awful," "hurt," "afraid," and "very upset."

Information management   Wolfe and her associates explain that another way in which privacy functions to manage social interaction is by regulating the disclosure or nondisclosure of personal information. For example, people are attuned to complex situational demands in determining the types of personal information they are willing to disclose to other persons (Cozby, 1973). In some situations, such as a job interview, an individual may disclose some personal information while at the same time employing a protective and defensive style of responding. The advent of advanced computer technology and the use of social security numbers for many business transactions have made it increasingly difficult for individuals to manage the disclosure of information about themselves (Rule, 1974).

Alan Westin (1967), in elaborating on the information-management function of privacy, explains that the selective disclosure of personal information meets the individual's need to share confidences with trusted others. Nondisclosure helps the individual to maintain an appropriate psychological distance from other persons in the many situations in which limited communication is appropriate and required.

Maintaining group order   Another function of privacy is to help in the maintenance of group order. Barry Schwartz (1968) has pointed out that the ability to withdraw into privacy can help to make life with a difficult person possible. Similarly, Westin (1967) explains that periods of privacy allow the individual to achieve an emotional release from the accumulated tensions of everyday life.

Privacy also helps to maintain the status divisions that are essential to effective group functioning. In the armed forces, for example, enlisted men and women live in dormitories, noncommissioned officers have their own rooms in the barracks, and officers of high rank warrant a residence separated from the barracks. Similarly, in the business world, the higher people rise on the organizational ladder, the more personal privacy they are ensured and the greater protection they are provided from unwanted intrusions.

Schwartz points out that privacy has a price; privacy is regularly bought and sold in hospitals, hotels, and transportation facilities. We are required to pay more for a private hospital room, for a quieter and larger hotel room, and for a private berth on a ship or train. Valerian Derlega and Alan Chaikin (1977) add that one characteristic of low-status, disadvantaged groups, such as welfare recipients, pris-

A Personal Tale of Privacy and Poverty

The following account of life in a poor Mexican community offers a poignant picture of how difficult it can be to achieve privacy in the midst of poverty. This passage is from Oscar Lewis's touching and sensitive book *The Children of Sánchez* (1961); it records the impressions of a Mexican woman of her childhood in a low-income environment. She recounts in richly personal terms the ways her feelings of privacy were affected by life in an environment where a family that at times included nine people lived in a single room.

It would have been a great luxury to be able to linger at the mirror to fix my hair or to put on make-up; I never could because of the sarcasm and ridicule of those in the room. My friends in the Casa Grande complained of their families in the same way. To this day, I look into the mirror hastily, as though I were doing something wrong. I also had to put up with remarks when I wanted to sing, or lie in a certain comfortable position or do anything that was not acceptable to my family.

Living in one room, one must go at the same rhythm as the others, willingly or unwillingly—there is no way except to follow the wishes of the strongest ones. After my father, Antonia had her way, then La Chata, then my brothers. The weaker ones could approve or disapprove, get angry or disgusted but could never express their opinions. For example, we all had to go to bed at the same time, when my father told us to. Even when we were grown up, he would say, "To bed! Tomorrow is a work day." This might be as early as eight or nine o'clock, when we weren't at all sleepy, but because my father had to get up early the next morning, the light had to be put out. Many times I wanted to draw or to read in the evening, but no sooner did I get started when, "To bed! Lights out!" and I was left with my drawing in my head or the story unfinished. . . . But these annoyances were insignificant compared to that of being scolded in the presence of everyone else. I often thought that if my father had berated me in private, I would not have minded so much. But everyone heard the awful things he said to me, even though they pretended not to, and it hurt and shamed me more. My sisters and brothers felt the same way. When one of us was scolded, the others felt equally punished. My father's words would build up little by little, until they covered us and made us fall in a crisis of tears.

[From O. Lewis, *The Children of Sánchez*, pp. 237–39. Copyright 1961. Reprinted by permission of Random House, Inc.]

oners, and mental patients, is that their privacy is limited or denied altogether. The lives of welfare recipients are constantly scrutinized; even the sex lives of welfare mothers are monitored by caseworkers in an effort to determine the recipients' eligibility for aid. Finally, privacy is a luxury that is associated with status, and is often inaccessible to the poor (Schwartz, 1968) (see box, "A Personal Tale of Privacy and Poverty").

ESTABLISHING A SENSE OF PERSONAL IDENTITY

Self-identity   Altman (1975) explains that in addition to managing social interaction, privacy serves the essential psychological function of helping the individual to establish a sense of personal identity. In fact, the interpersonal function of privacy, while important in its own right, also acts in the service of self-identity; self-identity is possible only after the individual has learned to define his or her personal boundaries and limitations. Altman points out that among children's first steps in becoming individuals is the ability to distinguish themselves from other persons.

This process of self-recognition and self-definition is dependent on the ability to regulate the nature and style of social interaction with other persons. If we find that we cannot control our interactions with other persons, we are provided negative information about our personal competencies. But if we observe that we can effectively regulate our interchanges with others, we are afforded positive information about our competency to deal with the world.

Self-evaluation   Altman explains that self-identity depends in part on our ability to evaluate ourselves in comparison with other persons. For example, by comparing ourselves with others, we are able to determine our personal strengths, our shortcomings, and our overall worth as persons. Westin (1967) also emphasizes the important role of privacy in allowing us to evaluate ourselves in comparison with the other persons in our social milieu. For individuals to assess their own worth, they must be able to step back from the activities of daily life and "take stock" of themselves. The self-evaluative aspect of privacy is similar to a religious retreat or the exile of a political leader. Altman adds that privacy allows us to be "offstage," an experience that is necessary if we are to experiment with and evaluate new social behaviors (see Goffman, 1959). He points out that children and teenagers will sometimes stand before a mirror and carefully practice facial expressions and body stances to convey various moods and feelings.

Personal autonomy   A major function of privacy is to help the individual develop and maintain a sense of personal autonomy (Westin, 1967). Autonomy involves a person's feelings of independence, conscious choice, and freedom from manipulation by others. In discussing the important role of privacy in the development of feelings of autonomy, Altman (1975) emphasizes that what is essential is not the inclusion or exclusion of others in itself; but rather the individual's sense that he or she can regulate interpersonal contact.

Similarly, Wolfe and her associates (Wolfe and Laufer, 1974; Laufer, Proshansky, and Wolfe, 1976; Laufer and Wolfe, 1977) contend that privacy enhances feelings of personal autonomy by affording the individual an experience of personal volition. The choice to be alone is a statement of the autonomy of the self. In their interviews with schoolchildren, they found that children's perceptions of experiences as private were tied to feelings of personal autonomy. In describing private experiences, children commonly said, "I felt independent"; "I could do what I wanted to do"; "I could have my own opinions" (Laufer and Wolfe, 1977:27).

Deindividuation   If privacy fosters positive feelings of personal autonomy, we might ask how individuals are affected psychologically by an invasion of their privacy. Some investigators (Beardsley, 1971; Gross, 1971) believe that invasions of privacy are psychologically harmful precisely because they destroy the feeling of personal autonomy, leaving people feeling unable to control their interactions with the social world. Altman (1975) emphasizes that the negative psychological effects of invasion of privacy are due more to the loss of control than to the exposure of personal information. While the issue has not been examined empirically, it seems possible that some of the psychological effects of a loss of privacy may be mediated by the experience of *learned helplessness* (see "Theoretical Perspectives on Environmental Stress," Chapter 6) associated with these feelings of reduced personal control in the social realm.

Ellen Berscheid (1977) discusses the psychological consequences of the loss of privacy in terms of "deindividuation"—a psychological state characterized by the loss of personal identity and a feeling of submersion in an anonymous group (see Festinger, Pepitone, and Newcomb, 1952). Some investigators (Diener, Fraser, Beaman, and Kelem, 1976; Zimbardo, 1969) have found that under conditions of such social anonymity, people are much more prone to behave in an antisocial fashion.

Especially strong negative feelings about the self are associated with the "social nakedness" experienced by mental patients and prisoners who are under surveillance at all times (Schwartz, 1968). Erving Goffman (1961) has poignantly described the almost total loss of personal privacy faced by mental patients. In the hospital he observed, patients were stripped of personal possessions, physical examinations and property inspections were conducted at the staff's will, many toilets were without doors, and patients' activities were constantly observed throughout the day and night.

Empirical evidence   A major limitation of the arguments concerning the psychological functions of privacy and the psychological effects of a loss of privacy is that they are based more on speculation than on empirical evidence. We need empirically based studies of the psychological functions of privacy in people's lives. Until such empirical support is available, we must maintain some degree of caution toward the observations that have been offered. The same sort of reservation is appropriate as we turn now to a consideration of theoretical pespectives on privacy. While theoretical models of privacy are exciting and creative, they should be viewed with some caution until necessary supporting empirical evidence is available.

THEORETICAL PERSPECTIVES ON PRIVACY

Although privacy represents a relatively new area of research in environmental psychology, some important theoretical views of privacy have begun to emerge in recent years. Our analysis of theoretical views of privacy will be based primarily on the model of privacy developed by Irwin Altman (1974, 1975, 1976). Stephen Margulis (1977) points out that Altman's view is less a single theory of privacy than a broad orientation toward the theoretical analysis of privacy.

I emphasize Altman's model here precisely because it is broad enough to encompass other theories of privacy. In fact, Altman developed his model not only as a theory of privacy, but as a broader theoretical view of social interaction in general, with privacy as a central concept. Altman believes that the concept of privacy is central to understanding the relationship between environment and behavior. He argues that an adequate theoretical appreciation of privacy can serve to link together a wide range of social behaviors, including crowding, territoriality, and personal space.

A DYNAMIC MODEL OF PRIVACY

Dialectics   Altman (1974, 1975, 1976, 1977) proposes that privacy is a *dialectical* process; the oppositional qualities of being *open* or being *closed* to social interaction shift over time as social circumstances change. He notes that in the dialectics of privacy, openness and closedness work together in a unified and systemic manner. The varied strategies people employ to attain personal privacy work together "as an integrated system in much the same way as the instruments and sections of a symphony orchestra yield an integrated result" (1975:32).

This dialectical model of privacy is broader than theories that envision privacy solely as a withdrawal from social interaction. In effect, the dialectical model views privacy as a two-way street, sometimes involving a separation from other persons and sometimes involving social contact with others. A person's desired level of contact with others fluctuates over time, in accordance with changes in mood and in the surrounding social circumstances.

Altman points out that husbands and wives often work out an arrangement that allows each partner to achieve a desired level of social intimacy while also permitting each partner to attain necessary personal privacy. For some couples, this arrangement may involve private areas in the home to which one can retreat to read or to pursue a hobby. For other couples, personal privacy is attained through separate interests, such as sports or the theater, or through taking occasional separate vacations.

This dialectical model of privacy is similar to the theoretical model advanced by Wolfe and her colleagues (Wolfe and Laufer, 1974; Laufer, Proshansky, and Wolfe, 1976; Laufer and Wolfe, 1977). They propose that central to privacy is the degree of personal control an individual is able to exert. For them, privacy involves three types of control—control over choice, control over access, and control over stimulation. Privacy involves the freedom to choose when and where to be private. It also entails regulating access to oneself, such as by retreating to a private room behind a locked door. And it means controlling stimulation from other people, visual intrusions, and unwanted noise. Wolfe and her associates add that these aspects of control change as an individual progresses through the life cycle, and individual needs and social roles change. A somewhat similar view of privacy has been expressed by Proshansky, Ittelson, and Rivlin (1976), who contend that privacy involves freedom of choice. They propose that privacy allows people to maximize their behavioral options and to maintain a high level of personal control over their activities.

Boundary regulation   At the heart of Altman's dialectical model of privacy is the notion that privacy involves the regulation of interpersonal boundaries. He explains that people or groups use interpersonal *boundaries* or *barriers* to regulate their access to other persons. To describe these interpersonal boundaries, Altman uses the analogy of a cell membrane. The membrane regulates the cell's openness to the external environment just as interpersonal boundaries regulate our openness to the social environment. And just as the cell regulates its membrane in response to changes in both internal functioning and the surrounding environment, people regulate their interpersonal boundaries as social conditions change. Finally, the cell membrane regulates both *inputs* from the external environment to the cell and *outputs* from the cell to the surrounding environment. Similarly, interpersonal boundaries regulate both social inputs and social outputs. Privacy involves the regulation of inputs from other persons to the self, such as the choice of whether or not to answer the telephone. It also involves the regulation of outputs from the self to others, such as the decision to seek out other persons to discuss a personal problem or to wrestle with it alone.

One example of interpersonal boundary regulation in the service of personal privacy is the matter of self-disclosure. An important series of studies of self-disclosure, along with a discussion of how self-disclosure relates to privacy regulation, has been performed by Valerian Derlega and Alan Chaikin (Chaikin and Derlega, 1974a, 1974b; Derlega and Chaikin, 1977; Derlega, Wilson, and Chaikin, 1976). Referring to earlier work by Sidney Jourard (1966b, 1971), they define self-disclosure as "what one person tells another about himself/herself" (1977:103). They propose that self-disclosure be reconceptualized as a particular case of interpersonal boundary regulation in the maintenance of privacy.

Derlega and Chaikin explain that self-disclosure involves the regulation of two distinct boundaries. The outer boundary shown in Figure 8-2 is termed a "dyadic boundary," which closes off the leakage of personal information to uninvited third parties while permitting the self to be disclosed to chosen others. The inner boundary is a "self boundary," which regulates the disclosure of personal information to selected other persons.

As Figure 8-2 shows, for self-disclosure to occur, one must close the dyadic boundary (warding off outsiders) and open the self boundary (permitting the passage of personal information). Derlega and Chaikin, in agreement with Altman, add that the self boundary will be carefully regulated to permit the desired amount of self-disclosure in particular circumstances. Self-disclosure is higher among friends than among casual acquaintances and tends to be reciprocal; that is, we are

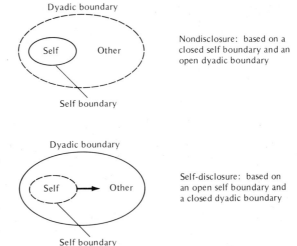

Figure 8-2

Self-disclosure as a function of self boundary and dyadic boundary adjustments.

From V. J. Derlega and A. L. Chaikin, "Privacy and Self-Disclosure in Social Relationships," Journal of Social Issues, *1977, 33:3 102–15. Reprinted by permission.*

likely to disclose more of ourselves to persons who have disclosed parts of themselves to us.

BOUNDARY REGULATION PROCESSES

Optimization   In his dialectical model of privacy, Altman emphasizes that people strive to attain an *optimum* level of privacy. In contrast to those theories that propose that the more privacy one has, the better off one is, Altman argues that either too little or too much privacy is unsatisfactory. Figure 8-3 plots the experience of privacy as a function of the match between the desired level of contact with other persons and the achieved or actual contact with others.

Figure 8-3 shows that a satisfactory match between desired and actual contact can occur when a person desires low contact and succeeds in achieving low contact, or when high contact is both sought and achieved. An unsatisfactory fit between desired and actual contact occurs when either low contact is desired and high contact is achieved, or high contact is sought and low contact is achieved. Altman offers the example of an individual working in an office. If the person seeks to be alone and a co-worker drops in to chat for fifteen minutes, the person experiences too much social contact. But if the worker desires social interaction, the same fifteen-minute conversation may be experienced as too little.

Multiple mechanisms   People employ a variety of behavioral mechanisms in their efforts to optimize personal privacy. These privacy mechanisms include verbal behavior, nonverbal use of the body, and environmental behavior. As we noted earlier, the diverse mechanisms people use to achieve personal privacy work together as a unified system, and change with personal and environmental circumstances.

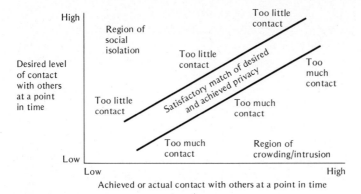

Figure 8-3

The dialectical and optimization properties of privacy.

From I. Altman, The Environment and Social Behavior, p. 26. Copyright ©
1975 by Wadsworth, Inc. Reprinted by permission of the publisher, Brooks/Cole
Publishing Company, Monterey, California.

Either the content or the structure of verbal messages can be used to achieve desired privacy. When we do not want to be disturbed, we tell other people, "Keep out" or "Please don't bother me now." When we desire social interaction, we say, "Come on in" or "I have something I'd like to tell you." Thus by the content of our messages—what we say—we tell people whether we want interaction or privacy. The *structure* of verbal behavior can be equally effective. For example, we might increase the loudness or raise the pitch of our injunction, "Keep out" if we are particularly determined not to be disturbed. Or we might emphasize the seriousness of our message by the combination of a strategic pause and inflection: "Keep . . . out."

People also attempt to attain the degree of privacy they desire through *nonverbal* behaviors. Altman explains that nonverbal behavior or "body language" involves the use of various parts of our bodies to communicate our desires. If we are in a public setting, such as a library or a study area, and do not wish to be disturbed, we may communicate this message by pulling in our arms and legs, orienting our bodies away from other persons, avoiding eye contact with passers-by, and assuming a serious facial expression. Altman also notes how people on a crowded elevator use nonverbal behaviors to communicate their intention not to intrude into one another's privacy. Familiar "elevator behaviors" include keeping our hands close to our sides, standing very still, staring at the floor-level indicator, or looking blankly ahead or at the floor.

Finally, Altman notes that people can strive to achieve a desired level of privacy by employing *environmental* features. People often use doors, fences, locks, and signs to convey their desire for privacy. Persons with sufficient space are also able to retreat to a private bedroom, study, or den to be alone. Those with less space may ingeniously employ room dividers, partitions, or even strategically arranged pieces of furniture to claim a private area. Even clothing can be used to communicate a desire for privacy: unless you know him well, the man in the three-piece suit tends to be approached, if at all, with a certain formality.

Cultural mechanisms for regulating privacy   Cultural norms and practices are also employed in the service of privacy. Mechanisms for regulating interpersonal boundaries are present in all cultures. Altman notes a dramatic mechanism to achieve privacy among the Tuaregs of North Africa. Only the eyes of Tuareg men are visible. They are heavily protected by a robe that reaches to the ankles, a turban, and a veil. The veil is worn continuously after a man reaches adulthood, even while eating and sleeping.

Even in cultures that at first appear to grant little personal privacy, careful observation will reveal the presence of subtle yet pervasive privacy mechanisms. Altman points out that in the Javanese culture, privacy initially appears to be minimal. People live in unfenced homes constructed with thin walls, often without doors. Yet closer observation reveals that the Javanese achieve personal privacy by means of "a wall of etiquette," involving restrained social contacts, the hiding of feelings, soft speech, and a high degree of decorum.

APPLICATIONS TO ENVIRONMENTAL PLANNING

DESIGNING FOR PRIVACY

While privacy plays an important role in enriching our personal and social lives, many aspects of today's architecture make it difficult to achieve. In fact, Jourard (1966b) contends that contemporary living environments are often so lacking in privacy that they leave us with the feeling that we are in a prison or highly impersonal dormitory. As Altman (1975) has explained, when designers do plan for privacy, conventional design solutions reflect only the "keep out" aspect of privacy. For example, a living environment might include an upstairs study where a family member can find solitude and a retreat from the bustle of family life.

Altman argues for a design philosophy that would reflect the dialectical nature of privacy. He proposes "responsive environments" that would allow for flexible movement between separateness and togetherness, meeting people's needs for various degrees of privacy. One common design feature, the door, provides an excellent example of such a flexible approach to privacy. When we are interested in social interaction, we can communicate our intention by leaving a door open; when we wish not to be disturbed, we can express this feeling by closing the door. Altman points out that the design philosophy of Japanese homes offers such a flexible approach to privacy. In Japan, living spaces are highly flexible environments; interior walls are movable, and can be shifted to meet a variety of personal and social needs.

Privacy and architecture   Of course, in order to develop a design philosophy that is responsive to people's need for privacy, we first need a model of privacy that is explicitly tied to physical environment variables. John Archea (1977) has pointed out that many theories of privacy have not adequately dealt with the ways in which the physical environment helps or hinders the attainment of a desired level of privacy. He proposes a model of privacy that specifically develops the link between privacy and physical settings. In Archea's model, the physical environment is seen as affecting privacy by regulating the degree of *visual access* and *visual* *exposure* individuals experience in particular settings. Visual access involves our ability to monitor our spatial surroundings by sight, while visual exposure concerns the extent to which our own behavior can be monitored visually by other persons.

In the same vein as Altman's dialectical model, Archea proposes that privacy involves both limiting exposure to prevent unwanted invasions and sufficient access to take advantage of social opportunities. Access and exposure control the distribution of social information on which interpersonal behavior depends. Features of the designed environment that can influence the distribution of social information include the position, size, fixity, color, and transparency of such architectural features as walls, doors, and corners.

OCCUPATIONAL PRIVACY

Sundstrom and his colleagues' (Sundstrom, Burt, and Kamp, 1980) study of privacy in Tennessee work environments adds further to our understanding of the relationship between privacy and architecture. In a correlational research design, they investigated the association between a variety of physical characteristics of the work environment and workers' experience of privacy on the job. They found that architectural features that allowed visual and acoustic isolation were consistently related to greater feelings of privacy in many work environments and a variety of occupational roles.

Administrative employees of the State of Tennessee rated work spaces as more private when they were partitioned from other work areas, had a door, and were not visible to neighboring co-workers. Similarly, female clerical employees in a hospital rated large rooms that held a large number of people as low in privacy. Finally, nonacademic employees at the University of Tennessee evaluated work settings as more private when they had at least some closed sides, had few neighbors, and were not visible to a supervisor.

These findings are of direct relevance to design decisions involving open-plan offices and may possibly be generalized to decisions concerning open-space classrooms. In Chapter 5 we saw that open-plan designs in occupational and educational environments have generally not increased performance and have often been evaluated negatively by environmental users. In a similar view, Sundstrom and his associates conclude that existing open-plan offices need to be provided with additional architectural features to enhance workers' feelings of privacy. They emphasize the need for design elements that limit visibility and audibility, such as room dividers, partitions, and buffer zones between work spaces. Consistent with some of the design recommendations for improved performance that we noted in Chapter 5, Sundstrom and his colleagues also encourage the use of acoustically treated ceilings or walls and sound-absorbent partitions. In all of the job categories they studied, employees preferred privacy over accessibility.

RESIDENTIAL PRIVACY

When Elizabeth Harman and John Betak (1974) questioned people in the market for a new home in Ontario, Canada, they found that people associated residential privacy with a single-family home on a large lot at some distance from neighbors. These people thus saw privacy primarily in terms of ability to regulate unwanted intrusions from neighbors by control of outdoor space.

Harman and Betak encourage designers to explore innovative strategies in residential design, such as the stacked housing units built for the Montreal Expo (see Safdie, 1966). These units attempted to blend the efficiency of multifamily housing with the amenities of single-family housing, including a private outdoor area for each unit. They add that families also need to be educated in ways of attaining private open space other than buying a single-family home. For example, outdoor privacy might be enhanced by the strategic arrangements of housing units, walls, or vegetation, or by attached housing that has a walled court area for each living unit.

Serge Chermayeff and Christopher Alexander, in their book *Community and Privacy* (1963), provide an approach to residential design that creatively meets people's needs for seclusion and social involvement. They emphasize the importance of an adequate boundary between the interior living space and the outside environment, especially noise and vehicular traffic. They also encourage designers to build interior living spaces that allow access to and an appreciation of the outside natural environment. And they suggest a combined work-and-ritual room, akin to the ancient family hearth, that would allow for a balance between separate interests and voluntary communality, and a balancing of the diverse interests of adults and children (Figure 8-4).

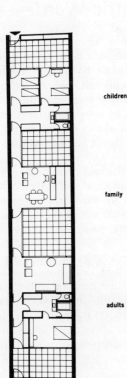

children

family

adults

Figure 8-4

This residential plan allows family members to choose between separation and communality. The central family area serves both as a social area and as a buffer between the realms of adults and children. The separate access to the children's rooms and the dressing-area buffer to the parents' bedroom add further to the experience of privacy.

From S. Chermayeff and C. Alexander, Community and Privacy, *p. 244. Copyright © 1963 by S. Chermayeff. Reprinted by permission of Doubleday & Company, Inc.*

INSTITUTIONAL PRIVACY

One area where planning for privacy is especially needed is the design of institutional environments; new approaches to institutional design must be sought in order to accommodate clients' right to and need for privacy. Erving Goffman's *Asylums* (1961) offers a poignant picture of the human costs of the lack of privacy in "total institutions." Humphrey Osmond (1957) argues that privacy for psychiatric patients is an essential component of a therapeutic program. Holahan and Slaikeu (1977) show that a lack of privacy in a therapeutic setting can negatively affect the rapport between a counselor and a client.

Studies have shown that private bedrooms, in contrast to multibed dormitories, can enhance the range and diversity of behaviors in which hospital patients engage (Ittelson, Proshansky, and Rivlin, 1970, 1976; Wolfe, 1975). Maxine Wolfe and Marian Golan (1976) also explain that when private or semiprivate bedrooms are available in institutional settings, doors should be provided for the rooms, and residents should decide whether these doors are open or closed. Institutional settings should also have spaces that allow for social intimacy. For example, small rooms or alcoves might be built where two residents can interact free from the surveillance of staff or the intrusions of other residents (see box, "Elusive Privacy on a Psychiatric Ward").

——Elusive Privacy on a Psychiatric Ward——

Robert Sommer and Bonnie Kroll (1979) conducted a survey to assess the ability of the physical environment of a California psychiatric hospital to meet the social and psychological needs of its patients and staff. Both patients and staff complained of the hospital's lack of privacy. Patients felt that the lack of privacy in their sleeping areas and bathrooms was among the worst features of the hospital. Particular complaints concerned the lack of shower curtains and toilet doors in the bathrooms. Patients were also upset about a bathroom design that placed tubs, showers, and open toilets in the direct line of sight from the bathroom entrance.

Hospital staff also complained about lack of privacy: the lack of separate staff toilets and staff sinks on the wards, the absence of individual staff offices, the inadequacy of the work surfaces and even of room to sit down in the shared work areas. Many wards were without lounges where staff members could retreat for private conversations, and the few lounges that were available were too small to allow for sufficient privacy.

Fortunately, environmental standards and requirements for psychiatric hospital design are beginning to improve. In fact, Sommer and Kroll note that six months after the completion of their survey, the psychiatric hospital they studied was stripped of its accreditation by the Joint Commission on the Accreditation of Hospitals in California. The commission emphasized the hospital's environmental deficiencies, and referred specifically to the absence of doors and partitions in the bathrooms, the lack of privacy in the sleeping areas, and the shortage of individual storage space.

Finally, residents in total institutions should be allowed to maintain control of some personal areas, possessions, and clothing, which are essential to a sense of personal identity. Wolfe and Golan add that residents in an institution that has not adequately planned for personal privacy will resort to drastic personal solutions in their quest for it. Some children in a psychiatric hospital they studied reported feigning emotional upset in order to be sent to a seclusion room. The hospital staff viewed the seclusion room as a form of punishment, but it became attractive to some residents because it was the only means available to achieve personal privacy.

TERRITORIALITY

It may be surprising to learn that environmental psychology can help to explain some major conflicts between people, such as international confrontations. In fact, when people fight over land or property, whether a backyard, a corner office, or an entire country, they are exhibiting aspects of what environmental psychologists call *territoriality*. The fences around backyards, the nameplates on desks and offices, and the identification tags on stereos and bicycles also reflect territorial behavior.

Some popular books and articles have discussed human territorial behavior in the light of territoriality in animals. This popular literature has argued that territorial behavior in humans is part of our genetic heritage, and that aggression between people over land and property is biologically determined. Environmental psychologists have been interested in learning whether these biological theories of human territoriality do in fact provide an accurate appraisal of human behavior.

Research on human territorial behavior represents a relatively new area of study in environmental psychology. Yet, because territoriality is related to some dramatic forms of behavior, it offers an exciting and challenging field of study. Research on territoriality has been conducted in a wide range of naturalistic settings, including college dormitories, the inner city, and even a battleship. Environmental psychologists have discovered that territoriality plays an important role in our lives, and is relevant to such issues as interpersonal aggression and social status.

DEFINING TERRITORIALITY

Territoriality has been defined in a variety of ways, and a commonly agreed-on definition is needed (Edney, 1974; Kaufmann, 1971). Altman (1970, 1975) has attempted to identify the common themes that run through the various definitions of territoriality. His analysis provides a good guide to the definitions of territoriality that have been proposed by investigators in this area.

Some researchers have defined territoriality in terms of a geographical area that is personalized and defended from encroachment (Becker, 1973; Becker and Mayo, 1971; Sommer, 1969; Sommer and Becker, 1969). An alternative definition views territoriality as the use and defense of a spatial area by a person or group that regards it as their exclusive preserve (Pastalan, 1970). A third group of investigators has envisioned territoriality as the exclusive use of an area or objects by persons or groups (Altman and Haythorn, 1967; Altman, Taylor, and Wheeler, 1971; Sundstrom and Altman, 1974).

Altman discerns several common features in these definitions. He notes that they typically specify that territoriality involves *places or geographic areas,* and they characteristically emphasize *ownership of the place.* Further, such definitions often refer to *personalization* of the space, and commonly consider a territory to be the province of either an *individual* or a *group.* Many definitions refer to *defense* against territorial intrusions. Julian Edney (1974) has pointed out, however, that definitions of territoriality vary considerably in the emphasis they place on defense; most earlier definitions stressed defense, but some recent definitions avoid the notion of defense entirely. On the basis of these common themes identified by Altman, I propose the following definition of territoriality: *Territoriality is a pattern of behavior associated with the ownership or occupation of a place or geographic area by an individual or group, and may involve personalization and defense against intrusions.*

TYPES OF TERRITORY

Altman cautions that any proper definition of territoriality is necessarily complex because the concept encompasses more than one type of territory. Territories differ in terms of how close or central they are to a person's or a group's life, and in accordance with the duration of the users' claims to the area. With these distinctions in mind, Altman identifies three types of territory: primary territories, secondary territories, and public territories. Some recent evidence (Taylor & Stough, 1978) has provided empirical support for Altman's typology.

Primary territories   Primary territories are typically under relatively complete control of their users for a long period of time. They are central to the lives of their occupants, and their ownership is clearly recognized by other persons. Sidney Brower (1965) calls such territories "personal" territories. A person's home is a primary territory. Invasion of a primary territory by an outsider without permission is serious, and may present a threat to the owner's self-identity.

Secondary territories   Secondary territories have some degree of ownership, though this ownership is neither permanent nor exclusive. Secondary territories are less central to the lives of their users, and have a semipublic quality. Social clubs and neighborhood taverns are secondary territories. There are usually some formal or informal limitations on who is free to use secondary territories, but because they have a semipublic quality and the rules governing their use are often not clearly stated, they are open to encroachment by a variety of users. Stanford Lyman and Marvin Scott (1967) have labeled those secondary territories that allow a relatively high degree of control "home" territories (a social club), and those where control is relatively fragile as "interactional" territories (a group of people at a party).

Public territories   Public territories are open to public occupancy on a relatively temporary basis. They are not central to their occupants' lives. Parks, public transportation systems, restaurants, and telephone booths are public territories (Figure

8-5). Research on public territories has also included an investigation of territorial spacing on a beach (Edney and Jordan-Edney, 1974) and of temporary work spaces or "jurisdictions" on a navy ship (Roos, 1968) (see box, "Territoriality on a U.S. Navy Warship"). These territories are open to almost anyone, though users are typically required to conform to standard rules and customs. A restaurant may evict patrons who fail to follow conventional rules of behavior, for example, and bathers may be required to leave a beach after sunset.

RESEARCH METHODS FOR STUDYING TERRITORIALITY

Altman (1975) and Edney (1974) explain that the nature of territoriality requires most research in this area to be conducted in naturalistic field settings. The feelings of ownership and personalization that are essential to territoriality do not easily develop in artificial laboratory settings where subjects are present for only a short period of time.

Most field studies of territoriality have relied on *naturalistic observation*, the systematic observation of behavior in real-world contexts, such as hospitals, universities, and inner-city neighborhoods. When Sundstrom and Altman (1974) observed the territorial behavior of male juvenile offenders who lived in a cottage at a

Figure 8-5

Investigators have identified different types of territory. In this building, the restaurant on the first floor is a public territory, while the residence above is a private territory.

© *Jack Prelutski/Stock, Boston.*

——— Territoriality on a U.S. Navy Warship ———

Philip Roos (1968) offers a fascinating description of the complexity of territoriality in humans, based on his observation of territorial behavior on a navy warship on which he was stationed, which he calls the U.S.S. *Oswald A. Powers*. A unique type of public territory on the *Powers* consisted of *jurisdictions*—work spaces over which particular seamen held control for specific, limited periods of time. Roos describes how such jurisdictions operated during a period of cleaning and polishing the sailors referred to as "field day."

On field day, most normal shipboard work was brought to a halt to allow the seamen to devote full time to scrubbing bulkheads, shining the ship's fixtures, and polishing decks. Small groups of seamen were responsible for each area on the *Powers*, and during field day virtually every jurisdiction on the ship was roped off. Decks were a critical point; no one wanted his newly scrubbed deck walked on before it could be waxed and buffed. But because only one buffer was available to enlisted men, there was often a long wait between scrubbing and polishing.

Roos's own jurisdiction—a central deck area subject to heavy travel—presented a unique challenge. In order to allow necessary travel while at the same time protecting the freshly scrubbed surface from damage, he would wax and guard half of the area while awaiting the buffer, then wax the rest of the area and buff the entire area when the buffer became available.

rehabilitation center, they mapped the spaces the residents used throughout the cottage each day for ten weeks and discovered that the boys' territorial behavior was complexly related to their dominance in their social groups.

Some field studies of territoriality have employed *unobtrusive measures*. Edney and Jordan-Edney (1974), for instance, recorded the radii of the spaces that groups of people on a Connecticut beach marked off about themselves. They discovered that the longer that mixed-sex groups remained on the beach, the more spatial markers they deployed about them.

Some studies of territoriality have also used *survey research* methods, often in combination with naturalistic observation or unobtrusive measures. By combining survey research with unobtrusive measures in a suburban neighborhood, Edney (1972) found that residents of homes that were defended by "no trespassing" signs and fences had lived longer on the property and anticipated living there longer in the future than residents of homes that lacked such defenses.

Some studies of territoriality have used *experimental methods* in either field or laboratory contexts. Robert Sommer and Franklin Becker (1969), for instance, conducted a field experiment on the role of territorial markers in defense of places in a soda fountain and a library on a college campus. They discovered that markers not only protected a chair, but delayed use of the rest of the table as well, and that a personal marker (such as a jacket) was more effective than an impersonal one (such as a library journal). A few experimental studies of territoriality have been conducted in laboratory contexts, as when Altman and his colleagues studied territorial behavior in small experimental rooms where pairs of sailors were isolated for eight to ten days (Altman and Haythorn, 1967; Altman, Taylor, and Wheeler, 1971).

PSYCHOLOGICAL FUNCTIONS OF TERRITORIALITY

EVERYDAY FUNCTIONING

Territoriality helps to organize and manage the daily lives of individuals and social groups. Consider what life would be like without territoriality. As Edney (1976) points out, there would be no places for people to settle, and social life would resemble a huge, milling Grand Central Station. Daily activities that involve uninterrupted chains of behavior, from creative thought to preparing and eating a meal, would be open to constant and unmanageable interruptions. Our social life would be disrupted by our inability to know where to find particular people (and how to avoid others). Even sleep and lovemaking could not be ensured freedom from interruption.

Another psychological function of territoriality is to help us develop a cognitive "map" of the types of behavior that can be expected in particular places (Edney, 1976). Knowing what to anticipate in various locations allows us to plan and order our daily lives. The tie between particular places and associated activities helps us to organize discrete everyday behaviors into patterned and integrated behavioral chains. For example, a parent may be able to prepare and eat breakfast, read the daily paper, plan the day's activities, and feed and get the children off to school in a short time because of the predictable patterning of the activities in the kitchen.

Altman and his associates (Altman, Nelson, and Lett, 1972) have demonstrated some of the ways the organizing role of territoriality can facilitate day-to-day activities. They found that territoriality helps to order who sits where at mealtimes, since most families evolve fixed seating arrangements for meals. People who share a bedroom, such as parents or siblings, generally establish a mutually accepted system of territorial claims. People typically have their own dresser or dresser drawers, their own closet or side of a closet, and their own side of the bed.

In another series of studies, Altman and his colleagues (Altman and Haythorn, 1967; Altman, Taylor, and Wheeler, 1971) systematically observed pairs of sailors isolated together in a small experimental room for eight to ten days. They found that an important aspect of a pair's effective functioning was their ability to establish clear territories during the early part of the isolation period. Pairs of sailors who began by clearly establishing who would use which chair and how storage space would be divided proved most effective in their overall functioning.

SOCIAL ORGANIZATION

Another psychological function of territoriality is to develop and maintain social organization. Territoriality helps to order a social group in accordance with the relative social status or dominance of group members. In fact, the role of territoriality in the establishment and preservation of the dominance hierarchy in a group has been among the most heavily researched aspects of territoriality. While researchers' definitions of dominance have varied somewhat, it may generally be defined as the ability of one individual to influence another.

One series of early studies concerned with the relationship between dominance and territoriality was carried out by Aristide Esser and his colleagues at Rockland State Hospital in New York (Esser, 1968, 1973; Esser, Chamberlain, Chapple, and Kline, 1965). The pattern of results from these studies has been complex and somewhat contradictory. When Esser studied adult psychiatric patients on a hospital ward, he found that high dominance was associated with *low* territoriality; the most dominant patients roamed freely over the ward and did not claim personal territories. When he studied children in a psychiatric hospital, however, he found no relationship between dominance (defined here as "pecking order") and territoriality. Finally, when he observed institutionalized children in a residential cottage, he found that highly dominant youngsters demonstrated *high* levels of territoriality.

Later investigators have attempted to interpret these early findings. Altman (1975) notes that the "worth" of territories may be related to dominance; dominance may be associated with territoriality only when the space in question is viewed as desirable. In a similar vein, Sundstrom (1976) suggests that the varied spaces in the residential cottage Esser studied probably made some parts of the cottage more desirable than others. In the psychiatric wards Esser observed, in contrast, the more homogeneous environment may not have made "owning" any particular place desirable. Thus competition for territory would be expected to be strong only in the cottage, and we might anticipate, as Esser found, that the dominance hierarchy in this setting would influence the outcome of such competition.

Later studies of the relationship between territoriality and dominance have tended to support the view that territoriality is positively related to dominance, at least during periods of stable group structure. When Alton De Long (1970, 1971, 1973a, 1973b), for example, observed territoriality in seating arrangements among thirteen college students enrolled in a group dynamics seminar, he found that high dominance was associated with high territoriality in the whole class and in separate subgroups of the class that emerged when group leadership style was altered.

Sundstrom and Altman (1974) also found a positive relationship between territoriality and dominance in a ten-week longitudinal study of juvenile offenders in a residential facility. They point out, however, that territoriality and dominance were positively associated only during periods when the group structure was stable. When two new, highly dominant boys entered the group, the previous relationship between territoriality and dominance broke down.

A recent survey study by G. William Mercer and M. L. Benjamin (1980) suggests that the role of territoriality in facilitating social organization may be different for men and women. They administered a questionnaire concerning the use of their dormitory rooms to freshmen in double-occupancy rooms at a Canadian college. The investigators found that the men established larger personal territories (room areas reserved for their own exclusive use) than the women did, and that they used this personal territory as an insulator or place to retreat from their roommates. The women, in contrast, used their personal territory as a more aggressive social statement against roommates they typically did not like. Commonly shared territory in the room also served different social functions for the two sexes, with men using common territory as a neutral zone and women using it as a shared social area.

PERSONAL AND GROUP IDENTITY

Territoriality also serves as a basis for the development of a sense of personal and group identity. Edney (1976) remarks that a sense of group identity can emerge simply because individuals share the same locale. Sharing a territory gives people common knowledge and experiences. People who reside in the same neighborhood, for instance, tend to shop at the same stores, eat occasionally at the same neighborhood restaurant, attend the same schools and houses of worship, perhaps attend an annual block party. A dramatic example of group identity founded on shared territory can be seen in the behavior of teenage gangs in inner-city neighborhoods (see box, "Territorial Graffiti of the Teenage Street Gang").

Edney notes that the sense of group identity that emerges from common territorial claims can also foster social bonds between people. People may like and trust one another simply because they live in the same neighborhood. Residents may also be reluctant to leave a familiar neighborhood because they are unwilling to relinquish the sense of group identity that living in the local neighborhood has afforded them.

Territoriality can also help to foster and maintain a sense of personal identity (Edney, 1976; Lyman and Scott, 1967). An individual who has laid claim to a personal territory, such as a new home, a country cottage, or a private office, derives a feeling of distinctiveness from that area. Edney (1976) points out that people often rely on designations of personal territory to identify themselves and each other. For example, someone might be described as "living on Q Street" or "in the corner duplex." People can enhance the distinctiveness of a personal territory by personalizing the environment in such a way that it more clearly reflects something of themselves. For example, students may decorate their dormitory rooms with movie posters or travel posters that reflect their personal interests. William Hansen and Irwin Altman (1976) have explored some of the ways college students personalize their dormitory rooms. They photographed dormitory rooms at the end of the first quarter of the school year and compared the pictures with photographs they had taken of the same rooms at the beginning of the quarter. They found that students decorated their rooms in uniquely personal ways, reflecting individual interests and values, and that the amount of room decoration increased during the quarter.

In an interesting speculative paper, Clare Cooper (1974) has applied Carl Jung's (1969) psychological views in an effort to demonstrate how a house serves as a symbol of the self. Many people buy homes that they feel will bolster their image of themselves. A self-made entrepreneur, for instance, might choose a house that is large and somewhat ostentatious as a way of enhancing his or her new-found feelings of high social status. Cooper describes the house that Jung constructed for himself near Lake Zurich in Switzerland as a symbol of his own sense of self. He built the house of stone, and continued periodically to add new sections to it to represent the continuing growth of his personality. After the death of his wife, Jung added an upper story that he felt expressed the extension of personal consciousness that is achieved in old age. There is a small amount of empirical support for Cooper's speculations. Edward Sadalla and his associates (Sadalla, Burroughs, and Quaid, 1980) found that the inferences people drew about

Territorial Graffiti
of the Teenage Street Gang

David Ley and Roman Cybriwsky (1974), two geographers, conducted a fascinating study of teenage street gangs in Philadelphia. They discovered that territory was central to the group identity of a gang, and that the graffiti of inner-city neighborhoods were important indicators of local gang activity. Each gang occupied a fixed and permanent territory, and the turf that belonged to a particular gang was never in doubt to local youngsters or members of other gangs. In fact, most street gangs in Philadelphia took their names from intersections near the center of their turf, such as "45-R" or "28-Ox." One teenager commented:

It used to be terrible, man. It used not to be safe to walk beyond 41st Street without some guys getting at you. I live in 37th's territory but I was never with them. I know cats on Sutton and some on Richmond. Like I was between all three, and that's cool.   [P. 496]

Wall graffiti offered an accurate indication of a particular gang's turf; they tended to become more prevalent as one moved toward the core of a gang's territory. As Ley and Cybriwsky progressed from the territory of the 28-Ox gang to the adjoining territory of the 26-P gang, they observed that the proportion of 26-P to 28-Ox graffiti changed from 25 percent to 75 percent. They also discovered that the graffiti along the turf boundaries between two groups were aggressive statements directed at the opposing gang.

(From D. Ley and R. Cybriwsky, "Urban Graffiti as Territorial Markers," *Annals of the Association of American Geographers,* 1974, *64,* pp. 491–505. Reprinted by permission.)

homeowners' personalities from photographs of the interiors of their homes correlated significantly with the homeowners' ratings of their own personalities.

THEORETICAL PERSPECTIVES ON TERRITORIALITY

SOCIOBIOLOGICAL THEORIES

In recent years some social and behavioral scientists have become interested in analyzing human behavior and the mechanisms of human society from the vantage point of *sociobiology.* The sociobiological view contends that human behavior can be explained in terms of the biological and genetic heritage that human beings carry from their evolutionary past. Thus sociobiologists believe that we can understand much of human behavior by studying the behavior of the lower animals with whom humans share a common biological history. The publication of Edward Wilson's important and controversial book *Sociobiology: The New Synthesis* in 1975 brought the sociobiological perspective into sharp public and scientific focus. Donald Campbell's presidential address to the American Psychological Association in the same year surprised many psychologists by encouraging investigators to seek an understanding of human behavior in our evolutionary past.

A warning at a turf boundary.

© D. Clan Haun/Black Star.

The territorial imperative   Of all the aspects of human behavior that environmental psychologists have studied, territoriality has received the greatest amount of attention from a sociobiological view. Of course, the implications of the issues considered here may be generalized to other topics, such as personal space and crowding, as well as territoriality. The publication in 1966 of Robert Ardrey's popular book *The Territorial Imperative* brought the sociobiological view of human territoriality to the attention of a broad reading audience (although the term "sociobiology" had not yet been coined). This book, along with a few other widely read works (Ardrey, 1970; Lorenz, 1966, 1974; Tiger, 1969), argues that the social behaviors people engage in, particularly territoriality and interpersonal aggression, are rooted in our evolutionary history. Ardrey (1966) contends that territorial behavior in humans is instinctive, and that humans share with other animals a drive to claim and defend territory. Both Ardrey (1966) and Konrad Lorenz (1966) argue that humans are genetically predisposed to defend their territory from encroachment, and that such interpersonal aggression is a natural tendency among humans.

Territoriality in animals   Sociobiological theories of territoriality have drawn on the extensive literature on the nature and functions of territoriality in animals. For example, Eliot Howard (1948) offered a discussion of the nature and functions of territoriality in birds. C. R. Carpenter (1958) reviewed studies of territoriality in animals, and concluded that territorial behavior can be observed in all vertebrates.

He noted, however, that the specific expression of territoriality in animal populations varies according to several other factors, such as the particular species, its social organization, and its natural habitat.

Carpenter explained that the functions of territoriality among animals are varied and extensive, including population regulation, security, and the maintenance of dominance hierarchies. Vero Wynne-Edwards (1962, 1965) discussed territoriality in animals as part of a self-regulating system to control population size. More dominant individuals in animal populations excel in obtaining space, food, and mates, thus regulating reproduction in the group. Irenaeus Eibl-Eibesfeldt (1970) also proposed that territorial behavior helps to maintain stable social organization among a group of animals by reducing unnecessary competition for resources.

Territorial defense   Definitions of territoriality based on the sociobiological view emphasize active *defense* of territories against encroachments as a central aspect of territorial behavior (Edney, 1974). Lorenz (1969:xiii) defines territoriality as "the defense of a given area." The sociobiological view stresses that territorial defense is essential to territoriality in humans as well as in animals. Sociobiologists point out that humans and animals alike manifest defensive and aggressive behaviors, such as vocal warnings and postural threats, in response to territorial invasions. They also note that both humans and animals mark their territories in an effort to ward off and prevent encroachment. Ardrey (1966) proposes that the underlying biological need to defend against territorial invasion can explain the "patriotic" rush to arms in response to the invasion of national boundaries.

SOCIOCULTURAL THEORIES

Territoriality and culture   *Sociocultural* theories view territorial behavior as a function of social learning and cultural influences. Sociocultural theorists recognize that humans engage in territorial behaviors, but they contend that territoriality in humans is fundamentally different from territoriality in animals. They suggest that while biology and instinct can explain animal territoriality, territorial behavior in humans can be understood only in the context of the complex social organization and cultural diversity that characterize human societies. Popularizations of human territoriality as essentially similar to territoriality in animals, such as the work of Ardrey (1966), have been vehemently attacked by sociocultural theorists as gross oversimplifications (Elms, 1972; Klopfer, 1968).

Varied defenses   Sociocultural theories of territoriality recognize that defense may occur in response to encroachment, but emphasize that defense is not central to the definition of territoriality and does not inevitably follow from territorial invasion (Edney, 1974). Altman (1975) argues that people do not typically respond with aggressive physical acts to maintain their territory unless the invasion is deliberately malicious or territory is severely limited. An interesting area for future research might be territorial behavior under contrasting conditions of high and low spatial density. Sommer and Becker (1969), in their study of territoriality in university settings, noted that when demands for space are heavy, personal markers are more effective than impersonal ones in reserving spaces. Altman also notes that the implications of territorial encroachment vary in accordance with the type of territory that is invaded. People are inclined to treat an invasion of primary territory, such as their homes, far more seriously than an invasion of public territory.

Stanley Milgram (1977) describes a field study in New York City in which students asked people seated on the subway (a public territory) for their seats. The students offered no explanation for the unusual request, and appeared healthy and well adjusted. The overwhelming majority of subway passengers yielded their seats without a word of protest. Similarly, investigators found that people whose territory was encroached upon in a cafeteria (Becker and Mayo, 1971) and on a public beach (Edney and Jordan-Edney, 1974) failed to respond in defensive or aggressive ways. In a laboratory study Edney (1972) found that association with a territory led subjects in that area to *shorten* the distances between themselves. Finally, Edgar O'Neal and his colleagues (O'Neal, Caldwell, and Gallup, 1977) failed to find strong evidence of aggression in children (especially boys) whose play area was invaded.

Altman (1975) explains that people's responses to territorial encroachment vary as a function of both the *meaning* of the invasion and the types of *adjustive responses* that are available in the situation. Similarly, Stanford Lyman and Marvin Scott (1967) suggest that people typically rely on a range of subtle responses in coping with encroachment, and turn to active defense only as a last resort. People can place a barrier between themselves and an intruder, for instance. Such "insulation" may be either a physical obstacle (a fence or gate), a visible sign of ownership rights (a uniform), or a nonverbal body response (postural shifts or negative facial gestures).

People also employ spatial markers in a strategic manner to signal territorial ownership, and thereby *prevent* encroachment (Figure 8-6). A series of field studies in college cafeterias and libraries by Franklin Becker and his associates (Becker, 1973; Becker and Mayo, 1971; Sommer and Becker, 1969) is especially enlightening. In an initial study, Sommer and Becker (1969) demonstrated that physical objects, such as books, a sandwich, or a jacket, can be used to mark and defend a space during short absences. They also found that students in nearby seats would help to support a marked claim when they were asked to do so, although this willingness to defend someone else's place against an invader decreased the longer the claimant was away. Becker (1973) later showed that people in a college library preferred to sit at unmarked tables rather than marked ones, and that if they did choose a marked table, they spent less time at it than at an unmarked table.

Becker and Clara Mayo (1971) found that most claimants of marked spaces in cafeterias and libraries relinquished their places when the space was invaded. They suggest that because these spaces were not defended, they were not real territories. They speculate that markers served to identify comfortable interpersonal distances rather than to stake territorial claims. In fact, we shall see in Chapter 9 that many of the same responses that people use to cope with territorial encroachment can also be used to cope with an invasion of personal space. Edney (1974), however, points out that if active defense is not an essential aspect of the definition of territoriality, cafeteria and library spaces meet the criteria of public territories until the occupants leave. In fact, threats to someone's occupied space at a cafeteria

Figure 8-6
This student has arranged a variety of territorial markers to defend her library study space.

© *Lawrence Frank 1981.*

or library table probably involve aspects of both an invasion of territory (threatened loss of the occupied place) and an invasion of personal space (an uncomfortably close interpersonal distance).

Gilda Haber (1980) conducted a field experiment in classrooms at a Maryland college in order to examine people's reactions to a territorial intrusion that did not involve a simultaneous invasion of personal space. She had invaders sit in students' stably occupied classroom seats while the occupants were out of the room, and systematically observed occupants' reactions to the loss of their seats when they returned to the room. She discovered that 27 percent of the occupants defended their territory, first stopping and staring at the invader, then verbally claiming ownership of the seat. The longer the seat had been occupied or marked with personal property, the more likely it was to be defended. All of the students whose seats had been invaded (even those who did not defend their territory) arrived earlier than usual at the next class session and reoccupied their territory. Interestingly, Haber also found that students who occupied more central seats were more inclined to defend their territory than those who occupied seats at the periphery of the room.

Territoriality as a form of control   According to sociocultural theorists, territoriality in humans represents a form of control over the social environment. Edney (1972) suggests that in the context of human territoriality, the concept of control is broader and more meaningful than that of dominance. A person who holds territory gains control in three respects: (1) priority of access to a spatial area; (2) choice of the types of activity that will occur in the area; and (3) ability to resist the control of other persons in that area. The notion that territoriality provides a form of social control is also consistent with the theoretical models of Altman (1975), who views territoriality as a mechanism for regulating interpersonal boundaries, and Proshansky, Ittelson, and Rivlin (1976), who see territoriality as a means of enhancing one's freedom of choice. Both models propose that the control afforded by territory serves ultimately to help the individual achieve a desired level of privacy.

Some sociocultural theorists have suggested that the link between territoriality and control is particularly apparent in the context of people's social roles. Proshansky, Ittelson, and Rivlin (1976) point out that social roles often involve the control of particular places or objects, such as the boss's office, the executive conference room, or the teachers' lounge. Edney (1976) notes that the social organization function of territoriality is expressed in this tie to social roles. Altman (1975) adds that role-related territorial behavior in humans is applicable to other persons, objects, and ideas, as well as to spaces. The holding of copyrights and patents, for example, is a complex expression of territoriality.

Ralph Taylor and Roger Stough (1978) offer empirical support for the proposition that primary and secondary territories help to structure social roles by providing settings for social interactions that vary in formality and exclusiveness. They surveyed residents of Baltimore concerning the meanings of some *primary* (e.g., kitchen, living room), *secondary* (e.g., the sidewalk in front of one's house), and *public* (e.g., a nearby supermarket or shopping center) territories in their lives. They found that primary territories afforded residents privacy and solitude, an escape from neighbors, and a setting for controlled interaction with friends. Secondary territories offered settings for a wide range of informal contacts with neighbors among suburban residents. Public territories fulfilled a variety of routine functions, such as providing a place for the family to shop.

Arthur Patterson (1978) has extended the notion of territorial control to an analysis of how elderly people establish active mastery of their home environment. He surveyed homeowners aged 65 years or older in central Pennsylvania, and also recorded the number of territorial markers around the home, such as "no trespassing" signs, fences, and peepholes in doors. He discovered that homeowners with several territorial markers were less fearful of property loss or personal assault than residents with fewer territorial markers. This finding was especially pronounced among people who lived alone. Patterson speculates that the active creation of territorial displays may reflect an underlying sense of mastery of the environment, and this feeling of mastery may be associated with less fear of being victimized by crime.

A synthesis of views   Some sociocultural theorists have proposed that the sociocultural and sociobiological models of territoriality are not incompatible, and that a constructive synthesis of the two views is possible. These theorists suggest that the extensive evidence of territorial behavior in animals can help us to understand some aspects of human territoriality, but that the significance of biological influences on territoriality needs to be interpreted in the light of the complex social and cultural factors that shape human life. Altman (1975) and Edney (1976) suggest that their discussions of human territoriality in terms of complex social roles do not exclude the possibility that some aspects of human territoriality can be understood from an evolutionary perspective.

Aristide Esser (1972, 1976) has presented a speculative conceptual analysis of human territoriality that merges aspects of both the sociocultural and sociobiological views. He proposes that human spatial behavior, including territoriality, is hierarchically structured according to the levels of evolutionary development undergone by the human central nervous system. There are three levels of evolutionary development in the human central nervous system, and Esser suggests that each of these levels is associated with a corresponding level of spatial behavior (Table 8-3). The brainstem and limbic system in humans are shared with other animals. These parts of the central nervous system control self-preservation and elementary forms of social behavior, including territoriality and related dominance behavior. The neocortex, in contrast, is uniquely human, and controls complex forms of social behavior, such as social role playing. Esser concludes that while some elementary aspects of territorial behavior are shared with other animals, the more complex expressions of territoriality in social roles are the result of more recent evolutionary transformations in the central nervous system that are unique to human beings.

APPLICATIONS TO ENVIRONMENTAL PLANNING

DESIGNING FOR TERRITORIALITY

Edney (1974) notes that knowledge in the area of territoriality has generally not been applied to environmental design. He points out, however, that design applications can be generated from existing knowledge of human territorial behavior. Jacqueline Skaburskis (1974), writing for architects, underscores the need to consider territoriality as a design concern. She points out that when people occupy and use a place, they develop a sense of belonging and of emotional commitment to it.

Altman (1975) has advanced some general design principles that take account of territoriality. He encourages designers to build settings that create and clearly define the differences between various types of territorial settings. This admonition is particularly relevant to secondary and public territories, which are sometimes difficult to recognize. As examples of public settings that are poorly designed in terms of people's territorial needs, Altman points to restaurants that seat parties too close to one another and telephone booths that lack adequate sound shielding.

Altman points out that the challenge in designing for territoriality lies not in planning settings to meet particular functional needs, such as designing a restaurant where people can eat or a library where they can read and study, but in de-

Table 8-3. Spatial behavior associated with evolutionary levels of the human central nervous system.

Level of central nervous system	Evolutionary level	Sample spatial behavior
Brainstem	Shared with animals	Territoriality
Limbic system		Dominance
Neocortex	Uniquely human	Role playing

Source: Adapted from A. H. Esser, "A Discussion of Papers Presented in the Symposium Theoretical and Empirical Issues with Regard to Privacy, Territoriality, Personal Space, and Crowding," Environment and Behavior, 8: 117–25, © 1976 Sage Publications, Beverly Hills, with permission of the Publisher.

signing to meet people's changing desires in regard to control of social interaction and stimulation. Altman proposes that designs for secondary and public territories offer clear recognition of territory type, provide users with appropriate degrees of control, and indicate the degree of permanency of ownership. Designs that effectively meet these territorial needs will prevent unnecessary territorial intrusions, thus avoiding potential conflicts between users about territorial rights.

RESIDENTIAL TERRITORIALITY

Altman (1975) proposes that designing for appropriate degrees of territoriality is relevant to the design of residential environments as well. Research in New York City public housing conducted by Oscar Newman (1972) revealed that the level of crime in the housing project was related to poorly defined territorial boundaries. Although Newman's findings have been criticized on methodological grounds (see Adams, 1973; Kaplan, 1973), they do suggest some residential design problems related to territoriality. For example, Newman found that crime was high where secondary territories, such as lobby areas, hallways, and play areas, were not clearly designed as such, and functioned instead as public territories that could be easily entered by outsiders. Newman suggests that the creation of semiprivate entrance areas, the clustering of residential units, and the use of territorial markers would increase residents' territorial control and reduce crime. We shall return to Newman's design philosophy in Chapter 10.

TERRITORIALITY IN INSTITUTIONAL ENVIRONMENTS

Knowledge that control of territory helps individuals to develop a sense of personal identity and facilitates social organization suggests that enhancement of a sense of territoriality in institutional environments may produce psychological benefits. Not only are many total institutions designed without safe, personal territories where users can retreat from the bustle and activity of institutional life, but even articles of personal property may be confiscated to achieve a standardized institutional life. Russell Barton (1966) has argued that psychiatric patients would derive therapeutic benefits from having personal territories on hospital wards. The psychological costs of institutional life might be reduced further if residents were permitted to personalize such individual territories. Residents' sense of belonging, personal satisfaction, and personal usefulness might be enhanced if they could add their own decorations, personally meaningful knickknacks, or an individualized color scheme to their personal territories.

Holahan and Saegert (1973) carried out a field experiment in a psychiatric hospital in New York City that demonstrated how increased territoriality can foster a more positive therapeutic environment. The investigators planned and conducted the large-scale remodeling of a psychiatric ward in the hospital in an effort to improve its overall social-psychological atmosphere. A major component of the remodeling was the construction of partitioned, two-bed sections in the previously multibed dormitories.

Systematic observations on the ward six months after the remodeling had been completed revealed that the newly partitioned bedrooms provided important personalized territories. Before the renovation, patients had made no effort to personalize the space they occupied; after the change, books, magazines, towels, powder, and flowers appeared on the window ledges of the bedrooms. Comparisons of the remodeled ward with an unchanged control ward indicated that the remodeled ward had a more positive social climate and was better liked by patients.

Knowledge of human territorial behavior may also be applied to the design of classroom environments. A few years ago Edney (1976) encouraged research to evaluate the psychological benefits that schoolchildren might derive from being allowed access to personal territories. He speculated that children allowed access to individual territories might learn better than youngsters who circulate in an open-space classroom. Individual territories might help children to organize and integrate new material. Edney did not advocate a move from open classrooms to traditional classroom design (see Chapter 5), but suggested that within open-space settings children be given the option of entering individual territories when they feel the need or wish to do so. In fact, many new open classrooms do now include cubbyholes and time-out spaces where children can retreat from the larger common areas.

SUMMARY

Privacy has a variety of meanings in everyday speech, in the law and politics, and in behavioral science. These diverse meanings of privacy fall into two broad categories: *withdrawal* from other persons and *control* over personal information. Irwin Altman defines privacy as *the selective control of access to the self or to one's group.* The principal research method for studying privacy has involved self-report *survey* and *questionnaire* measures concerned with people's experience of privacy in a variety of real-world settings. A small number of investigators have also employed *naturalistic observation* and *unobtrusive measures* to study privacy.

An important psychological function of privacy is to regulate social interaction between a person or group and the social world. For example, privacy regulates the disclosure of personal information and helps to maintain group order. Privacy also serves the important psychological function of helping the individual to establish a sense of personal identity. Privacy helps us to define our personal boundaries, to evaluate ourselves in comparison with other persons, and to develop a sense of personal autonomy.

Theoretical models of privacy have been based on the notion that privacy involves the control of information between people and groups. Altman proposes that privacy is a *dialectical* process, in which the oppositional qualities of being open or closed to social interaction shift over time as social circumstances change. Thus the dialectical model views privacy as a two-way street, sometimes involving a separation from other persons and sometimes involving social contact. At the heart of the dialectical model is the notion that privacy involves the regulation of interpersonal boundaries, controlling both social inputs from other persons to the self and social outputs from the self to others. According to this model, people strive to attain an *optimal* level of privacy; either too little or too much is unsatisfactory. In order to optimize personal privacy, individuals employ multiple behavioral mechanisms, including verbal behavior, nonverbal use of the body, and environmental behaviors.

Knowledge of the nature and psychological functions of privacy has been applied to environmental planning in the areas of occupational, residential, and institutional design.

On the basis of features common to a number of definitions of territoriality, we may define territoriality as *a pattern of behavior associated with the ownership or occupation of a place or geographic area by an individual or group, and may involve personalization and defense against intrusions.* *Primary* territories are typically under relatively complete control by their users for a long period of time, and are central to the lives of their occupants. *Secondary* territories are characterized by a degree of ownership that is neither permanent nor exclusive, and are less central to the lives of their users. *Public* territories are open to public occupancy on a relatively temporary basis, and are not central to their occupants' lives. Most studies of territoriality have relied on *naturalistic observation* or *unobtrusive measures* in real-world contexts. Several studies of territoriality have also used *survey research* methods, and some studies have employed *experimental methods* in either a field or laboratory context.

An important psychological function of territoriality is to help individuals and social groups to organize and manage their daily lives. Territoriality helps people to predict the types of behavior that can be expected in particular places, and to organize everyday behaviors into integrated behavioral chains. Another psychological function of territoriality is to develop and maintain social organization. Territoriality helps to order a social group in accordance with the relative social status or dominance of group members. While research findings concerned with the relationship between territoriality and dominance have been complex, evidence tends to support the view that territoriality is positively related to dominance during periods when the group structure is stable. A further psychological function of territoriality is to serve as a basis for the development of a sense of personal and group identity. People who share a locale tend to have common knowledge and experiences, and thus to share social bonds. Control of territory also affords an individual a feeling of personal distinctiveness, and a context for unique self-expression.

Sociobiological theories of territoriality propose that territorial behavior in humans is rooted in our evolutionary history, and that we share with other animals an instinctive drive to claim and defend territory. Sociobiological theories have drawn on an extensive body of literature on the nature and functions of territoriality in animals. These theories emphasize active *defense* of territory against encroachments as a central aspect of territorial behavior.

Sociocultural theories of territoriality, in contrast, view territorial behavior as primarily a function of social learning and cultural influences. They believe that while biology and instinct can explain animal territoriality, territorial behavior in humans can be understood only in the context of the complex social organization and cultural diversity that characterize human societies. Sociocultural theories propose that, while defense may occur in response to encroachment, defense is not central to the definition of territoriality, and does not inevitably follow from invasion. Sociocultural theories see territoriality in humans as a form of control over the social environment. Such control is often expressed through social roles, which may involve the control of particular areas or objects. Some sociocultural theorists have proposed a synthesis of the sociobiological and sociocultural views, suggesting that the significance of biological influences on territoriality needs to be interpreted in the light of the complex social and cultural factors that shape human life.

Research and theoretical knowledge in the area of territoriality have been applied to environmental planning in the areas of residential and institutional design.

9 Personal Space

When we discuss interpersonal relationships, our language relies heavily on spatial images. When we speak of positive feelings between people, we talk of "closeness"; we are "touched" by the mutual support given by members of groups that are "close-knit." Negative feelings between people are discussed in terms of distance: we feel "far apart" from people we dislike; we keep them "at arm's length." When we experience too much pressure from other people, we speak of needing "elbow room"; we dislike having someone "breathe down our necks" and being "forced into a corner." When we have had all we can take, we demand that the other person "get off our backs."

Actually, our language captures a very real and important aspect of our relationships with one another. In our own daily lives, each of us relies constantly on the use of interpersonal space as an essential, though often subtle, aspect of our interactions with other people. Reflect, for example, on the way we communicate love and anger, satisfaction and disappointment, trust and fear. Important aspects of these interpersonal feelings are translated

273

into a pattern of body movements and gestures—a step closer or farther away, searching eye contact or a shy, downward glance, a direct, face-to-face stance or a subtle turning away. We shall see in this chapter that this complex array of movements and gestures is neither random nor accidental. They represent a predictable, patterned, and meaningful dimension in interpersonal relationships. In fact, the use of space in social relationships serves a diversity of functions and needs in the daily contacts between people. The manner in which space is used varies in accordance with age and cultural background, between women and men, and in some respects according to personality.

It is surprising that we are generally unaware of this important role of space in our social relationships. Until just two decades ago, behavioral scientists, too, had been almost totally oblivious of the varied and exciting ways people use space in their interpersonal relationships. Today, however, the study of the use of space in social interchange has become one of the most heavily investigated areas of environmental psychology. Literally hundreds of research studies have been conducted on this topic.

THE NATURE OF
PERSONAL SPACE

DEFINING PERSONAL SPACE

Proxemics   Today's interest in the ways people use space in interpersonal relationships is based on the ground-breaking observations and speculation of Edward Hall, an anthropologist. In 1966 Hall published *The Hidden Dimension,* a book that summarized and extended his earlier work in this area (Hall, 1955, 1959, 1960, 1963a, 1963b). He coined the term *proxemics* to define the scientific study of space as a medium of interpersonal communication.

Hall's observations were based on earlier work in the field of *ethology,* a branch of biology that studies the adaptive behavior of animals. Heini Hediger (1950, 1955, 1961), an animal psychologist in Switzerland, had identified a series of spatial zones surrounding each animal of a particular species that systematically regulate interactions with other animals of the same or other species. Two types of distance zones control interactions with members of other species. *Flight* distance is the point at which an animal will flee from the approach of another animal of a different species; *critical* distance is the narrow zone between flight distance and the point at which the stalked animal will turn and attack the intruder. A captive lion, for example, will flee an approaching human until it reaches a barrier. If the person continues to approach and enters the lion's critical distance, the lion will reverse direction and begin to stalk the person (Hall, 1966).

Two additional distance zones regulate interactions between animals of the same species. *Personal* distance is the space that is normally maintained between animals that have no intimate relationship to each other. *Social* distance is the point at which an animal begins to feel uneasy because it is out of touch with its own group. Thus personal distance is based on the notion of separation, social distance on the idea of containment. Hall suggests that although flight distance and critical distance have generally been eliminated from interaction between humans, personal and social distance still exert a regulatory influence on human interaction.

Interaction distances in humans   One of Hall's major contributions to the psychological study of spatial behavior is his identification and description of four distance zones (each with a near and far phase) that regulate social interactions between human beings (Figure 9-1). *Intimate* distance is the zone from the point of physical contact to eighteen inches from an individual, and is the area reserved for lovemaking, comforting, and physical contact sports, such as wrestling. *Personal* distance is the area from eighteen inches to four feet from a person, and is appropriate for interactions between very close friends and personal conversations between acquaintances. (This concept is related to Hediger's notion of personal distance in animals.) *Social* distance, which extends from four to twelve feet around the individual, is used for business contacts, with more formal and distant business restricted to the far extreme. Finally, *public* distance is the zone from twelve to twenty-five feet or more beyond a person, and is reserved for very formal contacts, such as those between a public speaker or an actor and an audience.

Personal space   Spurred by Hall's classic studies in proxemics, an extensive body of research and scholarship has developed in the area of what has come to be called "personal space." Personal space is defined as *the zone around an individual into which other persons may not trespass.* It has been compared to a bubble surrounding the individual, creating an invisible boundary between the person and potential intruders. Unlike a real bubble, the bubble of personal space is highly variable, and will shrink or expand in accordance with individual differences, changing circumstances, and the nature of particular interpersonal relationships.

While personal space has often been referred to as circular, some recent evi-

Figure 9-1
People carefully regulate the spatial distance between themselves and other people.

© *Mike Mazzashi/Stock, Boston.*

dence (Hayduk, 1975) suggests that personal space may not be a perfect circle (see box, "The Shape of Personal Space"). And while the notion of a bubble emphasizes spatial distance between people, we shall discover that behaviors other than distancing, including eye contact and body orientation, are also employed to maintain personal space. It is important to recognize that personal space is a product of forces toward both approach and avoidance, and, as such, identifies an *appropriate range* for specific types of interactions, rather than simply a defense against intrusion. As Robert Sommer has colorfully commented: "Like the porcupines in Schopenhauer's fable, people like to be close enough to obtain warmth and comradeship but far enough away to avoid pricking one another" (1969:26).

The most current models of personal space view it as a complex pattern of related behaviors that are adjusted systemically to changing circumstances. This theoretical position emphasizes that personal space is maintained by a range of interrelated behaviors in addition to interpersonal distance, including eye contact, head position, and body orientation. In discussing invasions of personal space, we shall see that a person whose personal space is invaded responds with a complex variety of patterned behaviors.

These systems models of personal space (and of spatial behavior generally) have drawn on a broader theoretical framework in the human and physical sciences known as *general systems theory* (see Boulding, 1968; von Bertalanffy, 1968). Essential to systems models is the notion that social and biological systems (e.g., an industrial organization, the human body, a forest ecosystem) consist of a variety of interlocking variables that function so as to maintain an overall state of equilibrium in the system over time. For example, the various organs in the human body function together in a complex interplay of mutual influences to maintain a steady body temperature while external conditions of temperature and humidity may vary greatly.

Personal space, privacy, and territoriality   We must distinguish personal space from two related though distinct concepts we examined in Chapter 8—privacy and territoriality. In distinguishing personal space from privacy, we must keep in mind that personal space always has a *spatial* referent—the distance between two people. Although, as we shall see, the physical distance between people may be less important in itself than the manner in which it regulates cues in interpersonal communication, the spatial referent is invariably part of the definition of personal space.

Privacy, in contrast, refers more broadly to the control of access between the self and others, and involves multiple mechanisms, such as verbal messages and type of clothing, in addition to spatial cues. Yet there is an important link between personal space and privacy: personal space provides one mechanism that can be used to achieve a desired level of privacy (see Altman, 1975). For example, a person who wants to keep other people from claiming his or her attention while studying in a college library (desires a high degree of privacy) might choose to sit at some distance from other persons in the area (to increase the personal space zone).

Personal space must also be distinguished from territoriality. Sommer (1969) notes that often the defense of personal space is so enmeshed with the defense of

The Shape of Personal Space

Researchers have generally referred to personal space as circular, yet Leslie Hayduk (1978) has pointed out that personal space is probably more complex than the image of a circle suggests. The notion that personal space is circular is based on a view of the personal space zone from above the individual's head; personal space is seen as a circle surrounding the individual, with the individual at the center. This circular view of personal space accounts for only two spatial dimensions and completely neglects the third.

Hayduk proposes a three-dimensional model of personal space, as shown in the accompanying figure. The X and Y axes at the base of the figure define the horizontal plane where the circular view of personal space is typically represented. The Z axis adds a third, horizontal dimension to the personal space zone. What originally appeared as a circle from a top view now appears as a cylinder of the individual's height.

The three-dimensional personal space zone may not be of uniform shape. For example, the dashed lines in Figure 9-2 show a personal space zone that is circular at all heights, is of uniform shape above the individual's waist, but tapers below the waist. Hayduk's three-dimensional model of personal space is still speculative; systematic empirical research is needed before we can fully understand the shape of personal space. Yet the notion that personal space involves three dimensions is compelling, and future research will need to examine its vertical as well as its horizontal dimensions.

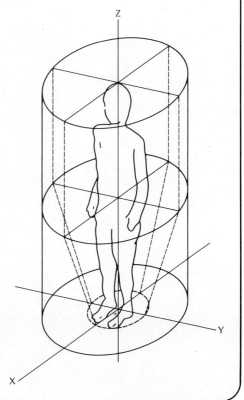

A three-dimensional model of personal space.

L. A. Hayduk, "Personal Space: An Evaluative and Orienting Overview," Psychological Bulletin, 1978, 85:117–34. Copyright 1978 by the American Psychological Association. Reprinted by permission of the publisher and author.

an immediate territory that the two processes may seem to be identical. We should remember, however, that personal space is an *invisible* boundary that *moves* with an individual as he or she changes location. Territory, in contrast, is a *visible* area that has a *stationary* location. For example, your personal space bubble moves with you as you go from your house or apartment (your territory) to the home or apartment of a friend (your friend's territory).

CULTURAL DIFFERENCES IN PERSONAL SPACE

Nationality   One of the most interesting findings of personal space research has been the observation that the manner in which people use personal space varies according to personal background. Among the most dramatic findings is evidence that differences in nationality operate as an especially strong influence on personal space. A major contribution of Hall's (1966) classic work in this field was to explain how individuals from different national backgrounds use personal space. In fact, such culturally based differences in the use of space may sometimes lead people of different cultural backgrounds inadvertently to offend each other by appearing coldly "stand-offish" or overly "pushy" (Figure 9-2).

Figure 9-2
Finding a comfortable interpersonal distance can be difficult when people are of different cultural backgrounds.

© *Peter Menzel/Stock, Boston.*

On the basis of his own observations, Hall identifies several stereotypic differences in personal space between Northern European and British cultures on the one hand and Mediterranean and Arab cultures on the other. Germans are characteristically more formal and distant in their use of personal space, and more offended than Americans by perceived violations of spatial norms. This cultural difference is even reflected in the doors of public and private buildings in Germany, which are considerably more solid and more heavily soundproofed than American doors. And while Americans see nothing wrong in adjusting one's chair to the needs of a particular situation, in Germany the shifting of a chair is considered a social infraction.

The English also show greater reserve and formality than Americans in their use of personal space. The British, however, rely heavily on body orientation, eye contact, and voice level to maintain personal distance. Hall notes that the British consider it rude to be overheard in conversation, and will so carefully modulate their voices in public that they sometimes appear "conspiratorial" to Americans. With similar reserve, the British will acknowledge that they are following a conversation by a blink of the eyes rather than a nod of the head or a verbal expression.

In Mediterranean cultures, in sharp contrast, people interact more closely than Americans, and with a higher level of sensory involvement. For example, the crowded sidewalks and public transportation of southern France involve more physical contact than is characteristic in America. Even more dramatic is the use of space in the Arab world, where pushing and shoving in public settings are commonplace. The most pronounced difference in the use of personal space between the American and Arab worlds is in the intense involvement across several sensory modalities that is characteristic of social interaction in the Middle East. In contrast to the American obsession with the elimination of any trace of body odor, the Arab deliberately accentuates natural body odors and considers it desirable to breathe on a friend while conversing.

Studies by environmental psychologists have lent empirical support to Hall's observations (Altman and Vinsel, 1977). When investigators compared interpersonal behavior in discussion groups composed of either Americans or Arabs, they found that Arabs not only maintained closer interpersonal distances than Americans, but also engaged in more physical contact and more direct eye contact, and spoke more loudly (Watson and Graves, 1966). Similarly, when Kenneth Little (1968) asked subjects to place small replicas of human figures on a field, he found that subjects from Northern Europe (Sweden and Scotland) placed figures farther apart than Americans did, while subjects from Mediterranean countries (southern Italy and Greece) placed figures closer together than Americans did. And in a cross-cultural study of people's ratings of the intimacy of a variety of seating arrangements, Robert Sommer (1968) found considerable similarity in the ratings of people from the United States, England, and Sweden, but some differences between American, Dutch, and Pakistani subjects.

Ethnic background   Since differences in personal space have been demonstrated in individuals from different national backgrounds, we may ask whether similar differences exist in different subcultural groups within a particular society. In fact, several researchers have addressed themselves to this question. Initial studies (Baxter, 1970; Thompson and Baxter, 1973; Willis, 1966) tended to indicate that blacks used more personal space than whites. Later research by John Aiello and Stanley Jones (Aiello and Jones, 1971; Jones, 1971), however, demonstrated that differences in personal space related to ethnic background were extremely complex, and tended to interact with other factors, such as age and sex. The most recent work in this area (Patterson, 1974; Scherer, 1974) suggests that personal space is more strongly affected by socioeconomic background than by ethnicity, and that what appeared to be ethnic differences in earlier studies were in fact due to underlying differences in socioeconomic status. Shawn Scherer (1974) found, for example, that persons from middle-class backgrounds tended to use more personal space than people from lower-class backgrounds among both black and white samples, and that when social class was controlled, no differences in personal space emerged between blacks and whites.

OTHER INDIVIDUAL DIFFERENCES IN PERSONAL SPACE

Development   Although most research studies of personal space have been restricted to adults, a small amount of work has been done with children. Some studies have investigated the developmental stage at which consistent personal space boundaries emerge, and have demonstrated that even preschool children show a systematic use of personal space in their interpersonal behavior (Castell, 1970; Duke and Wilson, 1973; Eberts and Lepper, 1975; King, 1966; Smetana, Bridgeman, and Bridgeman, 1978). Researchers have also speculated as to how personal space may change as a function of developmental stage. When Carol Guardo and Murray Meisels (1971b) asked school-aged children to place human figures on a field, they found that systematic figure placements were more evident among children in the seventh through tenth grades than among younger children. In general, evidence indicates that as children grow older, they demonstrate a preference for greater personal space (Aiello and Aiello, 1974; Jones and Aiello, 1973; Meisels and Guardo, 1969; Pedersen, 1973; Tennis and Dabbs, 1975), though there has been some inconsistency in research findings and more research is needed concerning age and the size of the personal space zone. Rae Carlson and M. A. Price (1966), who extended this developmental research among older children, contend that the development of stable social patterns in the placement of human figures continues on through adolescence and even into adulthood.

Sex   An especially interesting finding involves differences in personal space that are associated with sex. In fact, many of the other findings in this field tend to interact with sex, such as ethnic and developmental effects on personal space.

The personal space zone has been found to be greater for men than for women, even when the potential confounding influences of relative status and warmth have been controlled for (Wittig and Skolnick, 1978). Researchers who have observed interactions between two members of the same sex have consistently found that male-male pairs maintain greater interpersonal distance than female-female pairs (Aiello and Jones, 1971; Pellegrini and Empey, 1970; Sommer,

1959). Mixed-sex pairs have been shown to maintain closer spatial proximity than same-sex pairs (Duke and Nowicki, 1972; Hartnett, Bailey, and Gibson, 1970; Jourard, 1966a; Kuethe, 1962a, 1962b). Observations of spatial positioning between close friends on a college campus in South Africa reveals that the close proximity of mixed-sex dyads was due primarily to the spatial behavior of the women, who tended to move closer to men they liked (Edwards, 1972).

Research that has examined the relationship of ethnic background and developmental stage to personal space has shown a complicated picture of interactions with the sex of the subjects being studied. Aiello and Jones (1971) observed the behavior of first- and second-graders in a schoolyard, and found that, while white boys showed larger personal space zones than white girls, no sex effects were observed for black and Puerto Rican children. In a later study, Jones and Aiello (1973) reported that interactions between school-aged children in free discussion showed that, while black girls stood closer together than white girls, no ethnic effect was found among boys. Using a simulation technique, Guardo and Meisels (1971b) found that girls' figure placements showed relatively more stability than boys' between the third and eighth grades, but there were few sex-related differences in placements by the time students reached the ninth and tenth grades.

Schizophrenia   A particularly intriguing research issue has concerned the relationship between personal space and personality abnormality. In fact, some of the earliest research studies in the personal space field explored the effects of various forms of psychopathology on personal space. A series of studies by Mardi Horowitz and his associates (Horowitz, 1965, 1968; Horowitz, Duff, and Stratton, 1964) examined the use of personal space by schizophrenic patients. In one study, schizophrenics and normals were asked to approach another person until they felt uncomfortable. The interpersonal distances selected by the schizophrenics were found to be more variable than those of the normal group, sometimes closer to the other person than normal and sometimes farther away. When schizophrenics and normals were shown a drawing of a human figure and asked to draw a line around the figure to show the distance they liked to keep between themselves and another person, schizophrenics drew lines that established greater distances than normals did.

Horowitz also studied the personal space boundaries of psychiatric patients from the time they entered the hospital until they were discharged. He found that the personal space boundaries of both schizophrenic and depressed patients decreased as the degree of psychiatric disturbance lessened. Other researchers have reported similar findings. In some circumstances schizophrenics maintained greater interpersonal distances than normals (Sommer, 1959; Ziller and Grossman, 1967; Ziller, Megas, and Di Cencio, 1964), while in other situations schizophrenics were found to show greater variability in personal space than normals (Blumenthal and Meltzoff, 1967).

MEASURING PERSONAL SPACE

Because personal space behavior is an especially subtle aspect of interpersonal relationships and one of which people are typically unaware, the measurement strategies that have been developed to study personal space are quite varied. As we have seen, researchers have measured personal space by observing naturalistically how people use space in relating to one another, by asking people to approach one another in laboratory settings, and by simulating real-world situations (asking subjects to use small cutout figures to demonstrate their use of personal space). Let us examine these diverse strategies in more detail.

Naturalistic observation Personal space may be measured by systematical observation and recording of the distance between people involved in social interaction in real-world settings. Many investigators have used video tapes, movie films, or photographs to facilitate data collection in naturalistic settings. J. Smetana, D. L. Bridgeman, and B. Bridgeman (1978) employed naturalistic observation to measure personal space behavior in children in four nursery schools in Santa Cruz, California. They recorded all interactions between the children on video tape. Later they measured distances between interactants by "freezing" the tape at each interaction episode and using the 18-inch floor tiles in the setting as a measurement gauge.

Stop-distance procedure Personal space may also be measured in controlled laboratory settings by what has been termed the *stop-distance procedure*. Here the subject approaches or is approached by another person (often an experimenter), and stops the approach at the point where the subject begins to feel uncomfortable. Mardi Horowitz and his colleagues (Horowitz, Duff, and Stratton, 1964) used the stop-distance procedure in their study of personal space in hospitalized schizophrenic patients. While the greater artificiality of the stop-distance procedure makes its external validity lower than that of naturalistic observation, its greater experimental control and precision in measurement gives it greater internal validity.

Simulation measures Because it is difficult and time-consuming to measure personal space directly in either naturalistic or laboratory settings, many environmental psychologists have resorted to simulation measures. The use of simulation to study personal space was originated by James Kuethe (1962a, 1962b, 1964). In Kuethe's *figure-placement technique*, a subject was given two or more felt cutouts of human figures and asked to place them on a felt field in any manner the subject wished. Kuethe discovered that his subjects did not place the felt figures randomly or haphazardly; they responded according to particular response sets that determined both which figures belonged together and the degree to which they belonged together. These social response sets help to structure ambiguous situations, and individuals from the same cultural background tend to share very similar response sets. For example, subjects placed figures of a woman and a child nearer together than figures of a man and a child; but a figure of a dog was placed nearer to a man than to a woman (Figure 9-3).

To measure personal space with the figure-placement technique, the researcher asks the subject to imagine that one figure already on the field is a particular person, such as the subject's mother, father, or best friend. The subject is then

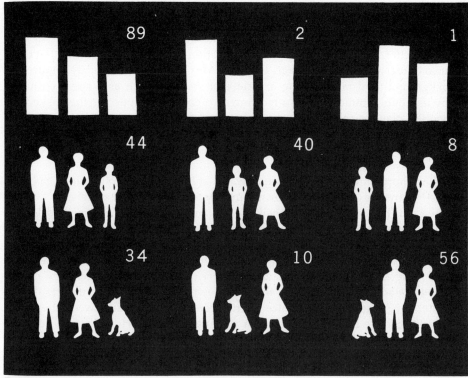

Figure 9-3
Frequencies with which 100 subjects ordered three groups of cutout figures.

From J. L. Kuethe, "Social Schemas," Journal of Abnormal and Social Psychology, 1962, 64:31–38. Copyright 1962 by the American Psychological Association. Reprinted by permission of the publisher and author.

asked to imagine that a second figure is him- or herself and to place that figure on the field in any position he or she chooses. The researcher assesses personal space by measuring the distance between the "self" and "other" figures on the field. Rae Carlson and M. A. Price (1966) used Kuethe's (1962) original set of human and nonhuman figures (see Figure 9-3) in their investigation of developmental trends in the way people employ space in interpersonal relationships. Alternative methods for simulating personal space include *paper-and-pencil* measures, where subjects are asked to mark a piece of paper to show the distance between themselves and another person (Duke and Nowicki, 1972), positioning small *dolls* on a field (Little, 1965), and expressing preferences for *photographs* showing people interacting in varied positions (Haase, 1970).

Validity of simulations   As we noted earlier in regard to other environmental behaviors, it is essential to ascertain whether subjects' responses to simulated conditions are similar to their behavior in real-world contexts; that is, the external validity of simulation techniques must be systematically explored. If simulation measures of personal space are to provide valid information, we need to be sure that there is a close correspondence between the way people place simulated human figures on a field and the way they use actual interpersonal space.

Kenneth Little (1965) examined this question by using both a simulation technique and real actresses on a stage. First he had subjects place simulated figures on a background field representing a variety of settings, such as a street corner, a lobby in a public building, and a location on a college campus. He then asked the subjects to play the role of a theater director, and to situate two real actresses on a stage representing the same environmental settings simulated earlier. The correspondence in interpersonal distance between the simulated figures and the actresses was remarkably close.

One shortcoming of Little's study was that in the real-life situations, subjects situated actresses rather than using interpersonal space themselves. Edward Gottheil and his associates (Gottheil, Corey, and Paredes, 1968) repeated Little's study in a manner that permitted a direct comparison between figure-placement distance and interpersonal distance used by subjects in a real-life setting. First they arranged an interview situation in which subjects were asked to place simulated figures representing themselves and the interviewer on a field. Later they photographed the actual distance from the interviewer selected by the subjects during the interview. The correspondence in interpersonal distance between the simulated and real-life situations was again remarkably close. Holahan and Levinger (1971), using a similar interview situation, also found a close correspondence between figure-placement distance and real interpersonal distance in an actual setting.

More recent evidence, however, has suggested that the correlation between simulation measures of personal space and actual personal space behavior may be lower than was previously assumed. After an extensive review of studies of personal space, Leslie Hayduk (1978), concludes that simulation measures do not provide a sufficiently strong index of the way people use personal space in real social settings. The chief problem is that simulation measures must rely on subjects' cognitive abilities. For example, in order to use figures to represent real interpersonal behaviors, subjects must be able to imagine a particular social and physical setting, to view themselves in interaction from a third-person perspective, and to transform the scale of real social relationships to the scale of small figures. The influence of cognitive ability on measures of personal space is of particular concern in testing children, where the range in cognitive ability may be considerable.

Some support for Hayduk's position comes from a recent study by Kathleen Love and John Aiello (1980). They asked pairs of female college students to have a discussion on a prearranged topic in an experimental setting. During their discussion, their interpersonal distance was unobtrusively recorded. The subjects were next presented with three traditional measures of personal space—two simulation measures (felt figure placements and doll placements) and the stop-distance procedure. Love and Aiello then explicitly asked subjects to place their figures or to stop the approach so as to reproduce the interpersonal distance they had themselves maintained during the discussion. The investigators found that the two simulation measures and the stop-distance procedure correlated poorly with the actual interpersonal distances in the discussion. They concluded that because personal space behavior occurs outside of people's awareness, it is difficult for them to duplicate actual interpersonal distances on simulation or stop-distance measures even when they are explicitly asked to do so.

On the basis of the accumulated research evidence concerning the external validity of simulation measures of personal space, we may conclude (1) that environmental psychologists interested in studying personal space should use naturalistic observation methods whenever possible; (2) that when naturalistic observation is not possible, the stop-distance procedure is preferable to simulation measures (Hayduk, 1978); and (3) that when simulation must be used (such as when a large sample of subjects is to be tested), the findings based on it should be accorded relatively less weight than findings based on naturalistic observation.

PSYCHOLOGICAL FUNCTIONS OF PERSONAL SPACE

SELF-PROTECTION

Environmental psychologists believe that an important function of personal space is self-protection. Personal space operates as a buffer against both physical and emotional threats from other persons. In fact, some researchers (Dosey and Meisels, 1969; Horowitz, Duff, and Stratton, 1964) have referred to personal space as a "body-buffer zone," thereby explicitly recognizing the self-protective function of the personal space boundary. Researchers have observed that when people find themselves in a threatening situation, they automatically enlarge their personal space zone in self-defense. For example, people maintained greater interpersonal distances when they were told that their physical and sexual attractiveness was being evaluated (Dosey and Meisels, 1969), and when they were given negative feedback about their performance on a task (Karabenick and Meisels, 1972).

Invasions of personal space   Consider for a moment how we might experimentally study the protective function of personal space. One procedure would be to intrude into another person's personal space zone and systematically to observe that person's reactions. In fact, the most widely used and most dramatic technique employed by environmental psychologists to investigate the protective function of personal space has been just such an experimental invasion of another individual's personal space. This *invasion technique* has been used with particular effectiveness by Robert Sommer and his associates (Sommer, 1969; Felipe and Sommer, 1966, 1972).

In one study, Sommer invaded the personal space of psychiatric patients in a 1,500-bed mental hospital in Mendocino, California. As subjects he selected male patients who were sitting alone on benches and who were not engaged in any particular activity. To invade the patient's personal space, Sommer walked over and sat beside him, without saying a word. He situated himself just six inches from the patient, and if the patient moved slightly away, Sommer moved also to keep the distance between the patient and himself at six inches. In order systematically to assess patients' reactions to the invasion of their personal space, Sommer selected a control group of patients who were also seated alone in the same general area, but whose space was not invaded. He reports that patients' reactions to the invasion were dramatic. Within two minutes, one-third of the invaded patients had fled their seats, while not one control patient had moved away. After nine minutes, half of the invaded patients had been driven away, while only 8 percent of the controls had left their seats (Figure 9-4).

In a second study, Nancy Felipe invaded the personal space of female students who were seated alone, reading or studying, in the study hall of a college library. Again control subjects were selected among other female students who were also seated alone in the study hall. In the most dramatic invasion situation, Felipe sat in the chair directly alongside the subject, and moved her chair as close as possible to the subject's chair without causing physical contact. After thirty minutes, 70 percent of the invaded subjects had retreated from their positions, while only 13 percent of the controls had left their seats. In a less serious invasion situation, however, when another chair or a table was situated between the invader and the subject, subjects showed little reaction to the intruder.

We might ask whether there are additional ways individuals cope with an invasion of their personal space in addition to simply fleeing the area. In fact, Sommer reports that in both invasion studies, subjects attempted initially to cope behaviorally with the invasion before fleeing. These behavioral adjustments to invasion were quite complex, and tended to vary from individual to individual. Some subjects altered their orientations toward the intruder by facing away or by subtly adjusting the angles of their chairs. Some subjects also adjusted their posture in a defensive fashion, by pulling in their shoulders, moving their elbows to their sides, or placing their chins in their hands. Other subjects used books and other objects as barriers against the invader. If these defensive maneuvers were unsuccessful, the subject then resorted to flight. Miles Patterson and his colleagues

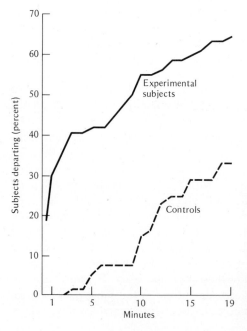

Figure 9-4

Percentage of psychiatric experimental patients who fled invasions of their personal space and of controls who changed seats, by number of minutes elapsed.

From R. Sommer, Personal Space: The Behavioral Basis of Design, p. 33. © 1969 by Prentice-Hall, Inc. Published by Prentice-Hall, Inc., Englewood Cliffs, NJ 07632. Reprinted by permission.

(Patterson, Mullens, and Romano, 1971) reported a similar pattern of complex behavioral responses to spatial invasion in a library. When we discuss theories of personal space, we shall see that the complex behavioral adjustments people make to such invasions are a central aspect in systems models of personal space.

Characteristics of the invader   Some studies have examined how characteristics of the person who invades someone else's personal space affect the response to the invasion. Research evidence has indicated that the invader's sex, age, and social status influence an individual's reaction to an invasion of personal space. Male invaders cause more movement on the part of an invaded party than female intruders (Bleda and Bleda, 1976; Dabbs, 1971). Interestingly, research has shown that men are also more disturbed than women by invasions of their personal space (Garfinkel, 1964; Patterson, Mullens, and Romano, 1971).

Anna Fry and Frank Willis (1971) have demonstrated that the age of the invader also helps to determine the victim's response. They had children stand less than six inches behind adults on a theater line, and found that while 5-year-olds were given a positive response, 8-year-olds tended to be ignored, and 10-year-olds were accorded the same negative reactions shown to intruding adults. Finally, David Barash (1973) found that the apparent status of the invader affected subjects' reactions in a library. When a male intruder was formally dressed in a suit and appeared to be a faculty member, students fled more precipitously than they did when the same intruder was casually dressed and appeared to be a fellow student.

The invader's perspective   So far we have examined the psychological effects of an invasion of personal space on the person whose space is invaded. We might also ask how the intruder is affected psychologically by the prospect of invading someone else's personal space. After all, someone who intrudes into another person's space is at the same time allowing an invasion of his or her own personal space. Environmental psychologists who have addressed this question have consistently found that people report discomfort and generally negative feelings about intruding into other people's personal space. For example, people tended to avoid drinking at a water fountain in a university building when another person (a confederate of the experimenters) was situated within five feet of the fountain (Baum, Reiss, and O'Hara, 1974). Interestingly, however, when the fountain was screened off (inserted into the wall rather than extending from the wall), people were more willing to stop and drink even when another person was nearby.

Nancy Thalhofer (1980) speculated that an individual's reluctance to drink from a water fountain when another person stands nearby might vary according to the overall number of persons in the area around the fountain. On the basis of an *information overload* model of crowding (see "Theoretical Perspectives on Crowding," Chapter 7), she reasoned that if people pay less attention to social cues in crowded conditions, they should feel less discomfort in violating another person's personal space when social density is high than when it is low. To test this hypothesis, she conducted a field experiment at a water fountain in the corridor of a university classroom building. The proportion of passers-by who drank from the fountain was observed under four experimental conditions: when the area around the fountain was typified by *high* and *low* social density, and (within each of these conditions) when an experimental *confederate* and when *no one* stood one foot from the fountain.

Table 9-1 shows the proportions of passers-by who drank from the water fountain under each of the four experimental conditions (300 subjects were observed under each condition). Consistent with Thalhofer's predictions, the confederate's personal space was violated more often when social density was high than when it was low. Thalhofer's study is especially interesting because it examined two social processes in the environment simultaneously—the invasion of personal space and coping with crowding. The joint effects of two or more psychological processes offer interesting possibilities for future research.

Additional research has demonstrated that people are reluctant to invade the space of two people who are interacting together—an indication that social groups are perceived to have a personal space zone comparable to an individual's. Research by James Cheyne and Michael Efran (Cheyne and Efran, 1972; Efran and Cheyne, 1973, 1974) demonstrated that individuals were reluctant to penetrate the personal space of two people who were conversing, but were relatively unconcerned about intruding when the pair was simply standing around. And when the two were more than four feet apart, intrusions increased. The sex of the pairs was also found to be important. Reluctance to intrude was greatest for mixed-sex pairs, intermediate for a pair of women, and least pronounced for male pairs. The behavior of the intruders revealed their own discomfort at invading other people's personal space. They tended to lower their heads, close their eyes, and mumble apologies as they passed through the interacting pair's space.

Eric Knowles (1973) reports that the size of the invaded group also affects an individual's inclination to intrude. He found that while people were generally disinclined to invade the personal space of a conversing group, the effect was more pronounced for a four-person group than for a two-person group. When we compare Knowles's findings with Thalhofer's, we should keep in mind that in Thalhofer's experiment the passers-by did not invade the space of a social group, but of an *individual* in the vicinity of other persons. Also, while Knowles's foursome was a coherent group of conversing persons, Thalhofer's condition of high social density consisted of a more disparate collection of individuals. It appears that when social

Table 9-1.   Proportions of passers-by who drank from a water fountain in the presence or absence of a person near the fountain, by social density.

Density	Person present	Person absent	Overall
Low	23/300 = 0.0767	55/300 = 0.1833	78/600 = 0.1300
High	47/300 = 0.1567	50/300 = 0.1667	97/600 = 0.1617
Overall	70/600 = 0.1167	105/600 = 0.1750	175/1200 = 0.1458

Source: Adapted from N. N. Thalhofer, "Violation of a Spacing Norm in High Social Density," Journal of Applied Social Psychology, 1980, 10:2: 175-83. Reprinted by permission.

density involves a coherent social group involved in conversation, the personal space of the group itself will be respected. Knowles also found that people were more reluctant to invade the space of a high-status group than of a low-status group, as reflected in the group's age and manner of dress. A further study (Knowles, Kreuser, Haas, Hyde, and Schuchart, 1976) found that pedestrians walked farther away from a group of people than from a single individual.

Anxiety, threat, and personal space   The self-protective function of personal space may be further understood if we examine personal space in regard to individuals who are anxious about social situations or who perceive other persons as threatening. Since personal space serves to protect an individual from physical and emotional threats, we would expect people who feel anxious or threatened in social situations to establish a larger personal space zone than people who do not. A series of research studies has in fact explored the relationship between personal space and anxiety. These studies, which have involved both the simulation method and observations of actual interpersonal behavior, have indicated that individuals who score high on questionnaire measures of anxiety maintain greater interpersonal distances than persons who score low on anxiety (Bailey, Hartnett, and Gibson, 1972; Karabenick and Meisels, 1972; Patterson, 1973b; Weinstein, 1968). Conversely, related research has indicated that shorter interpersonal distances are associated with high self-esteem, a more positive self-concept, and a clearer sense of body boundary (Frankel and Barrett, 1971; Frede, Gautney, and Baxter, 1968; Stratton, Tekippe, and Flick, 1973).

In a related line of investigation, Marshall Duke and Stephen Nowicki (1972) examined the relationship between personal space and personality differences in what researchers have called "locus of control" (Rotter, 1966). "Locus of control" refers to differences in the way individuals view the causation of events; "internals" are inclined to view outcomes as under the control of the self, and "externals" are inclined to view outcomes as under external sources of control. Duke and Nowicki, using a simulation measure, found that externals maintained larger personal space zones in the presence of strangers and authority figures than internals did. They explain their results in terms of a social-learning model of personal space, suggesting that externals have learned to feel more dependent on other people to meet their needs, and thus feel more threatened by strangers, whose responses are difficult to predict.

We may gain further insight into the relationship between perceived social threat and personal space by focusing on individuals who have been identified as emotionally disturbed or socially deviant (see box, "The Personal Space of Violent Prisoners"). An important line of investigation has examined the personal space of children and adolescents who have shown some degree of emotional disturbance. Laura Weinstein (1965), using the simulation method, found that boys between 8 and 12 years of age in a residential school for emotionally disturbed children placed greater distances between human figures than did a control sample of normal boys. Interestingly, the disturbed boys put the greatest distances between figures representing a woman and a child; presumably these placements reflected particularly negative relationships between the boys and their mothers and female

──The Personal Space of Violent Prisoners──

In an especially interesting study, A. F. Kinzel (1970) investigated the personal space of prisoners who had histories of violent behavior. In an experimental setting, an experimenter repeatedly walked toward the subject from various directions until the subject reported feeling uncomfortable. It was found that the average personal space boundary of violent prisoners was four times greater in total area than that of nonviolent prisoners. The greatest differences occurred when the experimenter approached from behind the subjects; in this situation the violent prisoners exhibited an especially distant personal space boundary. In fact, the violent prisoners were inclined to perceive the approach of the experimenter as a threat.

The behavior and comments of the subjects after the procedures supported the clinical observation that violent individuals tend to perceive nonthreatening intrusion as attack. Violent subjects frequently reported misperceiving the experimenter as "looming" or "rushing" at the subjects. Several mentioned that this was very much like the sensation they had had prior to assaults. . . . One subject repeatedly said nothing but moved from the center of the room with clenched fists each time he felt the experimenter was too close. Two subjects could not tolerate the experimenter behind them at any distance without looking at him for the first three trials. Another subject said, "If I didn't know you I might be ready for anything."   [Kinzel, 1970:63]

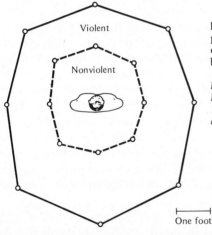

Personal space zones are greater for violent prisoners than for nonviolent prisoners, especially when they are approached from behind.

From A. F. Kinzel, "Body Buffer Zone in Violent Prisoners," American Journal of Psychiatry, 1970, 127:1:59–64. Copyright, 1970, the American Psychiatric Association.

teachers. Weinstein concludes that emotionally disturbed children perceive people, particularly adult women, as more punitive and less supportive and trustworthy than normal children do.

Rhoda Fisher (1967) found that schoolchildren in special classes for children with serious deportment problems placed more distance between toy human figures than normal children did. Other investigators who have studied the personal space of emotionally disturbed children (Du Hamel and Jarmon, 1971) and emotionally disturbed adolescents (Long, Ziller, and Bankes, 1970; Newman and Pollack, 1973) have reported a similar pattern of results. Emotionally disturbed youngsters consistently maintained greater personal space boundaries than their normal counterparts. In comparing these findings with those in regard to schizophrenics, we might speculate that the tendency of schizophrenics to have larger personal space zones than normal persons may also be related to underlying anxiety and perceived threat in social situations. The greater variability found in some schizophrenics' personal space zones may be associated with the severe thought disorder that characterizes schizophrenia.

Another study investigated the personal space of college students who showed some socioemotional disturbance. Jerry Higgins and his associates (Higgins, Peterson, and Lise-Lotte, 1969) used a simulation measure with a group of male undergraduates who were separated according to level of social adjustment. They found that students with poor social adjustment, like Weinstein's emotionally disturbed boys, placed more distance between figures representing a mother and son than socially well-adjusted students did. They speculate that this result reflects a more negative relationship between maladjusted students and their mothers.

INTERPERSONAL ATTRACTION

Friendship   Another social-psychological function of personal space is to regulate the expression of interpersonal attraction. In fact, some of the earliest studies in this field investigated personal space as a function of the degree of friendship between interacting partners. For example, Little (1965) found in both the simulation task and the positioning of actresses on a stage that good friends were placed closer together than casual acquaintances, and casual acquaintances were situated nearer to one another than strangers. Later studies in which the figure-placement technique was used with school-aged children have consistently shown that the closer the social relationship between people, the closer together subjects tend to put them (Bass and Weinstein, 1971; Guardo and Meisels, 1971b; Meisels and Guardo, 1969).

The finding that degree of friendship is an important determinant of interpersonal distance has also been confirmed by researchers who observed how people actually use space in naturalistic settings. When college students were observed in a campus setting, friends were found to approach closer to one another than acquaintances did (Willis, 1966). Similar results were found among schoolchildren in a classroom (Aiello and Cooper, 1972) and among psychiatric patients in a hospital cafeteria (Almond and Esser, 1965). Richard Heslin and Diane Boss (1980) naturalistically observed the nonverbal behavior of travelers and persons who were greeting or seeing them off at the Indianapolis airport. They found that the closer the relationship between two people, the more likely they were to be at the airport together and the more intimate their interaction was likely to be (e.g., an extended kiss or embrace in contrast to no physical contact or a simple handshake).

Attraction   Several researchers have directly investigated the relationship between personal space and the degree of interpersonal attraction between people. For example, researchers have asked whether interpersonal distance operates as a cue for interpersonal attraction; that is, does the spatial distance between two people provide information about how much they like each other? Findings based on simulated human figures (Goldring, 1967) and photos of people interacting (Haas and Pepper, 1972; Mehrabian, 1968a) have consistently shown that the shorter the distance between two people, the greater the attraction between them is perceived to be.

Other researchers have examined the relationship between personal space and interpersonal attraction in actual social situations. Howard Rosenfeld (1965) found that when subjects were asked to act in a way that would gain approval from another person, they positioned themselves closer to the other person than when they were not seeking approval. Albert Mehrabian (1968b) asked subjects in a laboratory context to sit as they would if they were communicating a variety of attitudes to a hypothetical addressee. He found that positive liking for another person was associated with decreased interpersonal distance along with increased eye contact and a more direct body orientation.

In a related line of investigation, researchers have asked whether a socially outgoing person might use personal space differently from someone who is reticent in social situations. Research in a variety of dicussion situations has indicated that extroverts—persons who are socially outgoing and affable—sit closer to other people than introverts—individuals who are socially shy and retiring (Cook, 1970; Patterson and Holmes, 1966). Similarly, Albert Mehrabian and Shirley Diamond (1971), using a laboratory task, found that individuals who scored high in need for affiliation or social contact with other people sat closer to other people than did individuals who scored low in need for affiliation.

Finally, research has shown that proximity tends to foster positive interpersonal feelings. Sidney Jourard (1970) found that subjects were more self-disclosing to an interviewer when the distance between them and the interviewer was decreased. Of course, on the basis of our earlier discussion of spatial invasions, we might expect that liking for another person would be facilitated at an *optimally* short distance, and that liking would decrease as the other person moves too close as well as too far away. In fact, Miles Patterson and Lee Sechrest (1970) have demonstrated that subjects showed more positive feelings toward an experimental confederate when they were situated at a distance of four feet than at either eight or two feet.

Dislike   If liking for another person is associated with proximity, we may ask if negative interpersonal feelings are reflected in greater interpersonal distances. In fact, a wide range of studies by environmental psychologists, based on both the simulation method and naturalistic observation, have demonstrated that negative feelings between people are associated with expansion of the personal space zone. For example, James Kuethe (1964) found that subjects who demonstrated social prejudice against black persons used a more distant manner of grouping black and white simulation figures than did less prejudiced individuals. Alexander Tolor and W. Ronald Salafia (1971) showed that simulation figures that were represented as having negative personal characteristics were placed at greater distances than figures that were seen in more positive personal terms. Carol Guardo and Murray Meisels (1971a) report that figures of a parent and child were placed farther apart when the parent was represented as criticizing the child than when the parent was seen as praising the child. In a study conducted in a naturalistic setting, children who were seen as unfriendly were not approached as closely by other children as were children who were considered to be friendly (King, 1966).

Similarity An interesting area of study is the relationship between personal space and the degree of perceived similarity between people. An extensive series of studies in this area has been conducted by Donald Byrne and his associates (Allgeier and Byrne, 1973; Byrne, 1961, 1971; Byrne and Buehler, 1955; Byrne, Baskett, and Hodges, 1971; Byrne, Ervin, and Lamberth, 1970). Byrne's initial research confirmed the observation that both degree of friendship and degree of interpersonal attraction are associated with shorter distances between people. Especially interesting was Byrne's finding that increased physical proximity is related to the degree of perceived similarity between people. For example, when female and male students were observed after a casual date, couples that were more similar in personality were found to stand closer to each other than couples whose personalities were dissimilar. Further research by Byrne indicates that degree of interpersonal attraction underlies the association between personal space and similarity: people are inclined to like another person more when they perceive similarities between themselves and the other person.

If similarity in personality is associated with shorter distances between people, we might ask whether other types of similarity between people result in a similar pattern. In fact, research has demonstrated that perceived similarity in several respects is associated with short interpersonal distances. Dale Lott and Robert Sommer (1974) found in a laboratory situation that individuals chose to sit closer to persons of similar status than to people whose status was either lower or higher than their own. R. Michael Latta (1978), who also used a laboratory situation, found that persons of discrepant status (defined by age differences) sat farther apart than did people of similar status. Dale Jorgenson (1975), using naturalistic observation with male employees at a series of training meetings, found that employees of equivalent status faced one another more directly while they were interacting than did employees of discrepant status, although he did not find significant differences in interpersonal distance associated with status differences. Students observed in a college campus setting were found to maintain shorter interpersonal distances as a function of similarity in age and ethnic background (Willis, 1966). Studies of the relationship between personal space and attitude similarity, however, have resulted in inconsistent findings. While an initial study (Little, Ulehla, and Henderson, 1968), based on the simulation method, indicated a tendency for individuals to place figures closer together when they were represented as sharing similar values, a later study (Tesch, Huston, and Indenbaum, 1973), involving actual seating choices, indicated no relationship between proximity and attitude similarity.

Stigma If similarity is related to proximity, we might ask whether perceived differences between people are associated with greater interpersonal distances. Research studies that have examined people's reactions to other persons who have a social disability, such as a physical handicap or an emotional problem, are relevant here. One study (Kleck, Buck, Goller, London, Pfeiffer, and Vukcevic, 1968), in which both the simulation method and naturalistic observation were used, found that people keep greater interpersonal distances from other individuals who are socially stigmatized, such as mental patients and amputees. Similarly, a study using simulation figures found that subjects placed more distance between "self" and "other" figures when the "other" figure represented either a physically handicapped person, a drug user, a homosexual, or an obese person (Wolfgang and Wolfgang, 1971). Ronald Comer and Jane Piliavin (1972), in an interview situation, also found that interpersonal distance grew when one person was physically disabled. Finally, Mary Worthington (1974) studied people's reactions to another person with a visible disability in a naturalistic setting. At a California airport, she had an experimenter ask passers-by directions to a nearby freeway. When the experimenter was confined to a wheelchair, subjects were helpfully responsive but stood at a greater distance than when the experimenter was seated in an ordinary chair.

Social influence   Since the distance we keep between ourselves and other people is so closely associated with our feelings toward them, we might speculate as to whether an individual's social influence varies in accordance with interpersonal distance. Is a person who sits close to you more or less persuasive than one who sits at a greater distance? A small body of research has been addressed to this question, and although the findings have been complex, these studies have generally supported the idea that there is an optimal interpersonal distance at which people are most persuasive.

Albert Mehrabian and M. Williams (1969) found that a communicator on film was perceived to be more persuasive when he was 4 feet from the camera than when he was 12 feet from the camera. Stuart Albert and James Dabbs (1970) found that subjects rated a persuasive communicator's distance from themselves as most appropriate at 5 to 6 feet—a distance of 1 to 2 feet was felt to be too short, 14 to 15 feet too long—but actual attitude change was greater at 14 to 15 feet than at either of the shorter distances. Subjects reported feeling pressured and irritated when a persuasive communicator was positioned inappropriately close to them (Dabbs, 1971). Some studies have examined the effects of an invasion of personal space on a request for assistance. Some evidence (Konecni, Libuser, Morton, and Ebbesen, 1975) suggests that people whose personal space has been invaded are unwilling to lend assistance when the invader appears to be in need, although additional work (Baron and Bell, 1976) has shown that the invaded party will offer help if the intruder directly asks for it (see box, "When an Invasion of Personal Space May Help in Obtaining Aid").

THEORETICAL PERSPECTIVES ON PERSONAL SPACE

PERSONAL SPACE AS A COMMUNICATION MEDIUM

Edward Hall's (1966) ground-breaking work on personal space was founded on a view of personal space as a medium of communication. More recently, Darhl Pedersen and Loyda Shears (1973) have presented a review of research in which per-

When an Invasion of Personal Space May Help in Obtaining Aid

How close should you stand to another person when you ask for help? Robert A. Baron (1978) conducted an interesting field experiment in a cafeteria at Purdue University to answer this question. Baron's findings help to clarify the complex results obtained by researchers who had investigated this topic earlier. He predicted that the effects of an invasion of someone's personal space on that person's willingness to lend a helping hand would be influenced by the apparent need of the invader. He reasoned that if the need seems to be great, the invasion of personal space may seem to be justified and may help the invader to obtain aid. But if the need appears to be small, the invasion may seem to be unjustified and may inhibit the aid the invader seeks.

To test these hypotheses, Baron had an experimental confederate approach students seated alone at tables in the cafeteria and request assistance in carrying out a course project. The request was made under four conditions: the degree of the requestor's need appeared to be either *high* (the project was presented as counting for half of the course grade) or *low* (the project was said to have nothing to do with grades), and (within each of these conditions) the requestor stood either *near* the subject (12–18 inches) or *far* from the subject (36–48 inches). The accompanying figure shows the mean number of hours of assistance volunteered by subjects in each of the four experimental conditions. Consistent with Baron's predictions, helping was facilitated by the invasion of personal space when the degree of apparent need was high, but hindered when the degree of need was low.

Mean number of hours of assistance volunteered, by interpersonal distance and requestor's apparent need.

From R. A. Baron, "Invasions of Personal Space and Helping: Mediating Effects of Invader's Apparent Need," Journal of Experimental Social Psychology, 1978, 14:304–12. Reprinted by permission.

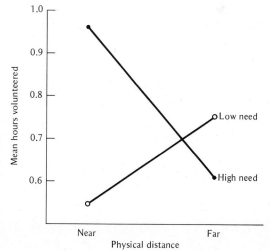

sonal space is viewed as a means of obtaining feedback on interpersonal feelings and attitudes. Our discussion of interpersonal attraction and personal space was consistent with a communication model of personal space. We found that when people like each other, for example, they communicate positive feelings by moving closer to each other, while they communicate feelings of dislike by increasing the distance that separates them.

When personal space is viewed as a medium of communication, we are including the use of interpersonal spatial cues among the broader category of behaviors that has been referred to as "nonverbal communication." In his book *Kinesics and Context* (1970), the anthropologist Ray Birdwhistell developed the theory that communication involves a very broad and complex array of body motions, in addition to verbal messages. Birdwhistell proposes a precise notational system for systematically recording nonverbal messages. While his categorization is broader than personal space behavior alone, his notational system can help us to appreciate the complex pattern of body motions and postural shifts that maintain personal space (Figure 9-5). We shall see that the notion that personal space is maintained by a broad array of interrelated behaviors is also central to systems models of personal space.

Hall (1966) speculated that personal space operates as a medium of communication by determining the amount and type of sensory information that passes between people. He emphasized that the communication associated with personal space involves more than one sensory system at a time. For example, variations in personal space affect our visual image of another person, our auditory perception of the other person's voice, our olfactory sense of the natural and artificial odors

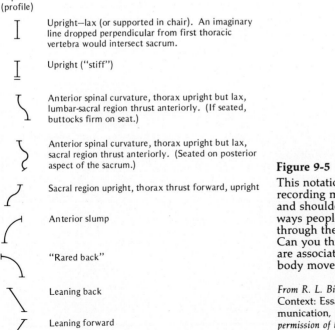

Figure 9-5

This notational system for recording movements of the trunk and shoulders shows some of the ways people can communicate through their body movements. Can you think of the messages that are associated with some of these body movements?

From R. L. Birdwhistell, Kinesics and Context: Essays on Body Motion Communication, 1970, p. 261. Reprinted by permission of the University of Pennsylvania Press.

associated with the person, and, in very close encounters, our ability to touch and feel the warmth of the other person's body. Hall's discussion of the four distance zones used by humans is based, in fact, on an underlying appreciation of how distance influences the communication of sensory information between people.

Consider Hall's discussion of the very different ways in which sensory information is communicated at an intimate distance and at a public distance. In the intimate distance zone (close phase), our perception of another person is dominated by intense sensory information. We can touch and smell the other person, and experience the heat of the person's body. Vision may be blurred, though when clear vision is possible, the visual detail is remarkable. The voice is much less important, and Hall notes that even a whisper tends to increase perceived distance. In the public distance zone, in contrast, our tactile, olfactory, and thermal senses are no longer in play. Vision is extremely important, but the fine details of the skin and eyes cannot be distinguished, the body appears flat rather than round, and eye color is imperceptible. Verbal communication is possible, but messages are more carefully phrased at this distance and tend to be more formal.

Fascinatingly, although personal space is an important aspect of human communication, much of the information we communicate by this medium is sent and received without our full conscious awareness. In fact, it is precisely because the cues associated with personal space are so complex and are communicated through such a wide range of sensory modalities that we sometimes misunderstand the messages we receive in this fashion. Hall gives the example of a job applicant of one cultural background interviewing for a job in a different cultural setting. He points out that the complex nonverbal messages encoded in the use of feet, hands, body, and eyes may cause even a highly motivated job applicant to appear nonchalant and uninterested to the prospective employer.

In addition, nonverbal body cues may assume different meanings to different observer groups. John Sterrett (1978) asked 160 managers in the insurance industry to rate the potential job effectiveness of a male job applicant shown in a videotaped interview. Sterrett's findings are summarized in Table 9-2. Male managers rated the applicant as more ambitious when nonverbal cues were of high intensity (e.g., strong eye contact and intense hand gestures). Female managers, in contrast, rated the applicant as more ambitious when nonverbal cues were low in intensity

Table 9-2.   Mean ratings of ambition by male and female managers, by level of intensity of nonverbal communication in a videotaped interview.

Intensity of nonverbal communication	Male managers	Female managers
Low	6.5	7.5
High	7.0	6.5

Source: Adapted from J. H. Sterrett, "The Job Interview: Body Language and Perceptions of Potential Effectiveness," Journal of Applied Psychology, 1978, 63: 388–90. Copyright 1978 by the American Psychological Association. Reprinted by permission of the publisher and author.

(e.g., weak eye contact and minimal hand gestures). It may be that both male and female managers perceived intense nonverbal cues as signs of aggression, but that men interpreted aggression as a good indication of ambition, while women saw it as a poor indicator of ambition.

THE STRESS MODEL OF PERSONAL SPACE

An alternative model of personal space is the stress model. Here the maintenance of adequate personal space is viewed as an adaptive mechanism that serves to reduce the level of stress that may be experienced by the human organism. Our discussion of the self-protective function of personal space was consistent with the stress model. Personal space provides a buffer or shield against a variety of stressors that may impinge on an individual when interpersonal distances become too short.

The stress model of personal space has been most fully developed by Gary Evans and his colleagues (Evans, 1974, 1979; Evans and Eichelman, 1976; Evans and Howard, 1973). Evans has pointed out that the personal space boundary serves to protect the individual from stressors that may be associated with proximity to other persons. One such stressor is a potential increase in aggression when other persons are physically close. Another is the information overload that may result from the excessive proximity of other persons. Visual cues, odors, and temperature are increased when personal space is invaded (Hall, 1966). Evans adds that this increase in the quality and quantity of sensory information at very short interpersonal distances may "overload" the human organism, leading to both stress and confusion.

If an invasion of personal space is perceived as stressful, we would expect to find physiological indications that the invaded person is experiencing a stress reaction. In fact, invasion of personal space has been shown to be associated with several indices of physiological arousal, including an increased galvanic skin response (Evans, 1972; McBride, King, and James, 1965) and a higher level of palmar sweat (Bergman, 1971).

Evans suggests that a simple stress model is not sufficient to explain the full complexity of human behavior in regard to personal space. Most important, the experience of stress in humans has a unique cognitive component. Evans contends that we need to know how an individual *interprets* an overload of sensory information before we can predict whether this information will lead to stress and arousal.

Evans has proposed that the stress model be broadened from a functional perspective that emphasizes the cognitive dimension of personal space and which also takes account of the underlying tension between the human need to form groups and the individual's need for autonomy. Throughout our evolutionary history, the tendency of human beings to join together in groups has promoted human survival; hunting and gathering could be carried out more effectively as collective enterprises than as individual pursuits. At the same time, however, constant close contact between people may eventually present a threat to the individual's sense of self and to personal ego boundaries. Evans notes that while the roots of personal space behavior may be found in our evolutionary history, the cognitive dimensions of humans' use of space introduces a qualitative difference between the adaptive functions of personal space in humans and in animals. (For a fuller discussion of the role of our evolutionary past in determining human spatial behavior, see "Theoretical Perspectives on Territoriality," Chapter 8.)

Miles Patterson (1976; 1978) has, in fact, recently proposed an *arousal* model of personal space that incorporates aspects of both a stress reaction and a cognitive dimension. Patterson explains that an individual's reaction to an invasion of personal space has two distinct stages. First, consistent with a stress reaction, comes a general state of arousal, accompanied by increased adrenaline secretion, heart rate, blood pressure, and skin conductance. Next, the arousal initiates a cognitive process, by which the individual tries to ascertain the nature and source of the arousal. This cognitive search for meaning is completed when the source of the arousal is labeled and the individual determines the appropriate response in such a situation.

Stephen Worchel and his associates (Worchel and Teddlie, 1976; Worchel and Yohai, 1979) have advanced a model of personal space and crowding that is similar to Patterson's arousal model. They suggest that the arousal associated with personal space invasions is the basis of the psychological experience of crowding. They explain that when an individual's personal space is curtailed, a state of arousal ensues; when this arousal is *attributed* to a violation of personal space, the psychological experience is one of being crowded.

In a series of creatively designed field experiments on the campus of Ohio State University, Robert Smith and Eric Knowles (1978, 1979) provided data that are consistent with Patterson's two-stage arousal model. While their research was initially designed to indicate whether an arousal process or a cognitive process is central to human reactions to invasion of personal space, their results indicate that both arousal and cognitive processes are involved. Their findings are also consistent with Patterson's model: they suggest that the initial reaction to an invasion of personal space is a general state of arousal, and that this arousal reaction is followed by a secondary, cognitive process.

Smith and Knowles found that when pedestrians at a crosswalk experienced an invasion of their personal space, they were less inclined to help another pedestrian (an experimental confederate who dropped a pen while crossing the street), whether or not the invasion appeared to be justified. The reluctance to help another pedestrian regardless of the particular characteristics of the situation was consistent with an initial general arousal reaction. Smith and Knowles also found, however, that the pedestrians whose space had been invaded were selectively helpful to the *invader* during other trials (also an experimental confederate who dropped a pen while crossing the street), depending on whether or not the invasion seemed appropriate. An artist who was sketching a street scene was helped more often than someone who was simply waiting to cross the street. This differential helping based on the particular characteristics of the situation is consistent with a secondary cognitive process, involving a labeling of the invasion of personal space as appropriate or inappropriate.

SYSTEMS MODELS OF PERSONAL SPACE

The earliest research based on a systems view of personal space was conducted by Michael Argyle and his associates (Argyle and Dean, 1965; Argyle and Ingham, 1972; Argyle and Cook, 1976; Argyle and Graham, 1976). Argyle contends that personal space behaviors function as a system to maintain an optimal level of intimacy between people. If one element of this behavioral system, such as physical distance, is altered so that intimacy becomes uncomfortably close, other components of the system, such as eye contact or body orientation, will be adjusted to *compensate* for the undesired intimacy. Argyle found that when subjects were asked to approach either another person or a photograph of another person, they gradually decreased their eye contact as their physical closeness to the stimulus person increased. At a distance of two feet, not only was eye contact sharply reduced, but the subjects tended to display compensatory postural shifts, such as leaning backward and facing away.

Later research by investigators who have observed visual and orienting behavior as a function of interpersonal distance has tended to concur with Argyle's compensatory model (Aiello, 1972, 1977; Goldberg, Kiesler, and Collins, 1969; Sundstrom, 1978a; Sundstrom and Sundstrom, 1977; Terry and Lower, 1979). In a comprehensive review of studies that bear on the compensatory model, Miles Patterson (1973a) concludes that there is strong support for the idea of compensatory relationships between interpersonal distance and both eye contact and angle of orientation. As interpersonal distance is decreased, both the overall amount of eye contact and the directness of the angle of orientation between individuals are decreased in a way that maintains the intimacy of the interaction at a constant level.

Systems models of social and biological functioning, like Argyle's compensatory model, have generally emphasized mutually *counteracting* influences among variables. For example, the processes by which the body maintains a steady temperature involve counteracting influences between heat-generating and heat-reducing forces. Some systems theorists, however, have emphasized that *amplifying* influences among variables are also essential to the functioning of living systems (see Buckley, 1967; Maruyama, 1963). Interpersonal attraction, for example, is often generated and maintained by amplifying influences between two parties.

Patterson's (1976; 1978) more recent *arousal* model of personal space may also be viewed as a systems model. He extends the Argyle model by including both mutually counteracting and mutually amplifying influences among variables that maintain personal space. Patterson notes that interpersonal closeness does not always lead to compensatory responses that lessen closeness, but may sometimes result in *reciprocal* behaviors that serve to enhance it. In this formulation, interpersonal closeness is seen as arousal producing, but the arousal may be perceived as either positive or negative depending on characteristics of the situation and of the relationship between the persons. While negatively labeled arousal will result in compensatory behaviors, positively labeled arousal will lead to reciprocal reactions.

Irwin Altman's (1975) model of spatial behavior (see "Theoretical Perspectives on Privacy," Chapter 8) also represents a systems model that encompasses personal space along with privacy, territoriality, and crowding. Central to Altman's systems view is the notion that personal space is a behavioral mechanism that may be used along with a variety of other mechanisms to achieve a desired level of privacy. In Altman's view, the maintenance of personal space is a dynamic process of moving toward or away from other persons by which the individual makes the self more or less accessible to others. This moving toward or away from others involves a variety of nonverbal behaviors, such as angle of orientation, as well as interpersonal distancing. Altman's model of personal space, like Patterson's, includes reciprocal as well as compensatory influences among variables. Most important, in Altman's model the systemic behaviors employed to maintain personal space are themselves embedded in a more comprehensive system of interlocking verbal, nonverbal, and environmental behaviors that form a coherent pattern designed to insure a desired level of privacy.

Paul Greenbaum and Howard Rosenfeld (1978) conducted an interesting field experiment that reflects a systems model of personal space. They were interested in learning how automobile drivers would respond to an encroachment on their personal space involving both interpersonal closeness and strong eye contact. They selected for their study a traffic intersection in a small city. An experimental confederate was stationed in a grassy area adjoining the intersection. The experimenters systematically manipulated the distance between the confederate and drivers who stopped at the traffic signal in the intersection and whether or not the confederate stared at drivers while they were stopped for the signal.

Greenbaum and Rosenfeld employed a pattern of behavioral indices to assess drivers' reactions to the invasion of their personal space. They recorded where drivers directed their gaze, the percentage of time they looked at the confederate, the time it took the driver to respond to the onset of the green light, and the time it took the driver to clear the intersection. They found that the predominant reaction to either a close or a staring confederate was to avoid the confederate. Many drivers avoided the normal stop position when the confederate was located close to the intersection, stopping instead either before or beyond the usual stopping point. When the confederate stared at them, drivers tended to avert their gaze or to drive away quickly or both. A small number of drivers intensified their gaze at the confederate, spoke to the confederate, and drove away more slowly. The investigators speculate that these approach responses probably reflected information-seeking behavior on the part of the drivers.

APPLICATIONS TO ENVIRONMENTAL PLANNING

ARCHITECTURE AND SMALL GROUP INTERACTION

Our appreciation of the social function of physical space can help us to design settings that facilitate rather than impede the natural patterns of social interaction between people (Evans, 1979b). In fact, Maxine Wolfe and Harold Proshansky (1974) have argued that no small group can function effectively if its physical setting conflicts with the normal patterns of social interaction in the group. Unfortunately, very often those physical settings that are designed to support contact among people, such as a dayroom in a hospital or a social lounge in a college dormitory, are built without a full appreciation of personal space behavior. Such settings may then function to impede the very social contact they were originally intended to encourage.

Architecture and personal space A small body of research has examined the effects of various architectural features on the size of the personal space zone. A general finding that has emerged from this research is that the more enclosed a setting becomes, the more personal space its occupants desire. For example, subjects observed during a seated conversation were found to increase their personal space as a function of decreasing room size (White, 1975). Another study, using the stop-distance procedure with male subjects, found that personal space became greater as ceiling height was lowered (Savinar, 1975). Similarly, James Dabbs and his associates (Dabbs, Fuller, and Carr, 1973; Tennis and Dabbs, 1975), who studied college students, prison inmates, and schoolchildren, report that people's need for personal space tends to be greater in a corner than in the center of a room. Finally, two studies (Little, 1965; Pempus, Sawaya, and Cooper, 1975), based on simulation and naturalistic methods, respectively, found that individuals use more personal space in indoor settings, where space is enclosed, than in outdoor settings.

Sociopetal spaces   Humphry Osmond (1957, 1959) classifies environmental settings according to whether they facilitate or impede normal social interaction. He defines as "sociofugal" such facilities as railway stations, jails, and hospitals, which tend to prevent or discourage the development of enduring social relationships (see box, "Airport Loneliness"). He defines as "sociopetal" such spaces as Indian tepees, Eskimo igloos, and Zulu kraals, which foster or encourage the growth of stable interpersonal relationships. Of course, environmental psychologists recognize that not all environments should function as active social settings; a study area in a dormitory and a reading room in a library, for example, have other functions. Hall (1966) suggests that the most desirable space is flexible, so that individuals may be socially involved or not, depending on the occasion.

Conversation   In an extensive series of research studies, Robert Sommer (1961, 1962, 1965, 1966, 1967, 1969, 1974) has empirically investigated the particular features of spatial settings that influence social interaction. He has studied what he has called "the limits of comfortable conversation." In one study he varied the distance between two sofas that faced each other in a lounge, and asked pairs of subjects to sit down and discuss various impersonal topics. He found that people preferred to sit across from each other, although at a slight angle, rather than side by side. When the two sofas were separated by five and a half feet, however, people preferred to sit side by side. In a similar study, Sommer used four chairs instead of two sofas. Again subjects preferred to face each other unless the face-to-face distance was greater than the side-by-side distance; in that case they tended to sit side by side.

Further research has examined the angle of orientation between individuals involved in social interaction (Table 9-3). When Sommer observed pairs of students conversing in a college cafeteria during nonmeal hours, he found that individuals seated at rectangular tables preferred corner-to-corner seating, and to a lesser extent opposite seating, rather than side-by-side arrangements. Similarly, Albert Mehrabian and Shirley Diamond (1971) demonstrated in an experimental setting that side-by-side seating was detrimental to conversation, while both face-to-face and corner-to-corner patterns facilitated social interaction. Some studies have indicated that when individuals perceive themselves to be in a competitive situation, face-to-face seating or other distant seating patterns are preferred to either corner-to-corner or side-by-side seating (Batchelor and Goethals, 1972;

Airport Loneliness

Robert Sommer (1974) studied the arrangement of space in the lounge areas of several American airports. He discovered that airport seating is typically arranged in a sociofugal manner that discourages comfortable conversation among airport patrons. Airport seating usually consists of institutional straight rows of chairs bolted to the floor; often rows of seats are arranged back to back or deployed classroom style, facing the ticket counter. Sommer notes that two movies—*The Loved One* and *The Graduate*—used the Los Angeles Airport to convey a feeling of loneliness and alienation. He argues that sociofugal seating is especially inappropriate because a great many travelers and visitors arrive at the airport with family, friends, or business associates.

Why, then, are airport lounges so consistently designed in an unsocial manner? After conversations with airport managers, Sommer concluded that the principal reason is economic. An airport's largest single source of revenue is its concessions—restaurants, cocktail lounges, shops. The underlying logic of uncomfortable, sociofugal seating is to drive travelers and visitors to the concessions, where they will spend money. Airport patrons "are regarded by the airlines as merchandise to be shipped elsewhere and by the concessionaires as sheep to be sheared" (p. 79).

Seating in many airport lounges is arranged in a particularly uncomfortable and unsocial manner.

© *Gabor Demien/Stock, Boston.*

Table 9-3.   Seating arrangements chosen by subjects for social activities in four conditions (percent).

Seating arrangement	Condition 1 (conversing)	Condition 2 (cooperating)	Condition 3 (co-acting)	Condition 4 (competing)
(X at top of table)	42	19	3	7
(X on both sides of table)	46	25	32	41
(X at left and lower-right of table)	1	5	43	20
(X at left and lower of table)	0	0	3	5
(two X stacked at left of table)	11	51	7	8
(X above and below table)	0	0	13	18
Total	100	100	101	99

Source: R. Sommer, Personal Space: The Behavioral Basis of Design, p. 62. © 1969 by Prentice-Hall, Inc. Published by Prentice-Hall, Inc., Englewood Cliffs, NJ 07632. Reprinted by permission.

Gardin, Kaplan, Firestone, and Cowan, 1973; Sommer, 1965, 1969). Other studies bear out a common stereotype: leaders gravitate toward and are seen as belonging at the "head of the table" or similar prominant positions (Hare and Bales, 1963; Howells and Becker, 1962; Sommer, 1961; Lott and Sommer, 1967).

APPLICATIONS IN THERAPEUTIC ENVIRONMENTS

Hospital settings   In a field research experiment, Sommer and Hugo Ross (1958) demonstrated the relevance of our knowledge of personal space to environmental planning in a psychiatric hospital in western Canada. They had been asked to try to increase social interaction between female patients on a geriatric ward that had been recently remodeled. Although the ward had been cheerfully painted and new furniture had been added in the dayroom, the dayroom was remarkable for its lack of social interaction. The experimenters noted that the chairs in the dayroom had been arranged in a highly sociofugal manner, most of them lining the walls. When the experimenters rearranged the chairs in a sociopetal manner, in groupings around small tables situated throughout the dayroom, the amount of social interaction among patients doubled.

Holahan (1972) replicated Sommer and Ross's findings in an experimental dayroom in a psychiatric hospital where extraneous environmental influences could be carefully controlled and patients could be randomly assigned to experimental conditions. In a sociofugal dayroom arrangement chairs were placed shoulder to shoulder along the walls of the room in the unsocial manner typical of psychiatric hospitals. In a sociopetal arrangement, chairs were situated around small tables in the middle of the room in an effort to facilitate social interaction among patients. Behavioral observations in the contrasting dayroom settings revealed that patients in the sociopetal arrangement engaged in significantly more social interaction and conversation than did those in the sociofugal arrangement. The most pronounced difference between the settings was in conversations that involved more than two persons. While such multiperson conversations were frequent in the sociopetal arrangement, they almost never occurred in the sociofugal setting.

Therapy settings   Research studies have also demonstrated the relevance of our knowledge of personal space to the design of settings for psychotherapy. Studies of photographs of a counselor and a client in a therapy setting demonstrated that as the distance between counselor and client is increased, the interaction is perceived as less warm and comfortable (Haas, 1970; Kelly, 1972). Intermediate distances, however, are less anxiety producing than interpersonal distances that are either too short or too long (Dinges and Oetting, 1972). Studies of actual behavior between an interviewer and a client have agreed that clients become less open and more uncomfortable when interpersonal distance becomes too great (Jourard, 1970; Lassen, 1973). Finally, using a variety of techniques, researchers have demonstrated that in counseling situations, people prefer corner-to-corner seating, and dislike arrangements that force them to interact across a desk or table (Haase and Di Mattia, 1970; Widgery and Stackpole, 1972).

APPLICATIONS IN EDUCATIONAL ENVIRONMENTS

Seminar rooms   Further research has demonstrated the relevance of our knowledge of personal space to the design of educational environments. When Sommer (1969) observed class participation by college students as a function of their seating location around a large table in a seminar room, he found that students who were seated directly opposite the instructor participated most often, while those seated at the side of the table were the next most frequent participants. Students who were seated adjacent to the instructor, shoulder to shoulder, tended not to participate at all. These findings are consistent with what has come to be called the

"Steinzor effect." Bernard Steinzor (1950) observed participation by members of small discussion groups seated in a circle, and found that interaction was greatest between those individuals who were seated opposite or nearly opposite one another. He speculated that people seated opposite each other have the greatest physical and expressive stimulus values for each other.

Classrooms   Sommer (1969) observed college students in a traditional straight-row classroom seating arrangement, and found that class participation was greatest among students seated in the front row. Similarly, Andrew Schwebel and Dennis Cherlin (1972), who observed schoolchildren in elementary classrooms, found that children who had been assigned to the front row were more attentive to the teacher than children seated in middle or back rows. Mele Koneya (1976) took account of students' inclinations to be verbal, and found that classroom seating position influenced the class participation of only those students who were moderate verbalizers; low verbalizers failed to participate no matter where they were seated.

Two studies have examined the relationship between classroom seating and college students' course grades. Franklin Becker and his associates (Becker, Sommer, Bee, and Oxley, 1973) found that students who sat at the front of three large classrooms at the University of California at Davis reported higher course grades than those who sat in the back of the room, with the highest grades in the front and center of the room (Figure 9-6). Students in the Becker study had been permitted

Figure 9-6
Environmental psychologists have found that the locations of college students' seats in a classroom can affect their course grades.

© *Frank Siteman/The Picture Cube.*

to select their own seats, however, and it is possible that self-selection may have influenced the study results (i.e., more highly motivated students may have selected seats at the front of the classroom).

Lloyd Stires (1980) tried to replicate Becker's findings while controlling for the influences of self-selection. He examined the actual course grades of students enrolled in two sections of a course in general psychology at Indiana University of Pennsylvania. In order to study the role of self-selection systematically, he permitted the students in one class section to select their own seats, while students in the other class section were seated alphabetically. He found that course grades at the end of the semester were, in fact, significantly influenced by the students' seating locations, whether they had chosen their seats or been assigned to them. The highest grades were obtained by students who sat in the middle of the room rather than at the sides. The classroom Stires studied was wider than it was deep, so that while seats in the middle of the room faced the instructor, those on the sides faced the center and forced students to turn their heads or bodies to look at the instructor.

SUMMARY

Personal space is the zone around an individual into which other persons may not lightly trespass. It identifies an appropriate range for social interactions that varies according to the individuals, the circumstances, and the nature of the interpersonal relationships. Four distance zones that regulate interactions between human beings have been identified: intimate distance, personal distance, social distance, and public distance.

Personal space varies from culture to culture. Northern Europeans and the British employ more personal space than Americans; people of Mediterranean and Arab cultures use less. Observed differences in the use of personal space by blacks and whites, once thought to be attributable to ethnic background, were later discovered to be a function of socioeconomic status rather than ethnicity. Persons with middle-class backgrounds have been found to use more personal space than people with lower-class backgrounds.

Developmentally oriented research has demonstrated that personal space behavior is evident by preschool age, though developmental changes in personal space continue into adulthood. Personal space becomes larger and more systematically used as the individual develops physically and mentally. Further research has shown that personal space varies by sex. Men use more personal space than women, and many of the other findings in this area, including reactions to invasion of personal space and ethnic and developmental effects, interact with the sex of the subjects being studied. Schizophrenics have been shown to maintain greater interpersonal distances than normal people in some circumstances; and in others to display more variability in personal space. Personal space has been measured by means of various strategies, including naturalistic observation in real-world settings, the stop-distance procedure in laboratory contexts, and techniques that simulate the manner in which people use interpersonal space.

Personal space serves several social-psychological functions. First, it provides self-protection, operating as a buffer against both physical and emotional threats from other persons. Research studies involving an experimentally controlled invasion of personal space have demonstrated that an individual whose personal space is invaded will respond with a variety of defensive maneuvers, including eventual flight from the area. Research has indicated that the invader's characteristics influence an individual's reaction to an invasion of personal space. Male invaders cause more movement than female invaders. Reactions to invasions also increase with age and social status. Additional research has shown that an invasion of someone else's personal space causes discomfort for the intruder, especially when the invasion occurs in the context of low social density and when the invaded party consists of more than one person, particularly a mixed-sex pair, and is engaged in conversation. Finally, interpersonal distance has been positively associated with anxiety, perceived social threat, and emotional disturbance.

A second function of personal space is the expression of interpersonal attraction. Proximity is associated with friendship, positive attraction between people, and perceived similarity between individuals. It is also positively associated with extroversion and need for affiliation. Finally, personal space serves to enhance social influence. People are perceived to be most persuasive at optimal interpersonal distances, neither very close nor very far.

Environmental psychologists have offered several theories of personal space. The *communication* model views personal space as a medium of nonverbal communication, determining the amount and type of sensory information that passes between people. An alternative *stress* model views personal space as an adaptive mechanism that protects the individual from a range of stressors (e.g., increases in aggression or an overload of sensory information) that might come into play when interpersonal distances become too short. The most recent theories of personal space are *systems* models, which envision personal space as a complex pattern of related behaviors that are adjusted systemically to changing circumstances. Researchers have addressed themselves to the types of physical spaces that tend to facilitate or impede social relationships, and have attempted to identify the spatial limits of comfortable conversation. Knowledge about personal space has been applied to the architectural design of hospital settings, therapy settings, college seminar rooms, and classrooms.

10 Affiliation and Support in the Urban Environment

I t may be surprising to hear that the physical environment can influence a person's selection of friends. Yet, consider for a moment your own best friends. Can you suggest some factors that may have affected your choice of these particular friends? Psychological explanations may come to mind first. You and your friends may share similar social and political attitudes, or enjoy the same sort of recreations. You may also discover, however, that aspects of the physical environment have played some part in your selection of these friends. You and your friends may be neighbors, or live in the same neighborhood, apartment complex, dormitory, or even be roommates.

Environmental psychologists have found that spatial factors in residential environments often exert an important influence on the development of patterns of friendship. For example, living near another person or sharing common facilities, such as a laundry room or parking lot, often provides the initial basis for the development of a friendship. Of course, other factors, such as a similarity in attitudes and interests, will also be important. Yet, remarkably, environmental psychologists have discovered that friendships that have been initi-

309

ated largely on the basis of environmental factors can provide a solid foundation for mutual help-giving and support and constitute an important element in an individual's sense of social identity.

In Chapters 6 and 7 we found that some features of the urban environment, such as excessive noise or a high degree of crowding, may tend to drive people apart socially. We saw that city residents may cope with an overload of social information by becoming less friendly with one another and less helpful to persons who need assistance. Although the research we discussed in Chapters 6 and 7 provides important information about some of the social-psychological costs of living in the city, it does not suggest that there are no positive social-psychological aspects to urban life. In this chapter we shall consider an important body of research that offers a picture of the positive social relationships that often develop in urban settings. We shall find that social life in the city can be friendly and supportive, and that many of the features of the city environment are conducive to positive social relationships.

Research findings by environmental psychologists who have studied positive social relationships in the urban environment are relevant to public policy decisions involving planned urban change, such as urban renewal and the design of public housing for low-income families. Environmental psychologists have discovered, in fact, that one reason for the failure of many urban renewal projects to achieve their social objectives has been policy makers' inadequate appreciation of the spatial dimensions of social bonds in the inner city. Public housing projects have often evolved into disastrous failures at a psychological level as well, because their physical design has been out of step with the natural patterns of affiliation among their residents.

Our discussion here will focus on affiliation and support in the urban environment. To date, research by environmental psychologists concerning social relationships in residential contexts has dealt overwhelmingly with urban settings, particularly working-class neighborhoods in the central city. Such research was initiated in response to pressing concern about the physical deterioration of the central city and to the social questions raised by the large-scale changes brought about by urban renewal. The relevance of the questions raised by these studies, however, is not limited to the central city, and in fact some research attention has been focused on socializing in the suburban environment as well. We may hope that as the field of environmental psychology continues to grow, increasing attention will be paid to the effects of suburban and rural residential environments on social relationships.

THE NATURE OF AFFILIATION AND SUPPORT IN THE CITY

A QUALITATIVE PICTURE OF THE CITY'S SOCIAL LIFE

A remarkable book by the architectural critic Jane Jacobs, *The Death and Life of Great American Cities* (1961), has had an important influence on the study of positive social relationships in the urban environment. Jacobs provides a richly personal, qualitative portrait of the positive social aspects of city life. Her book serves as a caution to scientists and professionals in the field of urban studies against discussing the city simply as a socially and psychologically negative habitat. It encourages practitioners in the field of urban studies to recognize and reinforce those features of the urban environment that contribute to the richness of its residents' social life (Figure 10-1).

The positive features of urban life have tended to be ignored or misunderstood because they typically do not conform to urban scholars' preconceived notions of social relationships. Urban designers have traditionally viewed the city's social life as taking place in formally designed social or recreational spaces, such as playgrounds or community centers. In fact, Jacobs points out, much of the vital social life of the central city occurs in public open-space settings that were originally designed for other functions. Public sidewalks offer one of the best and most frequently used settings for the city's social life. Jacobs describes the continuous and varied social uses of the city sidewalk during a typical day as a "ballet" in which the "dancers" reinforce each other and create a unified social whole.

Jacobs provides a rich image of the movements of the urban ballet on the sidewalks of Hudson Street, in her own neighborhood in New York's Greenwich Village. The sidewalk ballet continues from early morning till late at night, and attracts a varied group of players, including local residents, schoolchildren, visitors, and day and night workers in the neighborhood, from longshoremen and meat-market workers to business executives and communication scientists. Most impressively, the sidewalk ballet is able to accommodate the diverse behavioral needs of these disparate groups throughout the day. The sidewalk is simultaneously busy with children's street games, from roller skating to playing with bottle tops; with the daily errands of neighborhood residents engaged in household chores, shopping, and casual social banter; and with the needs of business people, from the arrangement of a storefront for the day's business to a dash to a local delicatessen for a quick sandwich.

Figure 10-1

Environmental psychologists now recognize that many aspects of the central city foster a rich and positive social life among residents.

© *Frances M. Cox/Stock, Boston.*

COMPONENTS OF THE CITY'S SOCIAL LIFE

Kinship   The history of social science research in urban environments has reflected a process of discovery that the central city is characterized by a more positive and robust social life than was once envisioned. Initial investigations by social scientists in working-class urban contexts revealed that the patterns of social interaction that are characteristic of middle-income communities, such as membership in formal organizations, are absent in the central city. These investigators concluded that the inner city lacked an established and meaningful social life. Later researchers who studied inner-city environments from the perspective of their essentially working-class residents, however, discovered that the inner-city neighborhood is in fact characterized by a highly developed social network, which is organized according to a pattern of interpersonal relationships that are congruent with social life in the working class. A central finding of this later research was that the social life of working-class urban neighborhoods was founded in large part on the ties of kinship (Dotson, 1951).

Boston's West End   An appreciation of the importance of kinship ties in working-class urban neighborhoods was gained by a series of research studies conducted during the late 1950s and early 1960s in Boston's West End, before the area was demolished as part of Boston's urban renewal program. The West End was an old section of Boston, situated in the heart of the city; its 7,000 residents were from working-class backgrounds, and of predominantly Italian heritage. The West End's narrow, winding streets were densely settled with three- to five-story walk-up apartments, built flush with the sidewalk. Its land uses were mixed, with many small shops and businesses interspersed among the aged tenements. Because many of the area's buildings were in poor repair, city planners slated the West End for urban renewal. Yet, as we shall see, the West End's narrow streets and mixed commercial and residential uses helped to support the area's social vitality (Figure 10-2).

   A penetrating and sensitive appreciation of life in the West End was afforded by the investigations of Herbert Gans, a sociologist and urban planner. Gans lived in the West End as a *participant observer* (a researcher who actually resides in the setting under study, while systematically gathering an in-depth record of life in that environment) for eight months in the late 1950s, just before the onset of redevelopment. He published his observations in a marvelous book titled *The Urban Villagers* (1962). Gans discovered that social relationships within the West End were the most important and central aspect of its residents' lives. Social relationships were founded primarily on kinship, and typically involved in-laws, siblings, and cousins. Gans coined the term "peer group society" to describe these primary social bonds, which were characteristically among peers— persons of the same sex, age, and life-cycle status. The peer group was dominant throughout the life of the West Ender, and its presence was felt from birth to death.

   "Sociability" was a primary theme in the life of the West Ender, and the most important part of residents' lives took place within the peer group. In fact, being alone for the West Ender brought feelings of isolation, discomfort, and fear. Gans drew a distinction between "object-oriented" and "person-oriented" styles of behaving. He speculated that West Enders, unlike most middle-income Americans, tended not to be object-oriented, that is, they did not strive to attain "things" such as career success, social status, or a high level of income. They were person-oriented in their style of behaving,

Figure 10-2
The narrow streets, the mixture of residential and commercial settings, and the active side-walks of this inner-city neighborhood are important ingredients of the area's social life.

© *Ellis Herwig/Stock, Boston.*

interested primarily in being a member of a group and in being liked and noticed by other persons in the group.

London's East End   Research evidence that supports the essential social-psychological role of kinship in working-class urban neighborhoods has come from studies in London's East End. The East End is environmentally similar to Boston's West End. It is an old section of London, densely settled with 54,000 working-class residents. Like the West End, it has a mixture of small stores among the many residences, which are walk-up apartments in old buildings.

Our knowledge of the role of kinship in the East End is based on a series of survey research studies conducted by Michael Young and Peter Willmott (Willmott, 1962; Willmott and Young, 1960; Young and Willmott, 1957), which compared social life in the East End and in two London suburbs. Like Gans, Young and Willmott found that kinship relations were a central and dominant aspect of these people's lives. Here, however, instead of horizontally structured kinship relations—relations among relatives of the same generation—as in the West End, kinship relations were structured vertically, through generations. Young and Willmott refer to this type of kinship group as an "extended family," and note that it commonly consists of a small cluster of families, made up primarily of three generations; grandparents, parents, and grandchildren.

The strongest ties in the extended families of the East End existed between mothers and daughters over three generations. The anchor of the extended family was the family matriarch, who was the oldest woman in a chain of descent and who was referred to as "Mum." Mum played an essential role in the daily functioning of the extended family. She offered assistance during illness or childbirth, babysat for grandchildren when mothers worked, and provided the locale where the extended family could regularly convene. In describing Mum's place in the family, Young and Willmott (1957:44) provide the following excerpts from an interview with a woman who lived in the East End:

> 'Then any time during the day, if I want a bit of salt or something like that, I go round to Mum to get it and have a bit of chat while I'm there.' If the children have anything wrong with them, 'I usually go round to my Mum and have a chat. If she thinks it's serious enough I'll take him to the doctor.' Her mother looked after Marilyn, the oldest child, for nearly three years. 'She's always had her when I worked. . . . Mum looks after my girl pretty good. When she comes in, I say "Have you had your tea?", and she says, as often as not, "I've had it at Nan's." '

Neighboring   While strong research evidence has demonstrated the central role of kinship in the social life of working-class urban neighborhoods, continuing research by environmental psychologists has indicated that patterns of affiliation between *neighbors* also provide a vital element in the social fabric of the central city. Marc Fried and Peggy Gleicher (1961) conducted a survey of almost 500 female residents of the West End before they were uprooted by urban renewal. On the basis of their survey results, they argue that previous research in working-class urban neighborhoods, by focusing almost exclusively on the social ties of kinship, tended to neglect the importance of alternative social relationships. Their own data reveal that the strongest relationship between residents' attitudes toward the West End and their social bonds there was provided by their feelings of psychological closeness to their neighbors (Table 10-1). Although the neighbor relationship may subsume kinship ties (one's relatives may also be neighbors), the relationship between feelings about neighbors and feelings about the West End was stronger than the relationship between feelings about the West End and feelings toward any group of relatives.

Table 10-1. Residents' attitudes toward West End, by feelings of closeness to neighbors.

Closeness to neighbors	Number of respondents	Feelings about West End (percent)			
		Totals	Strongly positive	Positive	Mixed negative
Very positive	78	100	63	28	9
Positive	265	100	37	42	21
Negative	117	100	20	39	41

Source: M. Fried, and P. Gleicher, "Some Sources of Residential Satisfaction in an Urban Slum," *Journal of the American Institute of Planners*, 1961, 27: 305–15. Reprinted by permission.

INFORMAL CHARACTER OF THE CITY'S SOCIAL LIFE

One of the reasons social scientists at first failed to appreciate the richness and depth of social relationships in the urban environment was that much of the city's social life is informal. In *The Death and Life of Great American Cities* (1961), Jane Jacobs provides a thoughtful analysis of the important social-psychological role of casual social contacts in the lives of urban residents. A vital aspect of urban social life consists of casual social contacts between people who meet one another accidentally, in outdoor public settings, while they are engaged in daily errands. While social relationships between relatives and friends are often carried on indoors, a great many social contacts among members of extended families and among neighbors take place on the street. Jacobs suggests, in fact, that such informal social contacts in public settings offer an ideal type of social interchange for city people, providing a necessary balance between a highly accessible form of social contact and the individual's need for privacy. Despite the casual nature of such street contacts, they form an extremely important part of the urbanite's social life, and casual "sidewalk acquaintanceships" can endure for many years and even decades.

Environmental psychologists have been concerned with developing a precise, empirical analysis of the city's outdoor social life that might complement Jacobs' impressionistic picture. One empirical study, relying on naturalistic observation, was conducted in Baltimore's inner city by Sidney Brower and Penelope Williamson (1974). They were interested in learning which of the outdoor spaces available in two low-income neighborhoods were actually used for recreational purposes. The two Baltimore neighborhoods that they studied were characterized by brick row houses, built flush with the sidewalk, and were settled by predominantly black low-income families.

Brower and Williamson's observations confirm that a significant amount of social interaction among residents occurred in the low-income areas' outdoor public spaces. In fact, like Jacobs, they discovered that most outdoor social recreation occurred in informal spaces, mostly along the street. Residents' choice of the street as a recreation area was deliberate; they congregated there even when alternative spaces, such as a yard, park, or playground, were available.

Another empirical study of the city's outdoor social life was conducted by Holahan (1976b) in a low-income section of New York City's Lower East Side. Using a naturalistic observation research strategy, Holahan systematically recorded outdoor social activity in a thirty-six-block neighborhood environmentally similar to the area Brower and Williamson studied in Baltimore. The neighborhood was multiethnic, with significant representations of Hispanic, black, and white residents. Fully 86 percent of outdoor activity in the neighborhood was social. This social activity was quite varied— people working or shopping together, active street games among children, and casual conversation, banter, and kibitzing among adults. As in the Baltimore neighborhoods, most outdoor socializing occurred along the street, in informal public spaces. In fact, 90 percent of outdoor behavior took place along the sidewalk, much of it in the vicinity of entrances to residential and commercial establishments.

Further research evidence, based on participant observation and survey research methods, has shown that the socializing among relatives and neighbors that occurs indoors in the central city is also largely informal. Gans's discussion of peer-group sociability in *The Urban Villagers* (1962) emphasizes that the West Enders' social life consisted essentially of informal conversation, in sharp contrast to the more formalized entertaining and party giving of the middle class. Although people arrived regularly for the informal gatherings of the peer group, they never received formal invitations or advance notification. Conversation tended to focus on casual local gossip about people familiar to the group, while group members dropped in and out of the gathering throughout the evening. Similarly, Marc Fried (1963) proposed that the cohesiveness of the social fabric in the West End rested on residents' ability to drop in on one another casually, without formality or advance warning.

THE ROLE OF THE ENVIRONMENT IN THE CITY'S SOCIAL LIFE

As environmental psychologists have become increasingly aware of the positive social features of urban life, they have come correspondingly to recognize the essential role played by the inner-city environment in the city's social life. Implicit in our review of the studies in Boston, London, Baltimore, and New York has been an assumption that the physical features of these urban districts helped to shape their vital social character.

The design of the *tenement houses* in these areas was especially congruent with their informal social life. The moderately low-rise design (three to five stories) of the tenements and their location flush with the sidewalk permitted residents to move easily between apartments and the outside street environment. Gans (1962) points out that residents going about their daily errands in the city were brought within easy range of the doors and windows of their neighbors. In addition, the stoops at the tenement entrances provided a natural place for residents to sit and participate casually in ongoing street activity (Hartman, 1963).

The *mixture of residential and commercial settings* in these inner-city districts also helped to foster their positive social character. Jacobs (1961) emphasizes that commercial sites, such as grocery stores, cleaners, and small restaurants, mixed among residential buildings attract residents to use the outdoor environment. The presence of people on the street naturally draws other residents to use the outdoor environment, and these "eyes along the street" help to discourage street crime. The entrances to inner-city commercial establishments also provide places where local residents may meet accidentally on their daily rounds and stop awhile to socialize (Holahan, 1976).

The relatively *narrow streets* in most inner-city neighborhoods also encourage their active street life. The reduced and slowed automobile traffic allows local youngsters to use the streets for recreational purposes, such as ball playing, bike riding, and roller skating (Brower and Williamson, 1974). The street activity generated by youngsters may then attract adult residents to sit by their windows or on tenement stoops to observe the ongoing street games (Jacobs, 1961). The narrow streets and alleys also permit casual socializing through open windows from building to building (Gans, 1962).

WHAT IS A SLUM?

The discovery that inner-city districts that are physically rundown may nevertheless support a rich and vital social life requires us to distinguish between a *physical* slum and a *social* slum. Not all urban environments that are physically deteriorated are necessarily similarly debilitated at a social-psychological level. There is strong research evidence to support the contention that although Boston's West End and London's East End showed signs of physical deterioration, both areas had a positive and beneficial social vitality that deserved to be preserved.

Gans (1962) reports that West Enders themselves did not consider their area a slum and strongly resented the city's definition of the West End as a slum. Although the exteriors of its tenement buildings showed evidence of physical decay, residents' apartments were well kept, clean, and up to date. Similarly, Marc Fried and Peggy Gleicher (1961), after studying potential sources of residential satisfaction in the West End, concluded that for the great majority of its residents, the West End was the focus of strongly positive sentiments. Residence in the area was highly stable, with minimal movement into and out of the area and little internal movement between dwelling units. Young and Willmott (1957:44) provide the following highly relevant comment by a resident of London's East End: "I suppose the buildings in [the East End] aren't all that good, but we don't look on this area as a pile of stones. It isn't the buildings that matter. We like the people here."

A realistic attitude   When we distinguish between physical and social slums, we must not lose sight of the fact that not all physically deteriorated urban environments are necessarily healthy social settings. Indeed, many urban settings that are physically deteriorated may also show evidence of serious social pathology, particularly if a large number of the area's residents are unemployed or only intermittently employed (Rainwater, 1966). As we shall see, a modern and attractive residential setting may actually function as a social slum if its design features are out of step with the characteristic social patterns of its residents. Neither does the fact that a central-city environment manifests a robust social life justify the neglect of essential physical repairs. But necessary physical renewal should be planned and carried out in a way that does not shred the fabric of social life in the region. Enlightened urban change facilitates rather than impedes the social-psychological vitality of the inner city. Gans (1962:16) eloquently summarizes this concern:

> Although it is fashionable these days to romanticize the slum, this has not been my purpose here. The West End was not a charming neighborhood of "noble peasants" living in an exotic fashion, resisting the mass-produced homogeneity of American culture and overflowing with a cohesive sense of community. It was a run-down area of people struggling with the problems of low income, poor education, and related difficulties. Even so, it was by and large a good place to live.

RESEARCH METHODS FOR STUDYING THE CITY'S SOCIAL LIFE

Jane Jacobs' views have encouraged urban researchers to examine the positive social forces at play in the urban environment. The environmental psychologist requires in addition, however, systematically collected empirical data in order to achieve a scientifically based understanding of affiliation in the urban environment.

The rich diversity of the city's social life has necessitated a varied battery of field research strategies. Thus far we have encountered three such strategies that have been employed to investigate the city's social life: naturalistic observation, survey research, and participant observation. The knowledge environmental psychologists have acquired concerning the nature of affiliation and support in the urban environment reflects the combined advantages of all three of these field research strategies. Each has its particular strengths and limitations (see also "Research Methods in Environmental Psychology," Chapter 1).

Naturalistic observation has been used when the investigator has wished to develop a behavioral portrait of social life in the setting without personally influencing the social activities under observation. The environmental psychologist systematically and unobtrusively compiles a behavioral record of public activities and social interaction in the setting. Because naturalistic observation involves minimal experimental constraints and is carried out in real-world contexts, its external validity tends to be high. At the same time, its lack of experimental control and the correlational nature of its findings tend to lower its internal validity.

The *survey research* strategy has been employed when the researcher has been concerned with environmental users' personal perceptions and evaluations of particular aspects of social life in the setting under study. The researcher asks residents to respond to a series of specific questions, presented either orally or in writing, which are designed to elicit their subjective impressions of social life in the area. Like naturalistic observation, the survey research strategy is generally higher in external validity (it is conducted in real-world contexts) than internal validity (its findings are correlational rather than experimental). Its external validity, however, may also be low if the survey questions asked are irrelevant to the respondents' real experiences or if they are worded in such a way that they seem confusing, artificial, or stilted to the respondents.

The *participant observation* strategy has been selected when the researcher has been interested in gaining an insider's qualitatively rich perspective on the city's social life. The participant observer actually lives in the community under study, and systematically compiles a rich, in-depth record of life there. Again, as in the other two strategies, external validity is higher than internal validity. The highly subjective nature of participant observation greatly limits its internal validity (it is both correlational and speculative), and may even threaten external validity if the personal biases of the participant observer color his or her perceptions and are not honestly and forthrightly stated.

PSYCHOLOGICAL FUNCTIONS OF NEIGHBORHOOD SOCIAL NETWORKS

MAINTAINING SOCIAL ORDER

Street-corner society   An important social-psychological function of neighborhood-based social networks in the inner city is to establish and maintain a sense of social organization and social control in the district. The first social scientists to investigate the social character of the central city failed to recognize the vital, informal social life of the city's working-class neighborhoods, and reached the conclusion that these neighborhoods were socially disorganized (Michelson, 1976). Social critics tended to see urban crime and social pathology as emanating from the "deviant subculture" that evolved when traditional middle-class forms of social organization and social control were lacking.

An exceptional book by William F. Whyte, *Street Corner Society* (1943), helped to reverse these negative attitudes toward the central city. Whyte's book was based on three and a half years of participant observation in a city slum district he called Cornerville. For many years Cornerville, which had been settled primarily by Italian immigrants, had been viewed as socially disorganized, as a problem neighborhood at odds with the rest of the community. Cornerville was known as the home of racketeers, corrupt politicians, poverty, and crime.

Whyte's extensive observations led him to conclude that Cornerville, far from being socially disorganized, was a closely knit society with a well-defined status hierarchy, shared rules of personal conduct, and well-established and enduring social traditions. Cornerville's social rules were informal, of course. No one had ever written them down; no one had to. Whyte describes in detail one street-corner gang that came together on Norton Street. The gang had no by-laws, no constitution, and never relied on parliamentary procedure, yet its meetings and rituals were highly stable and predictable, and group decisions and group functioning presented no problems (Figure 10-3).

Although Cornerville did not lack internal organization, it did have a problem in relating to the broader American society. The informal social structures of the local area generally failed to mesh with the social structure of the society around it. Whyte saw the local political style of Cornerville, which tended to fall prey to the influences of organized crime, as a product of this broader societal problem. He felt that a proper meshing with the surrounding society would be possible only when the residents of Cornerville achieved a fuller opportunity to participate in the larger American society.

Especially significant, in regard to our discussion of the social-psychological functions of neighborhood social networks, is Whyte's conclusion that Cornerville's social organization was founded on the area's rich, informal social life. For example, the social organization of the Norton Street gang was based on the habitual association of gang members over an extended period of time. The gang was characterized by a constantly high rate of social interchange, and its group structure was based on these informal interactions. The system of mutual social obligations that was essential to group cohesion arose from these daily interactions, and provided the basis of the group's well-defined

Figure 10-3

Environmental psychologists have found that street-corner gatherings in the central city can be an important part of residents' lives.

© Eric Kroll/Taurus Photos.

status hierarchy. The higher a member's status in the group, the more strictly he was expected to live up to his social obligations to other group members.

The relative status of groups in the Cornerville society, as well as the rules governing intergroup relations, were also founded on a system of informal personal relations that emphasized reciprocal social obligations. The "corner boys" were at the bottom of Cornerville's male social hierarchy; the "college boys" were at the top; and the "intermediaries" could participate in the activities of either group. A person at the bottom of the status hierarchy could approach someone at the top only through an intermediary. A corner boy had to establish a "connection" with an intermediary over time by performing small services for him, so that he developed a sense of mutual obligation. The intermediary, in turn, performed similar functions for the person at the top of the status hierarchy.

The social order of the slum   A similar picture of the social organization of an inner-city neighborhood has been provided by Gerald Suttles' book *The Social Order of the Slum* (1968), which was based on three years of participant observation on Chicago's Near West Side. Like Whyte's Cornerville, the Near West Side was considered by outsiders to be socially disorganized, its residents disadvantaged, inadequate, and culturally deprived. From the residents' perspective, however, the neighborhood was seen as highly organized, with a morality that demanded discipline and self-restraint. In contrast to the neighborhood Whyte studied, Chicago's Near West Side was multiethnic, with black, Italian, Puerto Rican, and Mexican residents, and much of the area's social organization was designed to structure relationships among these ethnic groups. Suttles points to the central role of informal neighborhood social networks in establishing and maintaining social order and social control in the area. The essential link in the neighborhood's communication network was the informal face-to-face social exchange of street life. The area's moral standards, like those of Cornerville, were highly personalistic; judgments of personal worth and social sanctions were based on past experiences of mutually honored commitments between residents.

The role of street life in socialization   Jacobs (1961) has pointed out that an important part of the socialization of children raised in the central city is accomplished by the informal and casual social contacts that are a natural part of the city's street life. On a city sidewalk that is lively and socially active, adults supervise the incidental play of neighborhood children, and help to accustom them to the social expectations and responsibilities of adult life. This remarkable socialization process goes on quite naturally while adults go about their daily affairs. In this way, children learn a cardinal rule of the social order of the central city: residents must assume a degree of responsibility for each other, even when they do not share the ties of kinship or intimate friendship. With some irony, Jacobs contrasts this natural socialization process with the common assumption that when children play in the "gutter," they are necessarily acquiring negative attitudes and antisocial behaviors.

The limitations of street-corner society   Although the social networks of the inner-city neighborhood have been found to possess a high degree of internal social organization, there are limitations to the social order that can be achieved through such informal networks. Elliot Liebow's *Tally's Corner* (1967) has provided a thoughtful analysis of the limitations of social order in the street-corner society. Liebow's book is based on his in-depth interviews and naturalistic observation of street life among a group of black men on a street corner in Washington, D.C.

The social organization of the street-corner group that Liebow observed was founded on face-to-face social contact between group members. Yet these street friendships tended to shift constantly in response to the economic, social, and psychological forces that operated on the lives of the low-income central-city residents. Transience was the most pervasive aspect of the street-corner group Liebow observed. Although the men on Tally's Corner claimed that their street friendships had endured for many years, most of these relationships were short-lived and lacked personal depth.

Richard Cloward and Lloyd Ohlin (1960), after studying central-city gangs, also concluded that in many slums the tentative efforts to evolve social order are limited by high levels of social mobility, changing housing and land uses, and economic instability. This finding does not indicate that the observations of Whyte and Suttles were inaccurate, but simply that while some urban slum districts are highly organized socially, others are not.

ESTABLISHING PERSONAL IDENTITY

A second social-psychological function of neighborhood-based social networks in the central city is to provide a frame of reference and a foundation for the individual's sense of personal identity. I am not referring here to a well-articulated feeling of "self-actualization," which, as Gans (1962) has suggested, is irrelevant to the life experiences of many low-income central-city residents. I am speaking rather of a basic sense of *belonging*—of being a part of a particular community and of sharing in the values and world view of that community. For a member of a central-city working-class community, this sense of belonging rests in large part on the individual's daily involvement in and attitudes toward the neighborhood social network.

Individuality   Gans (1962) has described the important function of the peer group in affording each person the opportunity to define his or her individuality. The peer group provides individuals with their only opportunity to express their personality, their unique character, and to establish a sense of separate identity distinct from that of the social group. Because one of the functions of the peer group is to provide a forum for each resident to forge his or her individuality, members of the group constantly jockey for status, power, and respect.

The inner-city neighborhood as home   The inner-city social network provides residents with a community that can be called home. For working-class residents of the central city, "home" goes beyond one's own apartment to encompass the sociospatial region that surrounds it. A series of research studies in the West End by Marc Fried and his colleagues (Fried, 1963; Fried and Gleicher, 1961; Fried and Levin, 1968) has provided an understanding of the sense in which local territory is perceived as home by its residents. The inner-city neighborhood, with its familiar faces and places, becomes home in a sense that is quite unfamiliar to most middle-income Americans, for whom the term "home" is reserved for one's own dwelling unit.

Chester Hartman (1963) has written that West Enders had a strong sense of being at home in the local neighborhood because they felt themselves to be personally rooted in a complex network of interpersonal ties and socially meaningful places. In fact, West Enders drew a very sharp social line between themselves and everyone else. A resident who (unaccountably) married someone from another neighborhood, for instance, was said to have "married outside the West End." In response to the question "How do you feel about living in the West End?" one resident said, "I love the West End—this is my country."

Group and spatial identity   Fried (1963) has also described how the inner-city environment provides the locus for an individual's sense of group and spatial identity. Drawing on earlier work by Erik Erikson (1946, 1956), Fried explains "group identity" as the feeling that one belongs to a larger human group. Essential to the sense of group identity is the feeling of shared human qualities and of communality with one's fellows. Spatial identity, in contrast, is based on a set of complex memories and associations of the familiar locale where one resides. In working-class inner-city neighborhoods, the senses of group identity and spatial identity are interrelated, since both are localized within the same residential area.

Personal values in the inner city   The neighborhood social network forms a basis for a sense of personal identity by providing the social context from which the individual's personal values are derived. William Ryan (1963) has identified two social values—friendliness and social expressiveness—that were strongly held by residents of the West End. When West Enders were asked to identify the major components of "good social standing," they listed "having a lot of friends" just below level of education and occupational standing, and above income level, social influence, and ethnic-group background. And West Enders valued a willingness to be involved in social relationships and an ability to derive enjoyment from these relationships.

SUPPORTING MENTAL HEALTH

Another social-psychological function of neighborhood-based social networks is to help the individual urban resident to maintain psychological health and adjustment. Environmental psychologists have found that people who are integrated into a neighborhood social network report fewer symptoms of psychological disturbance than individuals who are socially isolated. Leonard Duhl (1963) has pointed out the significance to mental health of a wide range of social relationships at the neighborhood level, including supportive relationships between neighbors, participation in neighborhood fraternal organizations, and even the informally supportive role played by the local bartender or grocer. Duhl cautions, however, that these natural supports that are available to urban residents in their own community can be destroyed by slum clearance and urban relocation projects (Figure 10-4).

Figure 10-4
The physical gains of urban renewal can sometimes be offset by the psychological costs that demolition imposes on residents.

© *Daniel S. Brody/Stock, Boston.*

Grieving for a lost home   The essential role played by neighborhood social bonds in the maintenance of mental health was brought home to environmental psychologists by a study conducted in Boston's West End by Marc Fried (1963). Fried interviewed more than five hundred West End residents both before their move from the West End and after they had been relocated to make room for urban renewal. He had anticipated that residents would report short-term psychological discomfort in response to the crisis of being forced to move. He discovered instead that they experienced a severe grief reaction, similar to the grief and mourning people experience for a lost loved one. And rather than being transitory, this grief reaction persisted for a long time after residents had resettled in a new neighborhood. In fact, almost 50 percent of respondents reported feeling sadness or depression up to a year after the move, and 25 percent of residents still felt sad or depressed two years after relocation.

Fried interprets this grief reaction as a response to the loss of the stable neighborhood social networks that had been a central ingredient in residents' community life. Dislocation from the old neighborhood had fragmented the established networks of easily accessible and familiar interpersonal contacts that were essential to the social life of the West End. Fried found that severe grief reactions were most common among residents who had the most positive feelings

——The Personal Costs of Relocation——

Marc Fried (1963) describes the personal costs of relocation incurred by one West End family—Mr. and Mrs. Figella—who were forced to leave their home when the West End was demolished to make way for urban renewal. The negative personal effects of relocation experienced by the Figellas were especially remarkable because they were able to buy a house in a Boston suburb. Mr. Figella kept his old job as a manual laborer after the move, and the family expressed satisfaction with the physical arrangements of their new home. What was lost was the old West End neighborhood itself, where both Mr. and Mrs. Figella had been born and raised.

When Fried interviewed the Figellas after their forced relocation from the West End, he discovered that the move had had severe psychological effects on the family. They expressed a strong sense of grief over the loss of their West End home. They were most acutely distressed by the loss of their friends in the West End, and by the difficulty in maintaining easy access to their relatives, who no longer lived in the immediate neighborhood. When Mrs. Figella was asked what she disliked about her new home, she replied, "It's in Arlington and I want to be in the West End" (p. 162). She spoke of the West End as "a wonderful place," where "the people are friendly" (p. 161). Mr. Figella remarked, "I come home from work and that's it. I just plant myself in the house" (p. 162). Fried concludes that despite the satisfactory physical arrangements in their new home, the loss of their familiar network of friends and their ready access to extended family in the West End imposed severe psychological and social costs on the Figella family.

about their West End neighbors, and most pronounced among persons who reported that their five closest friends had lived in the West End. When asked how they felt about being forced to leave the West End, typical resident responses were: "I lost all the friends I knew"; "I felt like my heart was taken out of me"; "I felt as though I had lost everything." The depth of this social loss could not be entirely remedied by a better apartment or even a home of one's own (see box, "The Personal Costs of Relocation").

THEORETICAL PERSPECTIVES ON NEIGHBORHOOD SOCIAL NETWORKS

PROPINQUITY AND FRIENDSHIP FORMATION

Implicit in our discussion of neighborhood-based social networks has been the assumption that the social relationships of central-city residents are based at least in part on their close physical proximity. In fact, the question of what role *propinquity* (proximity) plays in the development of social relationships has been the focus of considerable investigation and some controversy among environmental psychologists. In Chapter 9 we saw how spatial distances *within* settings, such as the seating arrangement in a classroom or hospital ward, can facilitate ongoing conversations between people in small groups. Here we shall consider how spatial distances *between* different settings, such as houses or apartment units, can influence the development of friendship between people in large groups or communities.

Student housing   Leon Festinger and his colleagues (Festinger, 1951; Festinger, Schachter, and Back, 1950) conducted a classic investigation of the effects of propinquity on people's choices of friends in a married-student complex at the Massachusetts Institute of Technology in the late 1940s. One complex consisted of seventeen two-story apartment buildings, with ten apartments in each building. The investigators, using a survey research strategy, interviewed all of the wives in these apartments concerning their closest friends in the complex. They discovered that close physical proximity played a dramatic role in friendship choices. Fully 65 percent of all friendships were among people who resided in the same building. Those friends who did not reside in the same building were likely to live in the same quadrangle. Especially remarkable was Festinger's finding that even within buildings the distance between apartments was closely related to friendship choices (Figure 10-5). The majority of friends (40 percent) were next-door neighbors. The percentage of friends decreased constantly as the distance between apartments increased, to 10 percent of friendships between those who lived at opposite ends of the same floor. Robert Priest and Jack Sawyer (1967) report similar findings in a study conducted in a large student residence hall at the University of Chicago. Students' liking for one another was closely related to the distance between rooms; students nearby were liked better than students whose rooms were situated farther away.

Interracial housing   Morton Deutsch and Mary Evans Collins (1951) conducted a survey study in four interracial housing projects in New York City and Newark, New Jersey, which offers further evidence that proximity influences people's choice of friends. Two of the housing projects they studied were totally integrated, with black and white families residing in adjoining apartments. Although the other

Figure 10-5

Percentage of possible friendship choices made by apartment residents, by physical distance between apartments in same building.

From L. Festinger, S. Schachter, and K. Back, Social Pressures in Informal Groups, *p. 41. Copyright 1950 by L. Festinger, S. Schachter, and K. Back. Copyright renewed 1978. Reprinted with permission of the publishers, Stanford University Press.*

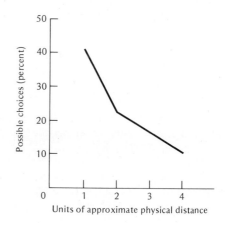

two projects had biracial populations, blacks and whites were housed in separate areas of the projects. Since the choice of whether a family was placed in an integrated or segregated project had been random, the housing situation permitted an opportunity to study the effects of residential propinquity on the development of friendships between black and white residents. Deutsch and Collins found that propinquity was a significant factor in friendship choice. Over 60 percent of the white housewives surveyed in the integrated project reported "friendly relations" with blacks, in contrast to less than 10 percent of white respondents in the segregated projects. In fact, in the segregated project, over 80 percent of the white women interviewed reported that they had had absolutely no contact with black residents.

Functional distance   We may ask at this point *how* propinquity affects friendship choices. Leon Festinger and his colleagues (Festinger, Schachter, and Back, 1950) have proposed that proximity leads to friendships because it fosters passive, unintentional contacts between people. In the process of going to and from the apartment and in working around it—carrying out the garbage, hanging out the laundry, and so on—residents accidentally meet one another, and such passive contacts are more likely to occur between people who live close to one another than between those who live farther apart.

The probability of passive contacts does not depend on spatial distance alone, but is also influenced by the *functional distance* between people. Functional distance arises from the arrangement and positioning of apartments, from paths for entering and leaving apartments, and from the sharing of common facilities, such as a stairway, a bank of mailboxes, or a laundry area. These functional connections between people increase the probability of passive contacts between them. Festinger (1951:159) presents a vivid example of the way in which functional distance can affect social relationships:

> In order to have the street appear "lived on," ten of the houses near the street had been turned so that they faced the street rather than the court area as did the other houses. This apparently small change in the direction in which a house faced had a considerable effect on the lives of the people who, by accident, happened to occupy these end houses. They had less than half as many friends in the project as did those whose houses faced the court area. The consistency of this finding left no doubt that the turning of these houses toward the street had made involuntary social isolates out of the persons who lived in them.

George Homans (1950) has developed a social-psychological theory that further helps to explain the link between proximity and liking. He suggests that the degree of liking grows as a function of increasing interaction. There are many ways in which people can reward one another, such as through material help, emotional support, or simply pleasant conversation, and these rewards are obtained most easily from people with whom we have frequent contact, such as neighbors or classmates.

Homogeneity   Although, as we have seen, some research evidence does suggest that propinquity affects friendship choices, investigation in this area has not been without controversy. While several researchers have agreed with Festinger and his colleagues that social relationships are strongly influenced by proximity (Caplow and Foreman, 1950; Ebbesen, Kjos, and Konecni, 1976; Kuper, 1953; Whyte, 1953, 1956; Rosow, 1961), others have argued that propinquity is relatively unimportant in friendship formation, and that the essential element in people's choices of friends is *homogeneity* (similarity in background and interests) (Gans, 1967; Lazarsfeld and Merton, 1954). Melvin Webber (1963) has contended that contemporary society is characterized by "community without propinquity": modern means of communication and transportation make it possible for people to maintain friendships over considerable distances with other persons with whom they share common attitudes and interests.

Locals and cosmopolitans   Recently environmental psychologists have avoided the "either-or" nature of the controversy surrounding the relative effects of propinquity and homogeneity, and have attempted instead to develop an integrative theory that can account for all of the diverse research findings in this area (Fischer, 1976). Researchers now believe that both propinquity and homogeneity play roles in the formation of friendships, and that their relative importance varies according to the characteristics of the individuals concerned. Several investigators, for example, have distinguished between what have been termed "local" and "cosmopolitan" urban lifestyles (Buttimer, 1972; Fried and Gleicher, 1961; Keller, 1968; Michelson, 1976):

> Some people are citizens of their local area, with limited horizons. Rich or poor, their interests lie within a limited range of territory. Their standards are local standards, and they evaluate others on a basis of idiosyncratic knowledge. Other people are very different. They are not bound to one place. They tune in to national or local media. They evaluate people and situations according to universal, objective criteria. The former people are locals; the latter, cosmopolitans.   [Michelson, 1976:87]

Those theorists who have drawn the local-cosmopolitan distinction have argued that the effects of propinquity on friendship formation are most pronounced among people with a local lifestyle, while homogeneity will exert stronger effects on friendship choices among people with a cosmopolitan lifestyle.

Neighboring in city and suburb   The distinction between local and cosmopolitan lifestyles helps us to understand some differences between patterns of socializing in the central city and in suburban communities. Herbert Gans (1961, 1967) has employed both survey and participant observation methods to study patterns of socializing in Levittown, New Jersey, and Park Forest, Illinois. Whyte (1956) had suggested earlier that propinquity influences friendship choice in suburbia; Gans found that friendships in these suburbs were influenced more by homogeneity than by propinquity. Some of the differences in city and suburban friendship patterns are attributable to underlying differences in social-class level. The central-city neighborhoods that have been most intensively studied have typically been working-class communities, while most suburbs today are middle-income communities. As people move up in social class, they tend to adopt more cosmopolitan lifestyles. Thus, in general, working-class residents of the central city are likely to develop local styles of socializing and to choose their friends among their neighbors, as the West Enders did, while middle-income suburbanites are more likely to exhibit a cosmopolitan style, and to choose as friends people who share their background and interests. In Levittown, New Jersey, where Gans lived for two years as a participant observer, 82 percent of the residents he interviewed reported "compatibility" as the basis for their selection of the neighbor they visited most frequently.

Gans notes that the site plan of suburban housing tends to discourage neighborly relations. In Levittown, where the streets curve, Gans found that constant, unplanned visual contact between next-door neighbors did not occur. Though Gans's data do not support the view that physical distance between neighbors affects their choice of friends, as Moos (1976) has pointed out, his findings are consistent with the notion that functional distance plays some role in friendship selection. Gans found that most socializing among neighbors in Levittown occurred between people who lived across the street from each other, and pointed out that front doors and driveways faced the street. In addition, the segregated land use that characterizes suburbia, along with the necessary heavy reliance on the automobile, strongly discourages the type of casual and passive social contacts that are typical of the central city.

Joint influences of propinquity and homogeneity   The extensive studies of social relationships in urban and suburban settings have made it clear that both propinquity and homogeneity affect people's choices of friends. As we have seen, individual differences between residents affect the relative importance of propinquity and homogeneity in individual cases. Propinquity and homogeneity serve different roles in the formation of friendship. Gans (1961) has suggested that propinquity helps people to initiate social relationships and to maintain less intensive relationships (to "be neighborly"), but propinquity alone cannot serve as a basis for deep and intimate friendships; for intensive relationships to develop, homogeneity in background and interests is essential. Thus, while proximity may initially bring residents together as casual acquaintances, only those relationships that are also homogeneous are likely to develop into lasting friendships. Gans concludes that on the basis of this view, we would expect a relationship between propinquity and friendship in a highly homogeneous setting, such as the student housing complex studied by Festinger and his colleagues.

Further evidence of the joint influences of propinquity and homogeneity on friendship choices has been provided by Robert Athanasiou and G. A. Yoshioka (1973), who studied friendship patterns among women in a housing development in Michigan. They found that friendship choices were influenced by close proximity and also by homogeneity in terms of life-cycle stage, social status, and political attitudes. Propinquity and homogeneity interacted in a complex manner. Similarity in life-cycle stage was important in the selection of friends regardless of the spatial distance between people, while similarity in social status was important only in the formation of friendships between next-door neighbors.

A THEORY OF DEFENSIBLE SPACE

Many high-rise housing projects for low-income families, instead of achieving the laudable social goals of city planners, have degenerated into places of crime and

Living with Fear in High-Rise Public Housing

One morning recently, Beatrice Smith dressed for work, then telephoned the maintenance man in her apartment building and asked him to escort her to the parking lot. When she returned home, she asked a security guard to accompany her from the parking lot to her sixth-floor apartment.

Mrs. Smith, a resident of Tyler House, a complex of high-rise apartments for low-income families, was not escorted simply because she is invalid or handicapped. Like most of her 1,300 neighbors in the massive building . . . she fears for her life.   [Whitaker, 1976:1]

This news item from the *Washington Post* goes on to note that Mrs. Smith's fears were not unfounded. In fact, police records indicate that during the last nine months of 1975, Tyler House experienced 17 burglaries, 11 larcenies, 6 stolen cars, 6 robberies, 4 rapes, 2 purse snatchings, a bomb scare, a drug arrest, and a murder. Residents throughout the building had installed heavy-duty locks on their doors. Many people refused to leave or enter the building unless they were accompanied by a family member or friend. Some residents even resorted to carrying concealed butcher knives on trips to the first-floor laundry room.

(From J. D. Whitaker, "Fear Stalks Tyler House Hallways," *Washington Post*, February 16, 1976, p. A1. Reprinted by permission.)

violence where people live in terror (see box, "Living with Fear in High-Rise Public Housing"). Shirley Angrist (1974) identified fear of crime and anxiety about physical security as major areas of concern in the lives of low-income residents of public housing projects. At a more qualitative level, Oscar Lewis's (1966) account of tenants who were moved from a slum to a housing project in Puerto Rico offers a vivid picture of the fear for personal safety that is endemic to life in a large public housing project.

Pruitt-Igoe   One of the most thoroughly investigated breakdowns of social order in high-rise public housing occurred in the Pruitt-Igoe housing project in St. Louis. The Pruitt-Igoe project consisted of approximately 40 eleven-story buildings, with a total of almost 3,000 apartments. From its inception, the project failed overwhelmingly to meet the social and psychological needs of its residents. Pruitt-Igoe became notorious for its rampant vandalism and constant sense of fear and distrust (Rainwater, 1966). The project, which had received architectural awards for its design, was razed to the ground in 1972—less than twenty years after its completion (Figure 10-6). Here is an example of an environment that was initially attractive, yet almost instantly functioned as a social slum.

William Yancy (1971) conducted a series of interviews with residents of Pruitt-Igoe in an effort to discover why the project had failed so totally as a human habitat. He found that Pruitt-Igoe lacked the cohesion, social order, and mutual support that have been found to characterize many central-city neighborhoods. When one resident was questioned about her neighbors, she lamented, "They are selfish. I've got no friends here. There's none of this door-to-door coffee business

Figure 10-6
Pruitt-Igoe's failure as a social environment led to its demolition less than twenty years
after it was built.

UPI.

of being friends here or anything like that. Down here, if you are sick you just go
to the hospital. There are no friends to help you" (p. 13). Another resident of
Pruitt-Igoe complained, "I used to watch the kids in this building. In the beginning
I tried to discipline them. I'd tell them every time I found them doing something
mischievous what was wrong and what was right. But kids don't like that; their
parents don't like it. . . . They put the blame on you. Watching children is dangerous" (p. 15).

Defensible space   When he analyzed this breakdown in natural social order and
social controls, Yancy concluded that the physical design of the Pruitt-Igoe project
had exerted an "atomizing" effect on the informal social networks that typify
many inner-city neighborhoods. At the core of the problem, he argued, was a lack
of adequate "defensible space"—physical space that is characterized by a high
level of social responsibility and personal safety. Defensible space is achieved
when the semipublic space between apartments is actively used by a large number
of residents, who assume some personal responsibility and interest in keeping the
area safe. The very presence of these persons in the area acts as a natural deterrent
to unlawful activities. This view is consistent with Jane Jacobs' (1961) contention
that if city streets are to be safe, there must be "eyes on the street," that is, the
street must be actively used by a large number of people, including residents and
shopkeepers, who are the natural proprietors of the street, and others who use the
area for a variety of reasons. Clarence Jeffery (1971) and Franklin Becker (1975)
argue similarly that planners can reduce urban crime by designing physical set-

tings that foster optimal social use and encourage residents to personalize the environment and assume responsibility for it.

Natural surveillance   Oscar Newman, in his book *Defensible Space* (1972), has developed a theory of how defensible space is achieved in urban neighborhoods (see also Chapter 8). Defensible space is found in housing environments that inhibit criminal activity by giving the impression of an underlying social order that can defend itself. Central to defensible space is the *natural surveillance* of the setting by its residents. Newman defines natural surveillance as the ability of residents to observe the public areas of the residential environment while they are engaged in their daily activities. Natural surveillance is encouraged when residential settings are designed in such a way that people cannot help viewing the common paths, entry and lobby areas, and outdoor areas as they go about their daily tasks. When natural surveillance is achieved, intruders quickly perceive that their activities are under observation by residents, and that unlawful activity will be easily recognized and dealt with. Later we shall consider some specific criteria for the design of urban high-rise housing that are consistent with a maximal degree of natural surveillance.

As we noted in Chapter 8, Newman's findings have been criticized on methodological grounds (see Adams, 1973; Kaplan, 1973); but some empirical support for the notion of defensible space has been provided by David Ley and Roman Cybriwsky (1974), who studied the distribution of stripped cars in Philadelphia's inner city. Using police crime statistics, they related the presence of stripped cars to the physical features of the immediate settings where the cars were located. Stripped cars were most frequently found at sites of institutional land use, alongside vacant buildings, and near the doorless and windowless sides of residential sites; they were rare in areas that could be observed easily from occupied resi-

Table 10-2.   Percentage of stripped cars observed at six Philadelphia settings and percentage of total street frontage occupied by each setting.

Setting	Street frontage (percent)	Stripped cars (percent)
Doorless flank (residential plus commercial use)	18.5	20.3
Institutional setting	16.7	22.5
Vacant house	7.9	20.3
Commercial setting (functioning and vacant)	14.3	6.5
Vacant/parking lot	1.6	13.0
Occupied house and apartments	41.0	18.1

Source: Adapted from D. Ley, and R. Cybriwsky, "The Spatial Ecology of Stripped Cars," *Environment and Behavior,* 6: 53–68, © 1974 Sage Publications, Beverly Hills, with permission of the Publisher.

dences (see Table 10-2). The authors conclude that structures that maximize occupant surveillance and foster local territorial control operate as the strongest environmental deterrents to crime.

Further empirical support for the relationship between urban crime and natural surveillance is provided by Patricia Bratingham and Paul Bratingham (1978), who studied the distribution of burglary rates in Tallahassee, Florida. They employed a mathematical model involving point-set topology to examine the relationship between burglary rates and particular environmental characteristics of neighborhoods. They hypothesized that natural surveillance would be greater in the interior of a residential neighborhood, where a stranger would be easily recognized, than at the neighborhood's boundaries, where strangers would be more common and less likely to be challenged. To differentiate between neighborhood interiors and boundaries, they examined the pattern of environmental variables, such as the average cost of housing, average rent, the percentage of single-family housing, and the percentages of small and large apartments. As they had predicted, the interiors of residential neighborhoods had lower burglary rates than the neighborhood boundaries. Burglary rates at the boundaries ranged from two to almost six times those at the interiors.

THEORIES OF SOCIAL SUPPORT

As we have seen, Marc Fried's (1963) research in Boston's West End demonstrated a link between residents' involvement in the local network of social relationships and their mental health. Fried's observations are consistent with those of a new and growing body of psychological research that has examined the link between social resources and mental health. Recent theoretical developments in this area have indicated that the psychological link between participation in a social network and mental health is the *social support* that involvement in a social network provides.

Support   M. Brewster Smith and Nicholas Hobbs (1966) have proposed that a breakdown in social support plays a causative role in the onset of mental illness. They argue that mental disorder is not the "private misery of an individual," but is closely tied to the failure of natural sources of social support—the extended family, the neighborhood, friends—to function effectively in the individual's life. A report by the President's Commission on Mental Health (1978) points out that people subjected to psychological stress typically turn first to informal networks of social support, including family, neighbors, and community organizations. Recently empirical evidence from a variety of sources has supported the contention that social support is positively associated with mental health. Correlational studies have shown that the more social support available to people, the fewer their symptoms of psychological disturbances (e.g., depression or anxiety) and psychosomatic illnesses (e.g., headaches or ulcers) (Holahan and Moos, 1981; Lin, Simeone, Ensel, and Kuo, 1979; Rabkin and Struening, 1976). Longitudinal research over periods of from one to five years has demonstrated that lack of social support is a good predictor of future psychological and psychosomatic complaints (Eaton, 1978; Gore, 1978; Holahan and Moos, 1981; Medalie and Goldbourt, 1976).

Recently psychologists have developed a conceptual model in which social support interacts with the overall level of stress an individual is experiencing to predict mental health. Researchers now believe that social support helps an individual to adjust to life stresses, such as moving to a new setting, suffering a financial setback, or having a serious illness in the family (see Caplan, 1974). In fact, research studies have indicated that persons who experience great stress are protected from harmful psychological consequences when the stress coincides with adequate levels of social support (Antonovsky, 1979; Caplan, 1974; Dean and Lin, 1977). From this perspective, the grief reaction that Fried observed in West Enders who were forced to leave their neighborhood is especially understandable, because the loss of their established sources of social support occurred simultaneously with the generally stressful situation of moving from a familiar setting to a new and unfamiliar residential environment.

Characteristics of support networks   Investigators have attempted to identify the features of urban social networks that provide positive social support to city residents. The *number of persons* in a support network has been shown to be an important predictor of the individual's perception of social support (Fischer, Baldassare, Gerson, Jackson, Jones, and Stueve, 1977; Wellman, Craven, Whittaker, Stevens, Shorter, Du Toit, and Baker, 1973). The *quality* of social support is also important; people obtain the greatest help from support networks that are cohesive and open to the expression of emotions (Holahan and Moos, 1980). Further research has demonstrated that *environmental factors,* such as length of residence in a neighborhood and a sense of belonging, determine the likelihood that people will have friends in their own neighborhood and the overall amount of social interaction they will have with neighbors (Biegel, Naparstek, and Khan, 1980). Finally, socially supportive urban networks are likely to be extremely *varied,* and may include family members, friends, and neighbors, as well as "natural care givers," such as the clergy, police, barbers, and bartenders (Kelly, 1964).

Mum   In the kinship network of London's East End, Mum provided a wide range of socially supportive services to her daughters (Young and Willmott, 1962). During pregnancy a woman would turn to Mum to discuss her doubts, questions, and fears. While she was in the hospital giving birth, or later during an occasional illness, Mum would be relied on to care for the house and family until the daughter recovered. When advice was needed on how to raise, train, or discipline a child, Mum was turned to rather than the free welfare clinic that was available in the neighborhood. Mum supported her daughter in countless ways, from helping her to find a new apartment to babysitting with the kids. When Mum grew old and required care herself, her daughter would return the support she had received over the years.

APPLICATIONS TO ENVIRONMENTAL PLANNING

The field of urban planning is one of the most challenging to the social and behavioral sciences. In no other public policy realm have the problems been more complex and the solutions more elusive. As we have seen, a part of the problem is to be found in some of the prevailing attitudes in regard to urban design. The assumptions of many policy makers that physically run-down sections of the central city are of necessity unhealthy human habitats and that the residents of these areas would prefer to live elsewhere have been shown to be overly simplistic.

We have seen, too, that often the well-intentioned planning decisions that have been based on such attitudes have done more harm than good to the social-psychological character of urban life. Investigators have argued that what is needed is an innovative design philosophy that is responsive to contemporary knowledge in the social and behavioral sciences and sufficiently focused to serve as a basis for practical design decisions (Gutman, 1966; Michelson, 1968; Wood, 1972). Here we shall consider some ways in which environmental psychologists' knowledge about the nature and functions of neighborhood-based social networks might be applied to urban planning. Our emphasis here will be on encouraging innovative planning and ideas that reflect a positive view of the city's social life. Admittedly, many of these ideas have not yet been tested in real urban settings. The impact of each of these ideas on the social-psychological fabric of urban life must be evaluated systematically before a final assessment is made concerning their value to efforts to solve the problems of our cities.

DESIGNING MULTIFUNCTIONAL OUTDOOR SPACES

In our discussion of the Pruitt-Igoe housing project, we found that the project failed to function effectively because of inadequate public space between apartments that could serve multiple functions. In fact, investigators have concluded that if urban residential environments are to function adequately at a social-psychological level, they must include a significant amount of multifunctional public space. Only this type of designed space provides the opportunity for the informal, casual contacts between people that are so important to social organization and control in the inner city.

Diversity   Jane Jacobs (1961) has argued that a critical challenge facing urban planners is the design of urban settings that encourage and sustain a broad *diversity* of uses by their residents. She believes that four conditions are indispensable to diversity in a city's streets and districts. First, the setting must serve *multiple functions*, so that people can use the same space for different purposes on different schedules. The "sidewalk ballet" on Hudson Street, for example, depended on a mixture of uses to sustain its movements, including the active street games of youngsters, the diverse daily chores of neighborhood residents, and the ongoing tasks and involvements of people who worked in the neighborhood.

Secondly, urban districts need *short blocks*, where opportunities to turn corners are frequent. When city blocks are long, residents are generally restricted to a single efficient path from one point to another; short blocks permit a variety of equally convenient paths between the same points (Figure 10-7). The availability of alternative routes, in turn, permits a mingling of the paths of many people who would otherwise be functionally isolated from one another.

Third, diversity requires a *mixture of buildings* that vary in age and condition. Jacobs emphasizes the importance of having some old buildings, where low and

Figure 10-7

While long city blocks (*top*) leave only one efficient route between two points, short blocks (*bottom*) permit a variety of equally efficient routes.

From J. Jacobs, The Death and Life of Great American Cities, *pp. 179, 181. Copyright 1961. Reprinted by permission of Random House, Inc.*

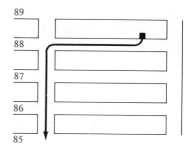

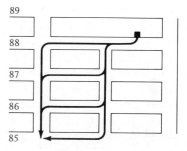

moderately priced commercial space is available. If an urban district has only new buildings, the high costs of new construction drive rents so high—often several times the cost of space in older buildings—that the only establishments that can afford to occupy them are chain stores, banks, and chain restaurants. The small shops and restaurants that are essential to the safety and public life of city neighborhoods can survive only with the low overhead of older buildings.

Finally, diversity requires a sufficiently dense *concentration of people.* Jacobs has drawn an important distinction between *outside density* (the number of dwelling units per acre) and *inside density* (the number of persons per dwelling unit). She proposes that high outside density is necessary to support both the amenities of urban life and the informal social life that characterizes the central city. William Michelson (1976) reached a similar conclusion after he reviewed the findings of research conducted in Boston's West End and London's East End. He suggests that the active social life of those areas depended on the easy physical access among residents afforded by high outside density.

In a similar vein, Christopher Alexander (1972) argues that cities should be designed in a way that recognizes and supports their natural diversity. While the city's functions naturally involve overlapping activities, urban designers have typically attempted to arrange cities in such a way that overlapping functions are ignored. The natural diversity, complexity, and richness of the city has been crippled by this kind of design. To illustrate his argument, Alexander points to urban college campuses isolated from the surrounding city environment. The typical isolated American campus is inconsistent with the flow and pattern of university life. Characteristically, student apartments and commercial establishments are interspersed in the area near the university; surrounding restaurants, bookstores, and movies always contain an intermingling of university and city life; and whole departments, such as a medical school with a hospital, may be actively involved in the life of the city.

Project spaces   On the basis of his observations of street life in New York City's Lower East Side, Holahan (1976b) has advanced several design suggestions to encourage behavioral diversity in the open spaces that surround housing projects. Even an aesthetically attractive environment is not sufficient to encourage outdoor socializing among adults unless it is able to support a range of activities. He suggests a design solution that would integrate the project with the natural patterns of urban life in the surrounding city environment.

Such a design philosophy might include both mixed functional uses of the outdoor space and appropriate design features to facilitate socializing. For example, the design of project space might encourage mixed uses—recreational and leisure activities, consumer behaviors, and task-oriented activities—that would attract project residents to use the available outdoor space. Then innovative design features in the outdoor setting, such as nooks, benches, tables, and recreational equipment, might be added to support the informal social contact likely to occur among residents who meet accidentally while pursuing diversified tasks in such multifunctional space.

Urban plazas   William H. Whyte (1972) spent two years observing how people used open-space plazas in New York City. He augmented his observations with

Figure 10-8
Many old city plazas have an active and varied social life.

© *Tyrone Hall/Stock, Boston.*

time-lapse and telephotography. Whyte concludes that most modern plazas in the city are underused because they are poorly designed as social settings. Such underuse of space tends to nullify the rich social life of the city and to attract street crime.

After observing older city plazas that functioned well as social settings (Figure 10-8), Whyte has offered suggestions for the design of new plazas that would function effectively as social settings. The most important factor in plaza use, Whyte believes, is "sittability." The best places to sit are often the simplest—ledges and steps, when they are broad enough. The most social plazas are those that are built in areas where they attract a varied clientele, including nearby residents, tourists, and office workers. Finally, in contrast to the prevailing view of many city officials, Whyte argues that the presence of sidewalk food vendors is

Innovative Design on the Streets of New York

Innovative planning ideas, such as those of Jane Jacobs and William H. Whyte, are beginning to influence actual urban planning decisions. In 1972 the New York City Planning Commission established a special Urban Design Group that was specifically interested in applying innovative, humane, and socially aware planning strategies in urban design. The Urban Design Group has been particularly concerned to encourage the design of socially effective plazas of the type discussed by Whyte. Wolf Von Eckardt (1978:26) has described how this new design philosophy works in New York:

> The new spirit of Jane Jacobism in New York is not just a matter of fashionable old house restoration and urban chic. . . . For old neighborhoods and new skyscrapers alike, it means finding the ingredients of healthy and pleasant city life and strengthening or creating them. It is done with zoning and building regulations and the bonuses a city can give a builder: We give you five rentable stories of office space more than the zoning calls for, if you give us a nice plaza in front of the building.

One example of the kind of positive difference the Urban Design Group can make may be seen in the neighborhood known as Little Italy. There the Urban Design Group worked with the local citizen group to draw up special legislation oriented toward preserving the unique ambience and charm of the neighborhood environment. The new regulations encouraged the design of new residential and commercial sites whose street fronts and building materials conformed to the existing neighborhood pattern. Further, developers who contributed to the improvement of neighborhood parks were permitted additional usable space. The arrangement proved a happy one for both the neighborhood and the developers. Fifteen new restaurants opened in the neighborhood in the first two years after the new regulations were established.

(From Wolf Von Eckardt, "Wolf Von Eckardt on Architecture: Double Manhattan," The New Republic, January 21, 1978, p. 26. Reprinted by permission of The New Republic, © 1978 The New Republic, Inc.)

essential to a plaza's social life (see box, "Innovative Design on the Streets of New York").

HUMANIZING THE DESIGN OF HIGH-RISE HOUSING

Despite the frequent failure of high-rise housing to function effectively as a social setting, the realities of contemporary architectural practice force us to recognize that such housing will continue to be built for the foreseeable future. The social-psychological problems associated with high-rise housing are not the result of high-rise living per se, but rather of the particular arrangement of space—especially the space between apartments—that has typified most high-rise design.

Semipublic space   William Yancy's (1971) analysis of the failure of the architectural design of the Pruitt-Igoe housing project to meet the social-psychological needs of its residents serves as a cautionary tale for the designers of high-rise housing for low-income residents. Yancy strongly criticizes the philosophy that encouraged a minimum of "wasted space," that is, semipublic space within Pruitt-Igoe's buildings and outside of individual apartments. An initial review of the Pruitt-Igoe architecture (*Architectural Forum*, 1951) praised the architects for limiting semipublic space between apartments. Yancy, however, argues for the reverse design philosophy. He suggests that the designers of multifamily housing for low-income residents, rather than seeing space between apartments as "wasted," should recognize that such space is appropriate and necessary. Such semipublic space and facilities offer areas that smaller and identifiable groups of residents can organize with a sense of personal "turf." Designers should minimize space in the residential setting that belongs to no one, and foster the development of spaces between dwelling units over which groups of residents will share a feeling of informal control.

A study of public housing in Baltimore (Wilner, Walkley, Pinkerton, and Tayback, 1962) demonstrates that housing that is designed with adequate semipublic space and common facilities can foster neighborliness, informal visiting, and mutual helping. Patrick Mullins and J. H. Robb (1977) found favorable resident reactions to a public housing project in New Zealand, which emphasized easy contact with neighbors and the localization of friends and kin. Lee Rainwater (1966) has proposed that public housing projects be designed so that the doors of several apartments open onto a common hallway, where common use of the shared space would foster positive social attitudes among residents as they cope with common problems. Daniel Amick and Frederick Kviz (1974) have shown that residents' sense of social isolation in high-rise housing is a product of the restricted interaction between individuals that occurs when the amount of public space between dwellings is inadequate.

Designing for defensible space   Oscar Newman (1972) has advanced some specific criteria for the design of urban high-rise housing based on the apparent relationship between defensible space (space that allows a maximum degree of natural surveillance) and the prevention of urban crime. First, on the basis of crime statistics, he strongly discourages the building of housing more than six stories high. Second, he encourages clustering project buildings in small building groups in order to divide the project into smaller and socially more manageable zones. Third, he suggests that corridor space within buildings be divided into distinct zones, each with its own small cluster of apartments. Unlocked swinging doors placed at strategic points along a corridor would serve the purpose. Finally, to optimize natural surveillance, he proposes that buildings be designed so that their lobbies face toward rather than away from the surrounding city streets, and so that several apartments are located near the lobby. The people who live in those apartments will be ideally situated to oversee activity in the lobby.

Elizabeth Wood (1962) has put forward a series of similar recommendations to designers of high-rise housing projects for low-income residents. She proposes:

1. Buildings should be designed to facilitate the visual identification of a family and its dwelling.
2. The physical design should encourage association and loitering on the building floor, in the lobbies, and on the grounds.
3. The design should also encourage the formation of informal social groups, such as by including shared facilities and well-planned outdoor benches.

Wood offers a few specific design proposals: Exterior or "balcony" corridors, when they are sufficiently wide, encourage a broader range of activities than interior corridors. In order to encourage social activities, building lobbies might include sitting areas and tables where residents could play chess, checkers, and other games. Finally, play-sitting areas for parents and preschool children should be built next to residential buildings, ideally in the form of paved areas just off the main sidewalk and close to the building entrance.

REDUCING THE PERSONAL COSTS OF URBAN RENEWAL

We have seen that the forced relocation that has been associated with urban renewal can lead to serious psychological consequences. The issue involved here is moral and political as well as psychological. Peter Marris (1963) has pointed out that urban renewal typically displaces the poorest of the city's population—cultural and ethnic minorities and immigrant groups. Because relocated families are usually poor, they cannot afford the new housing that replaces their old homes. Most relocated families move to neighborhoods that are similar to the neighborhood that was cleared, to live in substandard housing, and usually at higher rents.

Chester Hartman (1963, 1967, 1971, 1975) has addressed considerable attention over more than a decade to the human impact of the forced relocation associated with urban renewal. In a scholarly critique of federal housing policy, he concludes:

> . . . urban renewal . . . might more properly be labeled a de-housing program. This program was introduced in the 1949 Housing Act as "slum clearance," but was taken over at the local level by those who wished to reclaim urban land occupied by the poor for commercial, industrial, civic, and upper-income residential uses. Over half a million households, two-thirds of them nonwhite and virtually all in the lower income categories, have been forcibly uprooted. A substantial percentage of these persons were moved to substandard and overcrowded conditions and into areas scheduled for future clearance, at a cost of considerable personal and social disruption.   [Hartman, 1975:107]

Thus an area in which knowledge gleaned from research in environmental psychology is needed is planning for the relocation of people displaced by urban renewal.

Relocation   Since forced relocation presents a serious psychological threat to a resident's sense of group identity, we may encourage patterns of relocation that are oriented toward maintaining and reinforcing the relocatee's feeling of social identity. Young and Willmott (1957) encouraged planners to attempt to move residents one block at a time, as established social groups, to avoid destroying the social cohesiveness of the central city. Counseling for relocated families might be incorporated in the urban renewal program, in order to provide advice and suggestions on how best to cope with the socially disruptive forces of relocation. Marc Fried (1961) has offered some information relevant to such a counseling program. He identified three ways in which residents of the West End were able to cope effectively with relocation: (1) they made an effort to remain identified with the old neighborhood by relocating as close as possible to the West End; (2) they tried to maintain established social bonds by resettling among members of the extended family; and (3) they became increasingly involved in those social roles that did not change with relocation, such as the roles of spouse and parent.

Restricted demolition   A further way to diminish the negative social consequences of relocation might involve more restricted and controlled demolition of the established neighborhood. Fried (1961) has proposed that rather than demolishing an entire neighborhood at one time, planners might redevelop small parcels of the neighborhood in sequence. This strategy would help the neighborhood to retain its identity during the demolition period, and would allow relocated residents to resettle more easily within the old neighborhood.

Renovation   Chester Hartman (1975) argues that as an alternative to the demolition and rebuilding of inner-city neighborhoods, more effort might be directed toward rehabilitating the already existing housing. This strategy would offer residents the option of remaining in the inner city in decent housing conditions. More important, displacement and relocation stresses could be avoided, and a continued sense of neighborhood and community identity could be maintained. Hartman adds, however, that renovation typically requires more time and more money than planners anticipate, and that for a housing rehabilitation program to work, a range of carefully planned government aids and controls will be required. Similarly, D'Ann Swanson and her colleagues (Swanson, Swanson, and Dukes, 1980) encourage the development and systematic evaluation of publicly sponsored programs oriented toward neighborhood rehabilitation and preservation in an effort to minimize the human costs of forced relocation.

CITIZEN PARTICIPATION IN URBAN PLANNING

Some of the social-psychological costs of urban renewal, such as a sense of powerlessness and alienation, might be alleviated if residents were permitted to have some influence on the planning decisions that affect their neighborhoods. Citizens' perceptions of the impacts of planned changes on their lives can be of value in the evaluation of innovative and experimental urban planning programs. Since the 1960s we have seen important developments toward this end in planning strategies that have encouraged citizen participation in the development and conduct of urban renewal (Arnstein, 1969; Burke, 1968; Wilson, 1963). J. Douglas Porteous (1977) has suggested that on practical grounds citizen participation in urban renewal planning is likely to grow in the future because of increasing pressure from both citizens' groups and federal agencies. Urban decision makers are also becoming more aware that people do, after all, have a right to participate in the making of decisions that affect their lives.

Simulation games Some urban planning strategies that have evolved in recent years have been directed toward increased citizen participation in the planning process. One of the most interesting approaches to citizen participation is *simula-*

Simulating Urban Decision Making

A recent book, *Psych City: A Simulated Community* (Cohen, McManus, Fox, and Kastelnik, 1973), provides a simulation game appropriate for use by schools, community groups, and public agencies. The purpose of the game is to increase players' awareness of the political, psychological, and social forces that shape planning decisions. By actively playing the part of one of the individuals involved in the drama of urban change, such as a member of the city council, the president of the Chamber of Commerce, or a representative of a minority-group organization, participants learn how such people might think, feel, and behave in the planning situation.

Players may be required, for instance, to make a planning decision about the construction of low-income housing in the city. The problem is complex because the proposed housing would be built in an area of the city now occupied by many politically influential middle-income residents, who oppose low-income housing in their neighborhood. The situation is further complicated by the federal government's willingness to provide substantial funds for the new housing, as well as total funding for a desired downtown civic center, *only* if the low-income housing is constructed in the middle-income area of town.

Participants in the simulation game are asked to resolve the housing issue in a town meeting format, chaired by the mayor. Each player is assigned a particular role, and that player's words, actions, and viewpoints at the meeting reflect the role as the game defines it. One assigned role is that of a city councilmember from the middle-income area, who is defined as sharing his constituents' concern about the low-income housing but as willing to compromise on some issues. After participating in such a simulation game, players often report a broadened appreciation of the complexities of urban decision making and of the different needs and perceptions of various urban groups. We may hope that these increased sensitivities generalize to players' home communities, allowing them to play a more active and constructive role in their own community government.

tion gaming, in which players assume a range of planning roles in the development of a model city. A unique opportunity to learn and to broaden self-awareness occurs when urban designers and citizens are asked to reverse roles (Porteous, 1977). Jo Hasell and her colleagues (Hasell, Scavo, and Moore, 1980) discuss a simulation game that fostered the development of a productive partnership among representatives of local businesses, financial institutions, local government, and neighborhood organizations in the revitalization of targeted neighborhoods. Thomas Eisemon (1975) has described a simulation technique developed specifically to enable the residents of public housing projects to participate in actual housing plans (see also box, "Simulating Urban Decision Making").

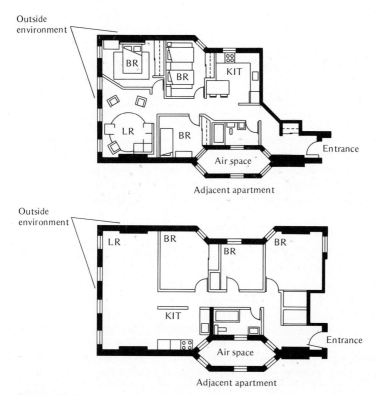

Figure 10-9
Contrasting design plans for a three-bedroom apartment for low-income Puerto Rican families. The upper plan incorporates information from observations of and discussions with future residents; the lower plan reflects the middle-class architect's preconceptions.

From J. Zeisel, "Symbolic Meaning of Space and the Physical Dimension of Social Relations: A Case Study of Sociological Research as the Basis of Architectural Planning," in J. Walton and D. Carns (eds.), Cities in Change: Studies on the Urban Condition. *Boston: Allyn & Bacon, 1973, pp. 258, 260. Reprinted with permission.*

Cultural sensitivity   Citizen participation in the urban planning process is of particular importance when a designer must choose among design alternatives in planning for a user group whose cultural or social-class background is different from the designer's. John Zeisel and Brent Brolin (Brolin and Zeisel, 1968; Zeisel, 1971, 1973) have demonstrated that information based on the views and needs of the particular user group can be effectively used in the design of multifamily housing for low-income urban residents. In one study they systematically applied more than 200 observations of behavior derived from Gans's *Urban Villagers* (1962) in developing guidelines for architects of multifamily housing for a similar population. In a later study, Zeisel designed two tenement buildings for Puerto Rican families in New York City on the basis of observations and discussions with the future residents.

For these Puerto Rican families, the kitchen tended to serve as a "turnstile" for people entering and leaving the apartment, with the person in the kitchen acting as the controller of entrances and exits. Zeisel designed the new setting so that the person in the kitchen would be the first to see anyone who entered the apartment and the last to see anyone who left. For the Puerto Rican residents, the kitchen and living room differed qualitatively as well as functionally; the kitchen was used for informal gatherings, while the living room served more formal functions. For these reasons, Zeisel clearly separated the kitchen and living room in his design. Finally, he found that the families used apartment windows for observation of and communication with ongoing street activities. And so he arranged the new apartment plan so that both the kitchen and living room had windows facing the street. Figure 10-9 contrasts Zeisel's new plan for a three-bedroom apartment with that of a middle-class architect whose design plan reflects his own preconceptions. The architect's plan allows visitors to pass the bedroom area before reaching the kitchen, has the kitchen and living room sharing an open space, and has no kitchen window facing the street.

Porteous (1977) has cautioned that, while citizen participation strategies offer hope of a more positive and fruitful role for urban residents in the planning process, some obstacles need to be overcome before citizen participation can achieve its social objectives. First, citizen participation can be unwieldy and time-consuming. In addition, identifying the "public" can be an almost impossible task, and frequently only the best organized and most articulate citizen groups are heard from. Porteous concludes that planners and decision makers must be committed to more than token participation; citizens must be permitted input at the time alternative plans are decided on, rather than simply being asked to adjust to a plan that has already been formulated.

SUMMARY

Environmental psychologists have discovered that many central-city environments are characterized by a more socially and psychologically positive social life than many people used to think. The social life of working-class urban neighborhoods is founded largely on the ties of kinship. Research in Boston's

West End showed the central role in West Enders' lives of the "peer group society"—typically relatives of the same sex and age group. In London's East End, research pointed to the essential role of horizontally structured kinship relations structured around mother-daughter relationships over three generations. Additional research, primarily in the West End, demonstrated that patterns of affiliation between neighbors also provide a vital element in the social life of the central city.

One of the reasons that social scientists initially failed to appreciate the richness and depth of social relationships in the urban environment was that much of the city's social life is informal. Social life in working-class neighborhoods consists largely of casual social contacts between people who meet accidentally but predictably in public, open-space settings. The physical features of the inner city that encourage its vital social character include low-rise tenement houses, a mixture of residential and commercial settings, and narrow streets. These findings indicate that we must distinguish between a *physical* slum (a district that is physically deteriorated) and a *social* slum (an area that is debilitated at a social-psychological level). The rich diversity of the city's social life has necessitated a varied and flexible battery of research strategies, including naturalistic observation, survey research, and participant observation.

A major function of neighborhood-based social networks in the inner city is to maintain social organization and control. Participant observers have discovered that inner-city neighborhoods can have a high degree of internal social organization, including a well-defined status hierarchy, shared rules of personal conduct, and enduring social traditions. The social order of such neighborhoods may be unstable, however, as a result of social mobility, shifting land uses, and underemployment.

Another social-psychological function of neighborhood social networks is to provide a foundation for the individual's sense of personal identity. The peer group provides individuals with an opportunity to express their personality, their unique character, and a sense of identity separate from that of the group. The neighborhood social network affords residents a broader setting they can call home. The central-city environment thus provides the locus for the individual's sense of group and spatial identity. The neighborhood network also forms the social context out of which the individual's personal values are derived, and helps the urbanite to maintain a positive psychological adjustment. People who are integrated into such a network have been found to have fewer symptoms of psychological disturbance than individuals who are socially isolated. Residents forced to relocate from the social network of the West End during urban renewal showed a severe grief reaction, including feelings of loss and depression, and psychosomatic complaints.

Theoretically oriented research has revealed that people's choices of friends are influenced by both proximity and homogeneity in background and interests. Proximity fosters passive, unintended contacts between people that tend to be mutually rewarding. Researchers have found that friendships tend to be based on propinquity when people have a "local" lifestyle, while homogeneity is more important to people who have a "cosmopolitan" lifestyle. Propinquity helps people to initiate social relationships and to maintain less intensive relationships, while homogeneity is essential for intimate relationships.

Environmental psychologists have theorized that the central reason for the breakdown of social order in many public housing projects has been a lack of adequate "defensible space." Defensible space is achieved when the space between apartments is actively used by a large number of residents, who assume personal responsibility for the area. Also essential to defensible space is the "natural surveillance" that occurs when residents can readily observe the public areas in a setting while they go about their daily activities.

The theoretical link between involvement in the neighborhood social network and mental health is the support that social relationships provide the individual. Family, friends, and neighbors are the first source of support sought in times of emotional crisis. Researchers believe that social support serves a particularly important function in enabling the individual to adjust in the face of highly stressful life events.

Environmental psychologists' knowledge of the social-psychological importance of neighborhood-based social networks can be applied to the solution of many planning problems. Environmental psychologists have demonstrated that public spaces in urban areas should be designed in such a way that they encourage and support multiple activities; that the design of high-rise urban housing must be humanized; that urban renewal programs must be more responsive to the social-psychological needs of central-city residents; and that citizens can and should participate in urban planning.

11 Environment and Behavior: A Unifying Framework

W e have seen that the physical environment plays a considerable role in the shaping of human affairs. Our behavior and experience are influenced by many aspects of the interior environments in which we live, work, and learn. Light, sound, temperature, amount of space, privacy, and territory influence the daily activities we pursue in designed settings. The physical structure and layout of architectural settings affect the nature and quality of the personal and social functions they house. Further, the broader physical contexts of human activities—neighborhoods, cities, natural landscapes—influence the character of human affairs.

We have seen, too, that the influences of the physical environment on our behavior and experience are diverse and far-reaching. Our perception of the world around us, our attitudes toward it, and our thoughts and ideas about the places where we live are all influenced by the physical settings of our lives. Our performance at school and on the job, our ability to cope with life's demands, and the quality of our relationships with other persons are also affected by the designed and natural environments that provide the stage for every human activity.

In sum, the influence of the physical environment on our lives is immense. In effect, all of our undertakings are shaped by the character of the physical settings—some immediate, others more remote—in which they take place. The interrelationship of environment and behavior is so complex that we must bring together the varied elements that make up the field of environmental psychology so that they can be seen to form an integrated picture.

AN ADAPTATIONAL VIEWPOINT

We saw in Chapter 1 that the field of environmental psychology has an *adaptational focus*. Environmental psychologists are especially interested in learning about the varied and complex *processes of adaptation* by which people cope with the demands of the physical environment. They take a *holistic* view of both the environment and the individual. An adaptational orientation envisions the individual as an *active* and dynamic participant in the process of coping with the environment.

The adaptational focus has been implicit throughout this book. Our discussion has been organized around the central psychological processes by which people cope with the environment—perception, cognition, attitudes, performance, problem-focused and emotion-focused coping, and dyadic boundary processes. So far as possible, we have taken a holistic view of human behavior and experience in dealing with the environment. When we discussed performance in learning and work environments, for example, we envisioned human performance as an integrated psychological and behavioral process. Similarly, in our analysis of environmental cognition, we viewed human information processing as a complex, integrated system.

Our review of the findings of environmental psychology has been based on an assumption that the person's role in the environment-behavior equation is an active and vital one. When we examined crowding and ambient environmental stressors, for instance, we emphasized the ability of a sense of personal control to mitigate the negative consequences of environmental stress. Finally, in exploring ways that research knowledge might be applied to environmental planning, we have repeatedly emphasized the importance of environmental users' inputs in the design process.

At the same time, however, the task of accumulating and organizing the vast amount of information in the field of environmental psychology has often made it necessary to emphasize discrete findings at the expense of a broader perspective. My purpose here is to adopt a broad view of environmental psychology, drawing explicitly on the adaptational viewpoint as a conceptual framework. In this chapter we shall focus on three characteristics of the adaptational viewpoint that are relevant to the goal of developing a unifying framework for understanding the relationship between environment and behavior.

We shall consider first the adaptational perspective's *holistic* model of the environment as a series of encircling contexts. This model will serve to draw together the various aspects of the environment we have considered in earlier chapters. We shall then discuss the adaptational view of the individual as a *total person* in relation to the environment. This aspect of the adaptational orientation will serve to unify the diverse psychological processes by which people deal with the physical environment. Finally, we shall view the person and the environment as involved in a *transactional* relationship. This transactional view will serve to link the environmental and personal variables we have considered in a single, comprehensive framework.

A HOLISTIC MODEL OF THE ENVIRONMENT

In considering the influences of the physical environment on human behavior and experience throughout this book, we have focused on a variety of features of the physical environment—interior design, the structure and configuration of buildings, geographic regions the size of a neighborhood or a whole city. Throughout this enterprise we have studied each level or aspect of the environment apart from other features of the environment. When we considered architectural interiors, we were unconcerned with building configuration, and when we examined buildings, we did not reflect on the surrounding neighborhood or regional setting. This approach has facilitated the task of accumulating and organizing the vast amount of information in this field. At the same time, however, a separate discussion of each feature of the environment may have conveyed the impression that each of these aspects operates independently of and in isolation from all the others. In fact, the various aspects of the environment exert overlapping, simultaneous, and interrelated influences on people's behavior. And a full understanding of how the physical environment shapes human behavior requires an appreciation of the fact that physical settings themselves are embedded in a broad social and cultural context. Here we shall consider a holistic model of the environment that incorporates the simultaneous influences of all aspects of the physical environment along with their social and cultural contexts.

Encircling contexts   Urie Bronfenbrenner (1976, 1977) has proposed a conceptual framework that is especially appropriate to our analysis. He envisions the environment that affects the functioning of an individual as an arrangement of *encircling contexts* that surround the individual. They may be thought of as rings that encompass smaller contexts and that are themselves surrounded by larger contexts. The *microsystem* consists of the immediate physical settings within which the person functions, such as the home, the school, and the workplace. The *exosystem* consists of the broader social structures, both formal and informal, that encompass the immediate settings in which the person functions—the neighborhood setting, agencies of government, communication and transportation facilities. The *macrosystem* is somewhat more abstract than the other systems, and represents the overarching cultural and subcultural patterns of which the microsystem and exosystem are concrete manifestations. For example, the macrosystem includes a cultural blueprint of a school classroom, so that classrooms within a particular culture tend to look and function in similar ways.

Bronfenbrenner's model may be adapted for our purposes as in Figure 11-1. First, the physical environment itself may be envisioned as a series of encircling contexts. For example, the most immediate aspect of the physical environment that affects an individual is the *interior design* of an architectural setting, such as room color, furnishings, and temperature. Next is the structure of the *building*, which forms the physical shell for the various aspects of the interior design. The building and interior design are some-

Figure 11-1
A holistic model of the environment.

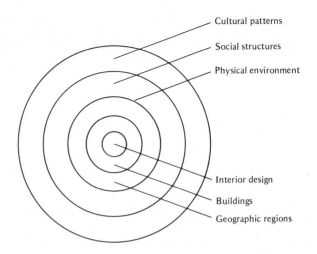

Cultural patterns

Social structures

Physical environment

Interior design

Buildings

Geographic regions

what similar to Bronfenbrenner's microsystem. At the next level, the building is encompassed by a broader *geographic region*, such as an urban district or a natural landscape.

All of these aspects of the physical environment are in turn surrounded by formal and informal *social structures*, such as organizational policies, a neighborhood association, or an extended family. These informal social structures are comparable to Bronfenbrenner's exosystem. Finally, the physical and social environments are encompassed by the *cultural* and subcultural *patterns* that dictate personal values and lifestyles, architectural trends, and the forms of social groups and organizations within particular societies. These cultural patterns are similar to Bronfenbrenner's macrosystem.

Let us consider an example of the simultaneous operation of these physical, social, and cultural domains of the total environment. A student studying in a college library may be affected simultaneously by the level of lighting in the study area (interior design), distracting noise from the floor above (building design), and the temptation to take off for the nearby hills (geograpic region). The student may also be responding to a stringent examination system at the university (social structure) and to a stereotype of the model college student (cultural pattern). The actual study behavior the student engages in reflects the complex interplay of these various influences.

The picture of the environment that emerges from this holistic model is complex. The model does, however, discourage a simplistic conception of environmental psychology that might suggest, quite incorrectly, that the influence of physical settings on people's behavior can be readily understood in terms of a simple causal model in which single causes produce single effects. As Rudolf Moos (1973, 1976) has argued, an adequate conceptualization of the role of the environment as a shaper of human behavior needs to take account of a broad array of organizational and societal variables as well as physical ones.

Although research in environmental psychology has tended to focus on physical environmental variables abstracted from the broader sociocultural context, some environmental psychologists have pointed to the need to incorporate the social and cultural levels of analysis in investigations of the human effects of the physical environment. Robert Bechtel (1976) has suggested that efforts to assess people's perceptions of the quality of the natural environment (e.g., perceptions of air pollution or water purity) should take account of the overlapping influences of sociocultural factors and the built environment. Similarly, Moos (1980) proposes an overall conceptual framework for understanding the behavior and experience of elderly residents of sheltered-care settings that accounts for the simultaneous effects of architectural features, institutional and organizational factors, and characteristics of the social environment.

THE TOTAL PERSON

In our review of the field of environmental psychology, we have examined a variety of psychological processes that characterize the individual's stance toward the physical environment. We have discussed how people perceive, evaluate, and form mental images of the physical world; how individuals function in normal settings and cope with stressful ones; and how people in social groups maintain personal space, achieve privacy and control territory, and engage in social behavior. In the course of our discussion of environment and behavior we have tended to focus on each of these psychological processes separately. For example, when we examined environmental perception, we were not concerned with how the perception of an area might affect the level of privacy an individual seeks to achieve there. When we discussed territoriality, we did not consider how territorial behavior in a setting might be altered by a person's mental map of that area. While this approach has facilitated discussion of each of these complex psychological processes, it may also have suggested that each process functions independently of the others. In fact, the psychological processes by which people deal with the physical environment are highly interrelated and function in a mutually supportive and complementary fashion.

THE PERSONALITY PARADIGM

The adaptational focus of environmental psychology emphasizes that the person functions as a *total, integrated entity* in the environment-behavior equation. In this respect, the adaptational view is similar to what Kenneth Craik (1976, 1977) has termed the "personality paradigm" in environmental psychology. The personality paradigm takes the whole person as the basic unit of analysis. In contrast to many schools of thought that have focused on isolated psychological processes, the personality paradigm is concerned with the total person as a *dynamically organized entity*. The personality approach recognizes that many complex and changing psychological processes are involved in person-environment relations, but stresses the unity and integration the functioning person brings to these varied processes.

A MODEL OF THE TOTAL PERSON

We may adapt this view of the total person to our model of environment and behavior, as in Figure 11-2. Information impinging on the individual from the environment is received, evaluated, and encoded by an interlocking network of psychological processes.

Figure 11-2
A model of the total person.

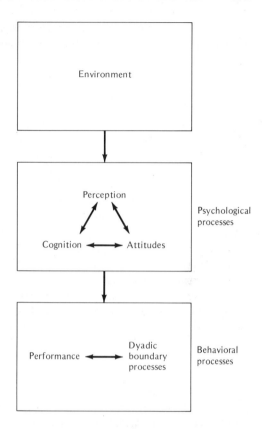

These processes include environmental perception, the development of environmental representations, and the formation of environmental attitudes. This "processed" environmental information then becomes the basis for the individual's decisions about how, when, and where to act in regard to the environment. The actions the individual engages in, whether individual behaviors (e.g., environmental performance) or social behaviors (e.g., behaviors involving personal space, privacy, territoriality, or affiliation) also operate as an interrelated system.

For example, an individual searching for a new house may see a home for sale (environmental perception) that he or she feels is attractively designed (environmental attitude) and located close to shopping, transportation, and recreational facilities (environmental cognition). On the basis of the interplay of these sources of information, the individual may go on to examine the house for structural strength, quality of insulation, and adequacy of electrical wiring (environmental performance). The individual may also call a friend for advice (social support), and seek out a place where they can talk confidentially (privacy). This complex sequence of interrelated environmental processes proceeds as a patterned and integrated whole, allowing the individual to cope effectively with the challenging task of purchasing a new home. The one-way direction of the arrow in Figure 11-2 between the psychological and behavioral processes is an overly simplified representation. For instance, on the basis of new information gleaned from an examination of the house or from the discussion with the friend, the individual might decide to take another look at the house and to reevaluate it, or to look at more houses before making a decision.

The model of the environment-behavior system that emerges from a consideration of the person as a total entity is complex. Yet this model is essential to an environmental psychology that strives to represent the way real people function in the real world. When our investigations focus on real people coping with the awesome challenge of today's environment and when the prospects of meaningful social application are so great, we must not allow overly simplistic methods to obscure the complex human processes under study. As Allan Wicker (1972) has noted, our research problems must dictate the strategies and measurement instruments we select, and not the reverse. Of course, this concern for external validity must be balanced by an effort to achieve a level of internal validity that enables us to draw sound inferences about the variables under study.

The model in Figure 11-2 bears some similarity to the ideas of some investigators we considered earlier. In our discussion of environmental cognition we considered Herbert Leff's (1978) model of information processing as a complex, integrated system (see Figure 3-8). In our analysis of performance in learning and work environments, we reviewed Ernest McCormick's (1976) model of human performance as an integrated psychological and behavioral process (see Figure 5-1). Both Leff and McCormick emphasize that an integrated pattern of psychological processes, including perceptual and cognitive components, form the basis for an individual's behavior in the environment.

Irwin Altman (1975, 1976, 1977) has argued cogently for a research orientation that is responsive to the complexity of the totally functioning person. Altman contends that traditional psychological research strategies, which have focused on isolated responses, have assumed a nonintegrated person whose various behaviors are unrelated to one another. He argues instead for research techniques that reflect the interrelationships among various behavior modalities, the multilevel nature of behavior, and complex patterns of simultaneously occurring activities. Many psychological processes occur simultaneously (e.g., perception, cognition, and emotional reactions) and are translated into several levels of overt behavior (e.g., territoriality, personal space, and verbal expressions).

Altman and Dalmas Taylor (1973) have proposed the notion of "social penetration" as an organizing conceptual framework for understanding how social relationships progress through the stages of stranger, casual acquaintance, close friend, and beyond. They contend that the social penetration process should be viewed as a system, which involves the simultaneous operation of behavior at various levels of response, including verbal, nonverbal, and environmentally oriented behavior. They emphasize that our understanding of the development of interpersonal relationships necessitates an analysis of the whole person, in contrast to isolated segments of behavior. Various levels of social behavior operate in unison, sometimes complementing one another and sometimes substituting for one another in sequence. This perspective is further developed in Altman's dialectical model of privacy, which we discussed in Chapter 8. Responsiveness to the total person does not mean that single psychological processes should never be studied, but that the total person should be studied in addition to the single processes traditionally emphasized. Synthesis should be pursued as energetically as analysis.

PERSON-ENVIRONMENT TRANSACTION

It is not enough to take a holistic view of the environment and of the person; the environment and the person together also form an integrated system. In order to predict a person's behavior accurately, we must know both the characteristics of the person and the nature of the environment that he or she is dealing with. Further, the relationship between person and environment is a *reciprocal* one; that is, not only does the environment affect people's behavior, but people, in turn, affect the environment.

INTERACTIONISM

$B = f(P,E)$ A guiding theme of the adaptational viewpoint is the belief that human behavior reflects an *interaction* between characteristics of the person and characteristics of the environment. This interactional model of behavior is captured in Kurt Lewin's (1936) classic dictum, $B = f(P,E)$—behavior is a function of both the person and the environment. This view, known as interactionism, synthesizes two conflicting theories of behavior. *Personologism* is a school of thought that posits stable, intraorganismic "traits" as the primary determinants of human behavior. *Situationism* is the antithesis of personologism, emphasizing environmental variables as dominant in the shaping of behavioral variation. *Interactionism* is the synthesis of these two extreme views, postulating that the essential source of variance in human behavior is the interaction of personal and environmental forces (Ekehammar, 1974).

Of course, the interactionist view does not imply that personal and environmental factors are of equal importance in every action. The relative ability of personal and environmental characteristics to predict behavior will depend on the specific environmental variable studied, the particular individual characteristics sampled, the type of behavior assessed, the nature of the subject sample, and the research paradigm employed (Bem and Allen, 1974; Mischel, 1973). The question of how much behavioral variance in a particular situation is a function of personal and environmental variables respectively is an empirical one, and has been the focus of an important body of environmental research since the early 1960s (see Ekehammar, 1974).

The conclusions of all of this research have been remarkably similar: in order to predict behavior we must take into consideration the many sources of variance, in both the person and the environment and particularly in the interaction of personal and environmental variables. When we examine the relationship between density and aggression, for example, we must take account of variance associated with degree of density (environmental variable), sex-related differences in aggression (personal variable), and the differential responses of men and women to various levels of density (person-environment interaction). After thirty years of often fruitless controversy, psychologists have rediscovered the essential wisdom of Lewin's (1936) basic equation, $B = f(P,E)$.

Reciprocal influences   Interactionist research in psychology has sometimes tended to view interaction in terms of two independent variables (personal characteristics and situational characteristics), in accordance with a unidirectional causal model of behavior. Norman Endler and David Magnusson (1976) have pointed out that much interactionist research has tended to see the person as a passive pawn of environmental forces. They argue that not only do environmental variables affect the behavior of individuals, but the individual is also an active participant in the shaping of environmental circumstances. They stress the need for models of human behavior that reflect multidirectional causality, and endorse Lawrence Pervin's (1968) suggestion that the term *transaction* be used in reference to the process of reciprocal causation.

In order to deal adequately with the active role of the person in environment-behavior transactions, we must develop a conceptual and linguistic framework capable of describing the person in terms that reflect the individual's capacity to act on the environment. Kenneth Craik and George McKechnie (Craik, 1970a, 1970b, Craik and McKechnie, 1977) suggest a merging of concepts from environmental psychology and personality theory in what they label *environmental dispositions.* While some personality characteristics have to do with the way a person relates to him- or herself (e.g., self-acceptance) and others with characteristic styles of relating to other persons (e.g., dominance), environmental dispositions are enduring styles of relating to the everyday physical environment. McKechnie (1974, 1977b) has developed the Environmental Response Inventory to identify and measure eight environmental dispositions: pastoralism, urbanism, environmental adaptation, stimulus seeking, environmental trust, antiquarianism, need for privacy, and mechanical orientation.

In an effort to understand the part played by the person in interaction with the environment, Walter Mischel (1973, 1977) has proposed a framework that combines concepts developed in cognitive psychology and social learning theory. He holds that individuals differ in the cognitive *construction competency,* or their ability to generate desired cognitions and response patterns, and in the way they *categorize* a particular situation. People also have varying *expectancies,* particularly with respect to the outcomes associated with specific response patterns and stimulus configurations. Finally, individuals place different subjective *values* on expected outcomes, and differ in the *self-regulatory systems* and *plans* they bring to a situation.

Holahan (1978) has applied the notion of reciprocal influences to a discussion of research knowledge and methodology in environmental psychology. Emphasizing the active role people play in dealing with the physical environment, he focuses on the adaptive processes by which people cope with environmental challenges. He has investigated in particular the positive and adaptive coping processes that enable people to deal with environmental demands in a high-rise university dormitory, a public housing project, and a psychiatric ward. He encourages the adoption of the view that the individual plays an active, creative, and problem-solving role in initiating environment-directed behavior. He advocates the use of research strategies that are sufficiently open-ended to measure a wide range of responses, including positive coping responses and problem-solving behaviors, so that a holistic picture of environmental behavior emerges.

Daniel Stokols (1976, 1977) has suggested that much research that reflects an underlying appreciation of people's active participation with the physical environment may be grouped under the concept of *human-environment optimization.* He defines human-environment optimization as the ways in which individuals and groups attempt to achieve optimal environments; that is, environments that maximally fulfill their goals and needs. There are three essential modes of human interaction with the environment: *orientation* (such as environmental perception and environmental cognition), *operation* (such as spatial behavior and coping with environmental stressors), and *evaluation* (such as environmental attitudes). These processes reflect the active way in which individuals and groups perceive, shape, and evaluate their environmental surroundings in response to personal needs. (For a similar view of the active processes people use in approaching the physical environment, see S. Kaplan, 1973.)

A TRANSACTIONAL MODEL

The transactional view of environment and behavior is presented schematically in Figure 11-3. This model, adapted from the work of Albert Bandura (1978), differs from traditional interactional models of behavior in two ways. First, as Bandura points out, traditional models have represented environmental and personal influences as having separate and unidirectional effects on behavior. Here the effects among all components—environmental, psychological, and behavioral—are reciprocal, with each component both acting on and being acted on, either directly or indirectly, by the other components in the model. Personal factors, such as an individual's expectations, influence that person's behavior; and the environmental changes brought about by this behavior in turn further alter or reinforce the individual's expectations in regard to future outcomes.

Second, traditional interactional models have typically depicted a simple direction of effect, with inputs at one end and outputs at the other. Environmental factors have generally been viewed as independent variables, psychological and cognitive factors as mediating variables, and the individual's behavior as a dependent variable. Here, in contrast, to emphasize the reciprocal effects among all components of the model, the variables are represented as interacting parts of an inte-

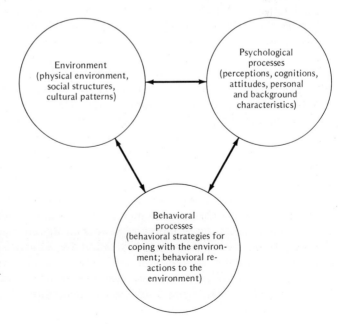

Figure 11-3

A transactional model of the relationships among the environment, psychological processes, and behavioral processes.

Adapted from Bandura, 1978.

grated system. Thus, as Bandura notes, each variable may be seen to function as an independent, mediating, or dependent variable, depending on the particular problem at hand and analytical focus selected.

Although the model shown in Figure 11-3 presents statistical challenges, some statistical procedures have recently been suggested for analyzing reciprocal causation. Lawrence James and B. Krishna Singh (1978) have proposed *two-stage least squares* for psychological research involving two or more variables that are interdependent. An application of the procedure to correlational data in a field setting demonstrated that outbreaks of violence at political protest rallies were caused by reciprocal influences between violence on the part of protesters and violence on the part of the police (Kritzer, 1977). Judith Howard (1979), who has specifically addressed the question of how to deal statistically with a model involving reciprocal influences among environmental, personal, and behavioral variables, suggests the applicability of two-stage least squares and of statistical procedures developed by K. G. Jöreskog (1970) to analyze covariance structures involving reciprocal influences over time.

TRANSACTIONAL STUDIES IN ENVIRONMENTAL PSYCHOLOGY

Environmental psychologists have recently applied aspects of a transactional model of person-environment relations to conceptualize research findings in several areas. A transactional perspective has been used to analyze students' behavior in educational environments, to investigate the psychological consequences of environmental stress, to study the environmental behavior of elderly persons, and to understand the human effects of the urban environment. These applications are directly relevant to psychological processes we have discussed in this book, including environmental performance, the process of coping with ambient environmental stressors and crowding, and the development of affiliative and supportive urban networks. They also suggest ways in which a transactional model may be used in other areas of investigation.

Educational environments   Rudolf Moos (1979) has developed a transactional model of the interrelationship between environmental and personal variables and stability or change in students' educational interests and behavior (Figure 11-4). Moos envisions the *environmental system* as including features of the physical setting, as well as organizational, aggregate, and social-climate factors. The *personal* system includes sociodemographic, expectation, personality, and coping skill variables. The environmental and personal systems influence one another, as when people seek out a particular setting or when an environment selectively admits members. Moos proposes that *cognitive appraisal* and *activation/arousal* mediate some of the effects of the environmental and personal systems on students' efforts to *adapt* and *cope* and the resulting influences on student *stability* or *change*. The mediating processes of appraisal and activation are influenced by both personal factors (a more talented student is more likely to experience a lack of educational challenge; some students are more easily motivated) and environmental factors (more challenging classes are often seen as competitive; different classes motivate students differently).

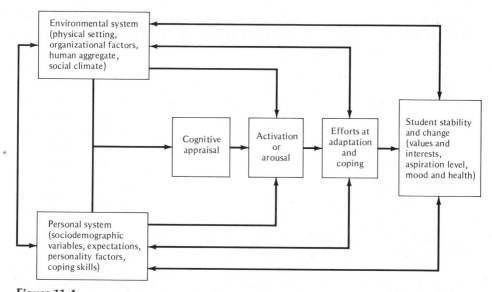

Figure 11-4

A model of the relationship between environmental and personal variables and student stability and change.

From R. H. Moos, Evaluating Educational Environments, *1979, p. 5. Reprinted by permission of Jossey-Bass, Inc., Publishers.*

Efforts to adapt are determined by both personal (students' coping skills vary) and environmental factors (settings differ in how they reward various coping skills). Ultimately, this complex interplay of environmental, personal, mediating, and coping variables affects students' personal values and interests, their level of aspiration, and their mood and health. This transactional model includes several reciprocal influences. For example, a student's coping efforts may alter the environmental system (creation of a new social group) or the personal system (participation in a new organization may change attitudes). Similarly, changes in students' educational interests and behavior may change either the environmental system (increased aspiration may make a setting more competitive) or the personal system (value changes may become integrated in an individual's self-concept).

Environmental stress   Daniel Stokols (1979) has applied a transactional orientation to environmental stress. He suggests that the level of environmental demand associated with a particular stressor can be assessed in terms of the degree of *controllability* (ratio of actual to ideal need fulfillment) and the degree of environmental *salience* (degree to which the environment is motivationally and perceptually significant) in the situation. Environmental demand (and the associated psychological distress) will be greatest when controllability is low and salience high. This model of stress views environmental stressors not as isolated stimuli, but as events embedded in a particular situation reflecting both personal needs and environmental conditions. Stokols and Raymond Novaco (in press) used a transactional model to investigate transportation stress (commuting in high traffic congestion), and found that people's reactions to transportation stress were mediated by personal variables (such as individual differences in coronary-prone behavior patterns) and the reciprocal influence of these variables on environmental contexts.

To describe human settings more broadly, Stokols and Sally Shumaker (Stokols, 1980; Stokols and Shumaker, 1981) have also proposed a general classification framework that reflects a transactional relationship between people (particularly social groups) and places. In this model settings are differentiated according to the composition and organization of their occupants and the dominant functional meanings associated with their physical milieus. Occupant characteristics can be either consistent with the functional orientation of a setting (e.g., people praying at a religious service) or inconsistent with it (e.g., a group joking and laughing at a religious service). This transactional view of settings can help to explain how settings emerge, persist, become modified, and eventually die out.

Environments and the elderly   M. Powell Lawton and Lucille Nahemow (Lawton, 1975; Lawton and Nahemow, 1973; Nahemow and Lawton, 1973, 1976) have applied a transactional model to the emotional and behavioral correlates of the aging process. They propose that the psychological functioning of the elderly needs to be assessed in terms of both environmental *press* (the demands the environment makes on an individual) and personal *competence* (the individual's ability to function effectively) (Figure 11-5). Highly competent older persons are able to function effectively in the face of a wide range of environmental demands, while less competent persons will function well only under limited environmental demands. They will function best at intermediate levels of environmental press that are moderately challenging. When the environmental press is either much greater or much less than the individual is accustomed to (*adaptation level* in the figure), emotional discomfort and maladaptive behavior will result.

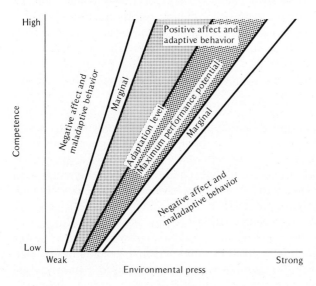

Figure 11-5

The behavioral and affective outcomes of person-environment transactions.

From M. P. Lawton and L. Nahemow, "Ecology and the Aging Process," in C. Eisdorfer and M. P. Lawton (eds.), The Psychology of Adult Development and Aging, *p. 661. Copyright 1973 by the American Psychological Association. Reprinted by permission of the publisher and author.*

Urban environment   William Michelson (1976, 1977) has presented a transactional framework that helps to unify conceptually the many variables involved in the "goodness of fit" between people and the urban settings in which they reside. The integrating aspect of Michelson's model is the notion of *intersystem congruence*, or the degree of match or mismatch between the built environment and the cultural, social, and personality systems. In general, the model proposes that "states of variables in one system [coexist] better with states of variables in another system than with other alternative states" (p. 26). Particular characteristics of urban physical settings seem to be congruent with and supportive of some social processes and incongruent with and disruptive of others. In the most recent formulation of his congruence model, Michelson has emphasized that it is the behavioral effects of the various systems, as reflected in an individual's lifestyle, that need to mesh with the designed environment, rather than the systems themselves.

APPLICATIONS TO ENVIRONMENTAL PLANNING

INTERDISCIPLINARY INVOLVEMENT

How might this transactional view of environment and behavior be applied to the planning and design of human settings? First, a holistic view of the environment, as a system in which physical features are only one of many interrelated elements, suggests the value of *interdisciplinary* involvement in the planning enterprise. As Harold Proshansky and Irwin Altman (1979) have pointed out, because the problems involved in human-environment relationships entail complex and interlocking factors that cut across the physical, psychological, societal, and cultural realms, this field is by definition interdisciplinary. The expertise required to accommodate the complex physical, human, and sociocultural demands implicit in the design of human settings necessarily crosses the boundaries of several disciplines.

Kenneth Craik (1970) has offered a thoughtful analysis of the plight of the designer in today's society. The designer, confronted with the challenge of designing complete environments for many millions of persons, is inundated by a tide of specialized information relevant to the design process. As contemporary design problems increase in scale and complexity, the social, cultural, and technological contexts of the design enterprise change even more rapidly. The scope and complexity of the contemporary design process tax the designer's competencies and intuitive skills to their limit.

Though we recognize the need for interdisciplinary involvement in the design process, we must not be naive about the difficulty of securing collaboration across disciplinary boundaries. Altman (1975) has described the difficult history of contact between designers and social scientists. The initial enthusiasm for collaboration during the early 1960s was replaced by a sense of disillusion on both sides by the end of the decade. Fruitful dialogue between designers and social scientists requires an honest recognition of and mutual respect for the underlying differences in their *approach* to environmental issues. As Altman points out, practitioners are typically *action-oriented*, while researchers are often interested in *understanding* a problem at an abstract level. While practitioners generally strive to *synthesize* diverse points of view and pragmatic concerns, researchers usually attempt to *analyze* complex phenomena in order to identify their discrete components. Only by learning to appreciate each other's style can designers and social scientists succeed in their efforts to tackle together the complexity of environmental planning.

Proshansky (1972) has discussed the implications of the interdisciplinary requirements of the environment and behavior field for graduate training in psychology. He argues for training models in psychology that would involve graduate students in working relationships with professionals in administrative and policy roles in a variety of community settings. Such a training philosophy would expose psychology students to the approaches, concerns, and concepts of other disciplines early in their education, and thus lay the groundwork for later professional collaboration. Proshansky's model of training is equally applicable to other social science fields, and to training in the design professions, where early collaboration with social scientists might facilitate later professional relationships. This model is not intended to make students generalists, but rather to produce individuals who are well grounded in a particular discipline without becoming too narrow in their focus.

PLURALISTIC PLANNING

The transactional model's emphases on the individual as a total entity and on the important contributions of individual differences (such as those associated with age, gender, and motivation) to environmental outcomes underscore the need for a *pluralistic* approach to the planning process. No isolated design element can meet the variety of psychological and physical needs that constitute the total person, and no single design solution can accommodate the diverse needs of a variety of environmental users. Rather, the planning process must be a pluralistic enterprise, in which a variety of patterned elements can meet the complex needs of the whole person and a diversity of environmental choices can match the requirements of a varied population of users. Ideal design solutions are in some degree flexible and modifiable, so that they can accommodate variations in user groups and shifts in individual needs and preferences over time.

Serge Chermayeff and Alexander Tzonis (1971) have offered a systems-oriented approach to urban design that emphasizes the goal of maximizing choice in the lives of urban residents. They encourage urban designers to create "exciting choices" for urbanites that represent real choices between alternative experiences, such as community and privacy or mobility and tranquility. They believe that urban planning should enhance social flexibility and be responsive to new and changing human aspirations. They envision the urban environment as the expression of a "dynamic order," characterized by a continuous process of interaction between growth and change.

Martin Krovetz (1977) encourages an approach to the design of learning environments that emphasizes pluralistic planning and a responsiveness to students' diverse educational needs. Each student possesses a uniquely personal learning style and a unique set of needs; no single educational environment can match the needs of all students. In reviewing research on individual differences in locus of control (see Rotter, 1966), Krovetz points out that students characterized by *internal* locus of control (people who tend to see reinforcements as contingent on their own behavior) typically require more information in approaching ambiguous tasks, are more active in attempting to control the environment, and display more achievement behavior than students characterized by *external* locus of control (people who are inclined to see reinforcements as controlled by external factors). He concludes that the design of educational environments should provide a variety of facilities and types of space that might accommodate new and diverse activities.

Lawton (1974) also calls for a pluralistic planning philosophy in the design of institutional environments for the elderly. He argues against an approach that lumps all old people in a single category: "Old people prefer . . ." The elderly differ as much among themselves as individuals in any other age group, and require *diversified* design solutions. The first rule in design for the elderly should be to maximize the congruity between the diverse personal characteristics of older people and a variety of design alternatives, rather than to seek an "ideal" design solution for all elderly persons. Lawton points out that institutional designers' penchant for leveling all individual needs to a single solution has been maintained because it is economically cheaper and intellectually less challenging than designing for pluralistic needs.

USER PARTICIPATION

The transactional model of the person-environment system, with its emphasis on the individual's influence on the environment as well as the environment's influence on the person, suggests the importance of *user participation* in the planning process. People bring natural motivation and competencies to the task of improving the physical environment. The energy, interest, and skill of environmental users constitute a valuable resource that can be tapped in the planning process to optimize the match between users' needs and design features (see Pogell, Balling, Passoneau, and Valadez, 1980).

Abraham Wandersman (1979) has developed a model of user participation in the design process that reflects a transactional orientation. In this model, *individual differences* among users—differences in personality, environmental preferences, perceived expertise, and sociodemographic characteristics—help to determine who participates and serve as moderating influences on the process of participation. The model also takes account of characteristics of *participation* (e.g., the nature of the environmental setting, technique of participation, and stage in the design process) and *feedback* from participation outcomes to the other variables in sequence. In addition, the model explores the role of personal *mediating* factors involving the individual's sense of control over the environment, personal needs and values, and cognitive coping factors (derived from Mischel, 1973, 1977), such as planning skills and expectancies about the effectiveness of individual action.

Robert Sommer (1972) has pointed out that decision makers sometimes argue against user participation because citizens may not be fully aware of the available options and trade-offs involved in planning decisions. Sommer has little patience with such arguments. Rather than exclude users from the design process on these grounds, he says, a more reasonable solution is to educate them. Design professionals and social scientists could conduct environmental workshops to educate users in the ways they are affected by physical settings and ways they might improve those settings. Such workshops should deal initially with people's immediate environments—homes, schools, offices, and neighborhoods—where the scale of the problem is more manageable and the effects of the individual's efforts are more apparent than in more complex settings. Sommer reports some fascinating and productive results from an environmental workshop he conducted in a hospital, where the staff members were induced to play the roles of patients so they could view the hospital environment from the patients' perspective:

> "As turn-on devices we used such prosthetic aids as crutches, wheelchairs, and gurneys. These produced some interesting perceptual experiences which were shared with the group at large. Distances seemed three times as long on crutches as they had previously. It took a very long time to go down the hallway in a wheelchair; when one person wheeled another, the speed of passage was very important. Wheeling a person at ordinary walking speed seemed much too fast; the person in the chair felt as if he were a bowling ball going down the alley. Tall men were particularly bothered by being looked down on as they sat in the wheelchair." [p. 44]

When a team of environmental psychologists set out to remodel a psychiatric hospital ward with the aim of making it more attractive and comfortable for patients and staff, they were forced to recognize and deal with the determination of users to participate in the design process (Holahan, 1976a, 1978). The researchers met with considerable resistance from the ward staff, who desired to play a larger role in the design decisions than they were originally accorded. After initially feeling frustrated by the staff's opposition, the investigators came to recognize that what they had perceived as "resistance" actually represented a natural and positive desire to participate in the remodeling on the part of a highly motivated staff. When the staff members were then encouraged to participate more actively in the project, their resistance lessened, and they made a variety of very constructive personal additions to the remodeling effort.

Citizens' sensitivity to and awareness of the environment has long-term consequences for the quality of human life. The responsibilities and trade-offs involved in the personal, professional, and corporate decisions of all citizens will determine the nature and degree of stressful environmental conditions faced by future generations. Practitioners and researchers in the field of environment and behavior can help to foster the kind of attitudinal and behavioral change necessary to improve those prospects. As René Dubos (1968) has cautioned, the issue is not *whether* we will survive, but the long-term physical and psychological *costs* of adapting to progressively more threatening environmental conditions. The responsibility to weigh the long-term consequences of today's environmental decisions cannot be abdicated without peril. Yet, in the history of the planet, the hour is reasonably early; opportunity remains for right-minded choices.

SUMMARY

In this chapter we have focused on three characteristics of the adaptational viewpoint that are relevant to the development of a unifying framework for the study of environment and behavior. First, we considered the adaptational viewpoint's *holistic* model of the environment as a series of encircling contexts. Second, we discussed the adaptational view of the individual as a *total person* who interacts with the environment. Third, we examined the adaptational viewpoint's emphasis on person and environment as involved in a *transactional* relationship.

A central theme of the adaptational viewpoint is a conception of behavior as nested in a series of encircling contexts. Urie Bronfenbrenner has identified the environmental systems that comprise the context of human behavior as the microsystem, the exosystem, and the macrosystem. We applied this notion of encircling contexts in developing a holistic model of the environment appropriate to our special concern with the physical environment (Figure 11-1). We envisioned the physical environment as consisting of three concentric rings—interior design, buildings, and geographic regions. The physical environment itself was viewed as encircled by two broader systems—social structures and cultural patterns. This holistic model of the environment emphasizes that the various aspects of the physical, social, and cultural environment exert overlapping, simultaneous, and interrelated influences on human behavior.

A chief emphasis of the adaptational model of environmental psychology is that the person functions as a total, integrated entity in the environment–behavior equation. Kenneth Craik explains that the "personality paradigm" in environmental psychology takes the whole person as the basic unit of analysis. In advancing a model of the total person, we applied this holistic view of personality to a model of environment and behavior (Figure 11-2). Information from the environment is received, evaluated, and encoded by the interlocking psychological processes of environmental perception, environmental cognition, and environmental attitude formation. This "processed" information then becomes the basis for the individual's behavior toward the environment, which operates as an interrelated pattern of individual and social behaviors. This model of the total person emphasizes that the psychological processes by which people deal with the physical environment are highly interrelated and function in a mutually supportive and complementary fashion.

A guiding theme of the adaptational viewpoint is the belief that people's behavior reflects an interaction between characteristics of the person and characteristics of the environment. This interactional model of behavior is best captured in Kurt Lewin's classic dictum, $B = f(P,E)$; behavior is a function of both the person and the environment. While some interactionist research has cast the person as a passive pawn of environmental forces, recent formulations have emphasized that person and environment have a transactional relationship, so that the individual actively shapes environmental circumstances while being influenced by the environment. This view of environment and behavior was incorporated in a transactional model adapted from the work of Albert Bandura (Figure 11-3), in which the effects among environmental, psychological, and behavioral variables are represented as reciprocal, with each variable both acting on and being acted on by the others.

The transactional perspective has been applied to the study of students' behavior in educational environments, the psychological consequences of environmental stress, the environmental behavior of the elderly, and the human effects of the urban environment. The transactional view of environment and behavior encourages interdisciplinary involvement in planning, a pluralistic approach to design, and user participation in the planning process.

REFERENCES

Abramson, L. Y., Seligman, M. E. P., and Teasdale, J. D. Learned helplessness in humans: Critique and reformulation. *Journal of Abnormal Psychology*, 1978, *87*, 49–74.

Acking, C. A., and Kuller, R. The perception of an interior as a function of its colour. *Ergonomics*, 1972, *15*, 645–654.

Acredolo, L. P. Frames of reference used by children for orientation in unfamiliar spaces. In G. Moore and R. Golledge (eds.), *Environmental knowing*. Stroudsburg, Pa.: Dowden, Hutchinson & Ross, 1976.

Acredolo, L. P. Developmental changes in the ability to coordinate perspectives of a large-scale environment. *Developmental Psychology*, 1977, *13*, 1–8.

Acredolo, L. P., Pick, H. L., and Olsen, M. Environmental differentiation and familiarity as determinants of children's memory for spatial location. *Developmental Psychology*, 1975, *11*, 495–501.

Acton, W. I. Speech intelligibility in a background noise and noise-induced hearing loss. *Ergonomics*, 1970, *13*, 546–554.

Adams, J. R. Review of defensible space. *Man-Environment Systems*, 1973, *3*, 267–268.

Aiello, J. R. A test of equilibrium theory: Visual interaction in relation to orientation, distance and sex of interactants. *Psychonomic Science*, 1972, *27*, 335–336.

Aiello, J. R., and Aiello, T. D. Development of personal space: Proxemic behavior of children six to sixteen. *Human Ecology*, 1974, *2*, 177–189.

Aiello, J. R., and Cooper, R. E. *The use of personal space as a function of social affect*. Paper presented at annual convention of the American Psychological Association, 1972.

Aiello, J. R., DeRisi, D., Epstein, Y. M., and Karlin, R. A. Crowding and the role of interpersonal distance preference. *Sociometry*, 1977, *40*, 271–282.

Aiello, J. R., Epstein, Y. M., and Karlin, R. A. Effects of crowding on electrodermal activity. *Sociological Symposium*, 1975, *14*, 43–57.

Aiello, J. R., and Jones, S. E. Field study of the proxemic behavior of young school children in three subcultural groups. *Journal of Personality and Social Psychology*, 1971, *19*, 351–356.

Ajzen, I., and Fishbein, M. Attitudinal and normative variables as predictors of specific behaviors. *Journal of Personality and Social Psychology*, 1973, *27*, 41–57.

Ajzen, I., and Fishbein, M. Attitude-behavior relations: A theoretical analysis and review of empirical research. *Psychological Bulletin*, 1977, *84*, 888–919.

Albert, S., and Dabbs, J. M., Jr. Physical distance and persuasion. *Journal of Personality and Social Psychology*, 1970, *15*, 265–270.

Alexander, C. *Notes on the synthesis of form*. Cambridge, Mass.: Harvard University Press, 1964.

Alexander, C. A city is not a tree. In G. Bell and J. Tyrwhitt (eds.), *Human identity in the urban environment*. New York: Penguin, 1972.

Allgeier, A. R., and Byrne, D. Attraction toward the opposite sex as a determinant of physical proximity. *Journal of Social Psychology*, 1973, *90*, 213–219.

Allport, F. H. *Theories of perception and the concept of structure*. New York: Wiley, 1955.

Allport, G., and Pettigrew, T. Cultural influence on the perception of movement: The trapezoidal illusion among the Zulus. *Journal of Abnormal and Social Psychology*, 1957, *55*, 104–113.

Almond, R., and Esser, A. H. Tablemate choice of psychiatric patients: A technique for measuring social contact. *Journal of Nervous and Mental Disease*, 1965, *141*, 68–82.

Althoff, P., and Greig, W. H. Environmental pollution control policy-making: An analysis of elite perceptions and preferences. *Environment and Behavior*, 1974, *6*, 259–288.

Altman, I. Territorial behavior in humans: An analysis of the concept. In L. Pastalan and D. H. Carson (eds.), *Spacial behavior of older people*. Ann Arbor: University of Michigan—Wayne State University Press, 1970.

Altman, I. *The environment and social behavior*. Monterey, Ca.: Brooks/Cole, 1975.

Altman, I. Privacy: A conceptual analysis. *Environment and Behavior*, 1976, *8*, 7–31.

Altman, I. Privacy regulation: Culturally universal or culturally specific? *Journal of Social Issues*, 1977, 33, 66–83.

Altman, I., and Chemers, M. *Culture and environment.* Monterey, Ca.: Brooks/Cole, 1980.

Altman, I., and Haythorn, W. W. The ecology of isolated groups. *Behavioral Science*, 1967, 12, 169–182.

Altman, I., Nelson, P. A., and Lett, E. E. The ecology of home environments. *Catalog of Selected Documents in Psychology.* Washington, D.C.: American Psychological Association, 1972.

Altman, I., and Taylor, D. A. *Social penetration: The development of interpersonal relationships.* New York: Holt, Rinehart & Winston, 1973.

Altman, I., Taylor, D. A., and Wheeler, L. Ecological aspects of group behavior in social isolation. *Journal of Applied Social Psychology*, 1971, 1, 76–100.

Altman, I., and Vinsel, A. M. Personal space: An analysis of E. T. Hall's proxemics framework. In I. Altman and J. Wohlwill (eds.), *Human behavior and environment: Advances in theory and research,* Vol. 2. New York: Plenum, 1977.

American National Standards Institute. *American National Standards specifications for making buildings and facilities accessible to and useable by the physically handicapped.* New York: American National Standards Institute, 1961.

APA Committee on Ethical Standards in Psychological Research. *Ethical principles in the conduct of research with human participants.* Washington, D.C.: American Psychological Association, 1973.

Ames, A., Jr. Visual perception and the rotating trapezoidal window. *Psychological Monographs*, 1957, 65, No. 324.

Amick, D. J., and Kviz, F. J. Density, building type, and social integration in public housing projects. *Man–Environment systems*, 1974, 4, 187–190.

Anderson, E. W., Andelman, R. J., Strauch, J. M., Fortuin, N. J., and Knelson, J. H. Effect of low-level carbon monoxide exposure on onset and duration of angina pectoris. *Annals of Internal Medicine*, 1973, 79, 46–50.

Anderson, J., and Tindall, M. *The concept of home range: New data for the study of territorial behavior.* Paper presented at annual conference of Environmental Design Research Association, Los Angeles, 1972.

Anderson, J. R. Arguments concerning representations for mental imagery. *Psychological Review*, 1978, 4, 249–277.

Andrews, H. F. Home range and urban knowledge of school-age children. *Environment and Behavior*, 1973, 5, 73–86.

Angrist, S. S. Dimensions of well-being in public housing families. *Environment and Behavior*, 1974, 6, 495–517.

Ankele, C. and Sommer, R. The cheapest apartments in town. *Environment and Behavior*, 1973, 5, 505–514.

Antonovsky, A. *Health, stress, and coping.* San Francisco: Jossey-Bass, 1979.

Appley, M. H., and Trumbull, R. *Psychological stress: Issues in research.* New York: Appleton-Century-Crofts, 1967.

Appleyard, D. Why buildings are known. *Environment and Behavior*, 1969, 1, 131–156.

Appleyard, D. Styles and methods of structuring a city. *Environment and Behavior*, 1970, 2, 100–117.

Appleyard, D. Notes on urban perception and knowledge. In R. M. Downs and D. Stea (eds.), *Image and environment: Cognitive mapping and spatial behavior.* Chicago: Aldine, 1973.

Appleyard, D. *Planning a pluralistic city.* Cambridge, Mass.: M.I.T. Press, 1976.

Appleyard, D., and Carp, F. The BART residential impact study: An empirical study of environmental impact. In T. G. Dickert and K. R. Domeny (eds.), *Environmental impact assessment: Guidelines and commentary.* Berkeley: University Extension, University of California, 1974.

Appleyard, D., and Craik, K. H. The Berkeley Environmental Simulation Project: Its use in environmental impact assessment. In T. G. Dickert and K. R. Domeny (eds.), *Environ-*

mental impact assessment: Guidelines and commentary. Berkeley: University Extension, University of California, 1974.

Appleyard, D., and Craik, K. H. The Berkeley environmental laboratory and its research programme. *International Review of Applied Psychology,* 1978, *27,* 53–55.

Appleyard, D., and Lintell, M. The environmental quality of city streets: The residents' viewpoint, *Journal of the American Institute of Planners,* 1972, *38,* 84–101.

Appleyard, D., Lynch, K., and Myer, J. R. *The view from the road.* Cambridge, Mass.: M.I.T. Press, 1964.

Archea, J. The place of architectural features in behavioral theories of privacy. *Journal of Social Issues,* 1977, *33,* 116–137.

Architectural Forum. Slum surgery in St. Louis. April, 1951, 128–136.

Ardrey, R. *The territorial imperative.* New York: Atheneum, 1966.

Ardrey, R. *The social contract.* New York: Atheneum, 1970.

Argyle, M., and Cook, M. *Gaze and mutual gaze.* New York: Cambridge University Press, 1976.

Argyle, M., and Dean, J. Eye-contact, distance and affiliation. *Sociometry,* 1965, *28,* 289–304.

Argyle, M., and Graham, J. A. The central Europe experiment: Looking at persons and looking at objects. *Environmental Psychology and Nonverbal Behavior,* 1976, *1,* 6–16.

Argyle, M., and Ingham, R. Gaze, mutual gaze and proximity. *Semiotica,* 1972, *6,* 32–50.

Arnstein, S. A ladder of citizen participation. *Journal of the American Institute of Planners,* 1969, *35,* 216–224.

Aronow, W. S., Harris, C. N., Isbell, M. W., Rokaw, S. N., and Imparato, B. Effect of freeway travel on angina pectoris. *Annals of Internal Medicine,* 1972, *77,* 669–676.

Ast, G. D. Moline, Illinois: Planning a barrier-free environment for the elderly and handicapped. In M. Bednar (ed.), *Barrier-free environments.* Stroudsburg, Pa.: Dowden, Hutchinson & Ross, Inc., 1977.

Athanasiou, R., and Yoshioka, G. A. The spatial characteristics of friendship formation. *Environment and Behavior,* 1973, *5,* 43–65.

Averill, J. Personal control over aversive stimuli and its relationship to stress. *Psychological Bulletin,* 1973, *80,* 286–303.

Azer, N. Z., McNall, P. E., and Leung, H. C. Effects of heat stress on performance. *Ergonomics,* 1972, *15,* 681–691.

Bagozzi, R. P., and Burnkrant, R. E. Attitude organization and the attitude-behavior relationship. *Journal of Personality and Social Psychology,* 1979, *37,* 913–929.

Bailey, K. G., Hartnett, J. J., and Gibson, S. W. Implied threat and the territorial factor in personal space. *Psychological Reports,* 1972, *30,* 263–270.

Baird, J. Studies of the cognitive representation of spatial relations. *Journal of Experimental Psychology: General,* 1979, *108,* 90–106.

Baird, J. C., Cassidy, B., and Kurr, J. Room preference as a function of architectural features and user activities. *Journal of Applied Psychology,* 1973, *63,* 719–727.

Baldassare, M., Knight, R., and Swan, S. Urban service and environmental stressors: The impact of the Bay Area Rapid Transit system on residential mobility. *Environment and Behavior,* 1979, *11,* 435–450.

Baltes, M. M., and Hayward, S. C. Application and evaluation of strategies to reduce pollution: Behavioral control of littering in a football stadium. *Journal of Applied Psychology,* 1976, *61,* 501–506.

Bandura, A. Analysis of modeling processes. In A. Bandura (ed.), *Modeling: Conflicting theories.* New York: Lieber-Atherton, 1974.

Bandura, A. The self system in reciprocal determinism. *American Psychologist,* 1978, *33,* 344–358.

Barash, D. P. Human ethology: Personal space reiterated. *Environment and Behavior,* 1973, *5,* 67–73.

Barker, M. L. Information and complexity: The conceptualization of air pollution by specialist groups. *Environment and Behavior,* 1974, *6,* 346–377.

Barker, M. L. Planning for environmental indices: Observer appraisals of air quality. In K. H. Craik and E. H. Zube (eds.), *Perceiving environmental quality: Research and Applications.* New York: Plenum, 1976.

Barker, R. G. Ecology and motivation. In M. R. Jones (ed.), *Nebraska Symposium on Motivation.* Lincoln: University of Nebraska Press, 1960.

Barker, R. G. (ed.). *The stream of behavior.* New York: Appleton-Century-Crofts, 1963.

Barker, R. G. Explorations in ecological psychology. *American Psychologist,* 1965, *20,* 1–14.

Barker, R. G. *Ecological psychology: Concepts and methods for studying the environment of human behavior.* Stanford, Ca.: Stanford University Press, 1968.

Barker, R. G. Wanted: An eco-behavioral science. In E. P. Willems and H. L. Raush (eds.), *Naturalistic viewpoints in psychological research.* New York: Holt, Rinehart & Winston, 1969.

Barker, R. G., and Associates. *Habitats, environments, and human behavior.* San Francisco: Jossey-Bass, 1978.

Barker, R. G., and Gump, P. V. *Big school, small school.* Stanford, Ca.: Stanford University Press, 1964.

Barker, R. G., and Schoggin, P. *Qualities of community life: Methods of measuring environment and behavior applied to an American and an English town.* San Francisco: Jossey-Bass, 1973.

Barker, R. G., and Wright, H. F. Psychological ecology and the problem of psycho-social development. *Child Development,* 1949, *20,* 131–143.

Barker, R. G., and Wright, H. F. *One boy's day.* New York: Harper & Row, 1951.

Barker, R. G., and Wright, H. F. *Midwest and its children.* New York: Harper & Row, 1955.

Barker, R. G., Wright, H. F., Barker, L. S., and Schoggen, M. *Specimen records of American and English children.* Lawrence: University of Kansas Press, 1961.

Baron, R. A. Aggression as a function of ambient temperature and prior anger arousal. *Journal of Personality and Social Psychology,* 1972, *21,* 183–189.

Baron, R. A. Invasions of personal space and helping: Mediating effects of invader's apparent need. *Journal of Experimental Social Psychology,* 1978, *14,* 304–312.

Baron, R. A., and Bell, P. A. Aggression and heat: Mediating effects of prior provocation and exposure to an aggressive model. *Journal of Personality and Social Psychology,* 1975, *31,* 825–832.

Baron, R. A., and Bell, P. A. Physical distance and helping: Some unexpected benefits of "crowding in" on others. *Journal of Applied Social Psychology,* 1976, *6,* 95–104.

Baron, R. A., and Lawton, S. F. Environmental influences on aggression: The facilitation of modeling effects by high ambient temperatures. *Psychonomic Science,* 1972, *26,* 80–83.

Baron, R. A., and Ransberger, V. M. Ambient temperature and the occurrence of collective violence: The "long hot summer" revisited. *Journal of Personality and Social Psychology,* 1978, *36,* 351–360.

Baron, R. M., Mandel, D. R., Adams, C. A., and Griffen, L. M. Effects of social density in university residential environments. *Journal of Personality and Social Psychology,* 1976, *34,* 434–446.

Barton, R. The patient's personal territory. *Hospital and Community Psychiatry,* 1966, *17,* 336.

Bartram, D. J. Comprehending spatial information: The relative efficiency of different methods of presenting information about bus routes. *Journal of Applied Psychology,* 1980, *65,* 103–110.

Bass, M. H., and Weinstein, M. S. Early development of interpersonal distance in children. *Canadian Journal of Behavioral Science,* 1971, *3,* 368–376.

Batchelor, J. P., and Goethals, G. R. Spacial arrangements in freely formed groups. *Sociometry,* 1972, *35,* 270–279.

Baum, A., Aiello, J. R., and Calesnick, L. E. Crowding and personal control: Social density and the development of learned helplessness. *Journal of Personality and Social Psychology,* 1978, *36,* 1000–1011.

Baum, A., and Davis, G. E. Spatial and social aspects of crowding perception. *Environment and Behavior,* 1976, *8,* 527–545.

Baum, A. and Davis, G. E. Reducing the stress of high-density living: An architectural intervention. *Journal of Personality and Social Psychology,* 1980, *38,* 471–481.

Baum, A., and Greenberg, C. I. Waiting for a crowd: The behavioral and perceptual effects of anticipated crowding. *Journal of Personality and Social Psychology,* 1975, *32,* 667–671.

Baum, A., Harpin, R. E., and Valins, S. The role of group phenomena in the experience of crowding. *Environment and Behavior,* 1975, *7,* 185–198.

Baum, A., and Koman, S. Differential response to anticipated crowding: Psychological effects of social and spatial density. *Journal of Personality and Social Psychology,* 1976, *34,* 526–536.

Baum, A., Reiss, M., and O'Hara, J. Architectural variants of reaction to spatial invasion. *Environment and Behavior,* 1974, *6,* 91–100.

Baum, A., and Valins, S. *Architecture and social behavior: Psychological studies in social density.* Hillsdale, N.J.: Lawrence Erlbaum, 1977.

Baxter, J. C. Interpersonal spacing in natural settings. *Sociometry,* 1970, *33,* 444–456.

Baxter, J. C., and Deanovich, B. S. Anxiety arousing effects of inappropriate crowding. *Journal of Consulting and Clinical Psychology,* 1970, *35,* 174–178.

Beardsley, E. L. Privacy: Autonomy and selective disclosure. In J. R. Pennock and J. W. Chapman (eds.), *Privacy.* New York: Atherton Press, 1971.

Bechtel, R. B. Perceived quality of residential environments: Some methodological issues. In K. E. Craik and E. H. Zube (eds.), *Perceiving environmental quality: Research and application.* New York: Plenum, 1976.

Bechtel, R. B. *Enclosing behavior.* Stroudsburg, Pa.: Dowden, Hutchinson & Ross, 1977.

Beck, R. J., and Wood, D. Cognitive transformation of information from urban geographic fields to mental maps. *Environment and Behavior,* 1978, *8,* 199–238.

Becker, F. D. Study of spatial markers. *Journal of Personality and Social Psychology,* 1973, *26,* 439–445.

Becker, F. D. The effect of physical and social factors on residents' sense of security in multi-family housing developments. *Journal of Architectural Research,* 1975, *4,* 18–24.

Becker, F. D. Children's play in multifamily housing. *Environment and Behavior,* 1976, *8,* 545–575.

Becker, F. D., and Mayo, C. Delineating personal distance and territoriality. *Environment and Behavior,* 1971, *3,* 375–381.

Becker, F. D., Sommer, R., Bee, J., and Oxley, B. College classroom ecology. *Sociometry,* 1973, *36,* 514–525.

Becker, L. J. Joint effect of feedback and goal setting on performance: A field study of residential energy conservation. *Journal of Applied Psychology,* 1978, *63,* 428–433.

Becker, L. J., and Seligman, C. Reducing air conditioning waste by signaling it is cool outside. *Personality and Social Psychology Bulletin,* 1978, *4,* 412–415.

Bednar, M. (ed.), *Barrier-free environments.* Stroudsburg, Pa.: Dowden, Hutchinson & Ross, 1977.

Bell, G., Randall, E., and Roeder, J. *Urban environments and human behavior: An annotated bibliography.* Stroudsburg, Pa.: Dowden, Hutchinson & Ross, 1973.

Bell, P. A., and Baron, R. A. Environmental influences on attraction: Effects of heat, attitude similarity, and personal evaluations. *Bulletin of the Psychonomic Society,* 1974, *4,* 479–481.

Bell, P. A., and Baron, R. A. Aggression and heat: The mediating role of negative affect. *Journal of Applied Social Psychology,* 1976, *6,* 18–30.

Bell, P. A., and Baron, R. A. Aggression and ambient temperature: The facilitating and inhibiting effects of hot and cold environments. *Bulletin of the Psychonomic Society,* 1977, *9,* 443–445.

Bem, D. J., and Allen, A. On predicting some of the people some of the time: The search for cross-situational consistencies in behavior. *Psychological Review,* 1974, *81,* 506–520.

Bem, D. J., and Lord, C. G. Template matching: A proposal for probing the ecological validity of experimental settings in social psychology. *Journal of Personality and Social Psychology,* 1979, *37,* 833–846.

Bennett, C. *Spaces for people: Human factors in design.* Englewood Cliffs, N.J.: Prentice-Hall, 1977.

Beranek, L. L. Noise. *Scientific American,* 1966, *215,* 66–76.

Berger, S. M. Conditioning through vicarious instigation. *Psychological Review*, 1962, *69*, 450–466.

Berglund, B., Berglund, U., and Lindvall, T. Psychological processing of odor mixtures. *Psychological Review*, 1976, *83*, 432–441.

Bergman, B. A. *The effects of group size, personal space and success-failure on physiological arousal, test performance and questionnaire responses.* Doctoral dissertation, Temple University, 1971.

Berkowitz, L. *A survey of social psychology.* Hinsdale, Ill.: Dryden, 1975.

Berlyne, D. E. *Conflict, arousal and curiosity.* New York: McGraw-Hill, 1960.

Berlyne, D. E. Arousal and reinforcement. In D. Levine (ed.), *Nebraska Symposium on Motivation.* Lincoln: University of Nebraska Press, 1967.

Berlyne, D. E. *Aesthetics and psychobiology.* New York: Appleton-Century-Crofts, 1972.

Berlyne, D. E. (ed.). *Studies in the new experimental aesthetics: Steps toward an objective psychology of aesthetic appreciation.* New York: Halsted Press, 1974.

Berscheid, E. Privacy: A hidden variable in experimental social psychology. *Journal of Social Issues*, 1977, *33*, 85–102.

Betak, J. F., Brummell, A. C., and Swingle, P. G. *An approach to elicit attributes of complex visual environments.* Paper presented at annual conference of the Environmental Design Research Association, 1974.

Beutell, A. W. An analytical basis for a lighting code. *Illuminating Engineering*, 1934, *27*, 5–16.

Bickman, L., Teger, A., Gabriele, T., McLaughlin, C., Berger, M., and Sunaday, E. Dormitory density and helping behavior. *Environment and Behavior*, 1973, *5*, 465–490.

Biederman, I. Perceiving real-world scenes. *Science*, 1972, *177*, 77–79.

Biegel, D., Naparstek, A. J., and Khan, M. M. *Determinants of social support systems.* Paper presented at annual conference of Environmental Design Research Association, Charleston, S.C., 1980.

Birdwhistell, R. L. *Kinesics and context.* Philadelphia: University of Pennsylvania Press, 1970.

Birren, F. *Color psychology and color therapy.* New Hyde Park, N.Y.: University Books, 1965.

Birren, J. E. The abuse of the urban aged. *Psychology Today*, 1970, *3*, (10), 37–38, 76.

Blackman, A. Scientism and planning. *The American Behavioral Scientist*, 1966, *10*, 24–28.

Blackwell, H. R. Development and use of a quantitative method for specification of interior illumination levels on the basis of performance data. *Illuminating Engineering*, 1959, *54*, 317–353.

Blackwell, H. R. Development of visual task evaluators for use in specifying recommended illumination levels. *Illuminating Engineering*, 1961, *56*, 543–544.

Blackwell, H. R. Further validation studies of visual task evaluation. *Illuminating Engineering*, 1964, *59*, 627–641.

Blackwell, H. R. A human factors approach to lighting recommendations and standards. In *Proceedings of the Sixteenth Annual Meeting of the Human Factors Society*, Santa Monica, Ca., 1972. (a)

Blackwell, H. R. A unified framework of methods for evaluating visual performance aspects of lighting. *C.E.I. Publication 19*, 1972. (b)

Blackwell, H. R., and Blackwell, O. M. The effect of illumination quantity upon the performance of different visual tasks. *Illuminating Engineering*, 1968, *63*, 143–152.

Blackwell, H. R., and Smith, S. W. Additional visual performance data for use in illumination specification systems. *Illuminating Engineering*, 1970, *65*, 389–410.

Blaut, J. M., McCleary, G. F., Jr., and Blaut, A. S. Environmental mapping in young children. *Environment and Behavior*, 1970, *2*, 335–349.

Bleda, P. R., and Bleda, S. E. *Effects of sex and smoking on reactions to personal space invasion at a shopping mall.* Paper presented at the meeting of the American Psychological Association, Washington, D.C., 1976.

Bleda, P. R., and Sandman, P. H. In smoke's way: Socioemotional reactions to another's smoking. *Journal of Applied Psychology*, 1977, *52*, 452–458.

Blumenthal, R., and Meltzoff, J. Social schemas and perceptual accuracy in schizophrenia. *British Journal of Social and Clinical Psychology*, 1967, *6*, 119–128.

Boggs, D. H., and Simon, J. R. Differential effect of noise on tasks of varying complexity. *Journal of Applied Psychology*, 1968, 52, 148–153.

Bolt, Beranek, and Newman, Inc. *Noise environment of urban and suburban areas.* Washington, D.C.: Federal Housing Administration, 1967.

Boulding, K. E. General systems theory: The skeleton of science. In W. Buckley (ed.), *Modern Systems Research for the Behavioral Scientist.* Chicago: Aldine, 1968.

Bourestom, N., and Pastalan, L. *Forced relocation: Setting, staff, and patient effects.* Final report to the Mental Health Services Development Branch, National Institute of Mental Health. Ann Arbor: Institute of Gerontology, University of Michigan, 1975.

Boutourline, S. The concept of environmental management. In H. M. Proshansky, W. H. Ittelson, and L. G. Rivlin (eds.), *Environmental psychology: Man and his physical setting.* New York: Holt, Rinehart & Winston, 1970.

Boyce, P. R. Age, illuminance, visual performance and preference. *Lighting Research Technology*, 1973, 5, 125–144.

Boyce, P. R. The luminous environment. In D. Canter and P. Stringer (eds.), *Environmental interaction: Psychological approaches to our physical surroundings.* New York: International Universities Press, 1975.

Bragdon, C. R. *Noise pollution: The unquiet crisis.* Philadelphia: University of Pennsylvania Press, 1970.

Brandt, J. A., and Chapman, N. J. *Student alterations of dormitory rooms: Social climate and satisfaction.* Paper presented at annual conference of Environmental Design Research Association, Charleston, S.C., 1980.

Bratingham, P. L., and Bratingham, P. J. A topological technique for regionalization. *Environment and Behavior*, 1978, 10, 335–353.

Bray, R. M., Kerr, N. L., and Atkin, R. S. Effects of group size, problem difficulty, and sex on group performance and member reactions. *Journal of Personality and Social Psychology*, 1978, 36, 1224–1240.

Brehm, J. W. *A theory of psychological reactance.* New York: Academic Press, 1966.

Briggs, R. Urban cognitive distance. In R. M. Downs and D. Stea (eds.), *Image and environment: Cognitive mapping and spatial behavior.* Chicago: Aldine, 1973.

Brill, M. Evaluating buildings on a performance basis. In C. Burnette, J. Lang, and D. Vaschon (eds.), *Architecture for human behavior: Collected papers from a mini-conference.* Philadelphia: Philadelphia Chapter/The American Institute of Architects, 1971.

Broadbent, D. E. Some effects of noise on visual performance. *Quarterly Journal of Experimental Psychology*, 1954, 6, 1–5.

Broadbent, D. E. Effect of noise on an "intellectual" task. *Journal of the Acoustical Society of America*, 1958, 30, 824–827.

Broadbent, D. E. *Decision and stress.* New York: Academic Press, 1971.

Broadbent, D. E. The current state of noise research: Reply to Poulton. *Psychological Bulletin*, 1978, 85, 1052–1067.

Brolin, B. C., and Zeisel, J. Mass housing: Social research and design. *Ekistics*, 1968, 158, 51–55.

Bronfenbrenner, U. The experimental ecology of education. *Educational Researcher*, 1976, 5, 5–15.

Bronfenbrenner, U. Toward an experimental ecology of human development. *American Psychologist*, 1977, 32, 513–531.

Bronzaft, A. L., Dobrow, S. B., and O'Hanlon, T. J. Spatial orientation in a subway system. *Environment and Behavior*, 1976, 8, 575–595.

Bronzaft, A. L., and McCarthy, D. P. The effect of elevated train noise on reading ability. *Environment and Behavior*, 1975, 7, 517–529.

Brookes, M. Office landscape: Does it work? *Applied Ergonomics*, 1972, 3, 224–236.

Brookes, M., and Kaplan, A. The office environment: Space planning and affective behavior. *Human Factors*, 1972, 14, 373–391.

Brower, S. N. Territoriality, the exterior spaces, the signs we learn to read. *Landscape*, 1965, 15, 9–12.

Brower, S. N., and Williamson, P. Outdoor recreation as a function of the urban environment. *Environment and Behavior*, 1974, *6*, 295–345.

Brunetti, F. A. *Open space: A status report.* Unpublished manuscript. School Planning Laboratory, School of Education, Stanford University, Stanford, California, 1971.

Brunetti, F. A. Noise, distracton, and privacy in conventional and open school environments. In W. J. Mitchell (ed.), *Environmental design: Research and practice.* Los Angeles: University of California, 1972.

Brunswik, E. *Perception and the representative design of psychological experiments.* Berkeley and Los Angeles: University of California Press, 1956.

Brunswik, E. The conceptual framework of psychology. In O. Neurath, R. Carnap, and C. Morris (eds.), *Foundation of the unity of science: Toward an international encyclopedia of unified science.* Chicago: University of Chicago Press, 1969.

Brush, R. O. Perceived quality of scenic and recreational environments: Some methodological issues. In K. H. Craik and E. H. Zube (eds.), *Perceiving environmental quality: Research and applications.* New York: Plenum, 1976.

Brush, R. O., and Shafer, E. L., Jr. Application of a landscape preference model to land management. In E. H. Zube, R. O. Brush, and J. Fabos (eds.), *Landscape assessment: Values, perceptions and resources.* Stroudsburg, Pa.: Dowden, Hutchinson & Ross, 1975.

Buckley, W. *Sociology and modern systems theory.* Englewood Cliffs, N.J.: Prentice-Hall, 1967.

Buechley, R., Van Bruggen, J., and Truppi, L. Heat Island = Death Island? *Environmental Research*, 1972, *5*, 85–92.

Burger, J. M., and Arkin, R. M. Prediction, control, and learned helplessness. *Journal of Personality and Social Psychology*, 1980, *38*, 482–491.

Burke, E. M. Citizen participation strategies. *Journal of the American Institute of Planners*, 1968, *34*, 290–291.

Burns, J. Development and implementation of an environmental evaluation and redesign process for a high school science department. In W. J. Mitchell (ed.), *Environmental design: Research and practice.* Los Angeles: University of California, 1972.

Burrows, A. A., and Zamarin, D. M. Aircraft noise and the community: Some recent survey findings. *Aerospace Medicine*, 1972, *43*, 27–33.

Burton, I., and Kates, R. W. The perception of natural hazards in resource management. *Natural Resources Journal*, 1964, *3*, 412–441.

Burton, I., Kates, R. W., and White, G. *The environment as hazard.* New York: Oxford University Press, 1978.

Buttel, F. H., and Flinn, W. L. The structure of support for the environmental movement, 1968–1970. *Rural Sociology*, 1974, *39*, 56–69.

Buttel, F. H., and Flinn, W. L. Social class and mass environmental beliefs: A reconsideration. *Environment and Behavior*, 1978, *10*, 433–450. (a)

Buttel, F. H., and Flinn, W. L. The politics of environmental concern: The impacts of party identification and political ideology on environmental attitudes. *Environment and Behavior*, 1978, *10*, 17–36. (b)

Buttimer, A. Social space and the planning of residential areas. *Environment and Behavior*, 1972, *4*, 279–318.

Byrne, D. The influence of propinquity and opportunities for interaction on classroom relationships. *Human Relations*, 1961, *14*, 63–69.

Byrne, D. *The attraction paradigm.* New York: Academic Press, 1971.

Byrne, D., Baskett, D. G., and Hodges, L. Behavioral indicators of interpersonal attraction. *Journal of Applied Social Psychology*, 1971, *1*, 137–149.

Byrne, D., and Buehler, J. A. A note on the influence of propinquity upon acquaintanceships. *Journal of Abnormal and Social Psychology*, 1955, *51*, 147–148.

Byrne, D., and Clore, G. L. A reinforcement model of evaluative responses. *Personality: An International Journal*, 1970, *1*, 103–128.

Byrne, D., Ervin, C. R., and Lamberth, J. Continuity between the experimental study of attraction and real life computer dating. *Journal of Personality and Social Psychology*, 1970, *16*, 157–165.

Cadwallader, M. Problems in cognitive distance: Implications for cognitive mapping. *Environment and Behavior*, 1979, *11*, 559–576.

Calhoun, J. B. A behavioral sink. In E. L. Bliss (ed.), *Roots of behavior*, New York: Harper & Row, 1962. (a)

Calhoun, J. B. Population density and social pathology. *Scientific American*, 1962, *206*, 139–148. (b)

Calhoun, J. B. Ecological factors in the development of behavior anomalies. In J. Zubin and H. F. Hunt (eds.), *Comparative psychopathology*. New York: Grune & Stratton, 1967.

Calhoun, J. B. Space and the strategy of life. In A. H. Esser (ed.), *Environment and behavior: The use of space by animals and men*. New York: Plenum, 1971.

Calvin, J. S., Dearinger, J. A., and Curtin, M. E. An attempt at assessing preferences for natural landscapes. *Environment and Behavior*, 1972, *4*, 447–470.

Cameron, P., Robertson, D., and Zaks, J. Sound pollution, noise pollution, and health: Community parameters. *Journal of Applied Psychology*, 1972, *56*, 67–74.

Campbell, D. E. Interior office design and visitor response. *Journal of Applied Psychology*, 1979, *64*, 648–653.

Campbell, D. T. On the conflicts between biological and social evolution and between psychology and moral tradition. *American Psychologist*, 1975, *30*, 1103–1126.

Campbell, D. T., and Stanley, J. C. *Experimental and quasi-experimental designs for research*. Chicago: Rand McNally, 1966.

Canter, D., Sanchez-Robles, J. C., and Watts, N. A scale for the cross-cultural evaluation of houses. In D. Canter and T. Lee (eds.), *Psychology and the Built Environment*, New York: Wiley, 1974.

Canter, D., and Tagg, S. K. Distance estimation in cities. *Environment and Behavior*, 1975, *7*, 59–81.

Caplan, G. *Support systems and community mental health: Lectures on concept development*. New York: Behavioral Publications, 1974.

Caplan, N., and Nelson, S. C. On being useful: The nature and consequences of psychological research on social problems. *American Psychologist*, 1973, *28*, 199–211.

Caplovitz, D. *The poor pay more*. New York: The Free Press, 1963.

Caplow, T., and Forman, R. Neighborhood interaction in a homogenous community. *American Sociological Review*, 1950, *15*, 357–366.

Carey, G. W. Density, crowding, stress and the ghetto. *American Behavioral Scientist*, 1972, *15*, 495–508.

Carls, E. G. The effects of people and man-induced conditions on preferences for outdoor recreation landscapes. *Journal of Leisure Research*, 1973, *6*, 113–124.

Carlsmith, J. M., and Anderson, C. A. Ambient temperature and the occurrence of collective violence: A new analysis. *Journal of Personality and Social Psychology*, 1979, *37*, 337–344.

Carlson, R., and Price, M. A. Generality of social schemas. *Journal of Personality and Social Psychology*, 1966, *3*, 589–592.

Carp, F. M. *A future for the aged*. Austin: University of Texas Press, 1966.

Carp, F. M. Housing and living environments of older people. In R. H. Binstock and E. Shanas (eds.), *Handbook of aging and the social sciences*. New York: Van Nostrand Reinhold, 1976.

Carp, F. M. Environmental effects upon the mobility of older people. *Environment and Behavior*, 1980, *12*, 139–156.

Carp, F. M., Zawadski, R. T., and Shokrkon, H. Dimensions of environmental quality. *Environment and Behavior*, 1976, *8*, 239–265.

Carpenter, C. R. Territoriality: A review of concepts and problems. In A. Roe and G. G. Simpson (eds.), *Behavior and evolution*. New Haven: Yale University Press, 1958.

Carr, S., and Schissler, D. The city as a trip: Perceptual selection and memory in the view from the road. *Environment and Behavior*, 1969, *1*, 7–36.

Carson, D. Population concentration and human stress. In B. F. Rourke (ed.), *Explorations in the psychology of stress and anxiety*. Don Mills, Ontario: Longmans Canada Limited, 1969.

Cassel, J. Physical illness in response to stress. In S. Levine and N. A. Scotch (eds.), *Social stress*. Chicago: Aldine, 1970.

Cassel, J. Health consequences of population density and crowding. In R. Revelle (ed.), *Rapid population growth: Consequences and policy implications*. Baltimore: Johns Hopkins University Press, 1971.

Castell, R. Effect of familiar and unfamiliar environments on proximity behavior of young children. *Journal of Experimental Child Psychology*, 1970, *9*, 342–347.

Chaikin, A. L., and Derlega, V. J. *Self-disclosure*. Morristown, N.J.: General Learning Press, 1974. (a)

Chaikin, A. L., and Derlega, V. J. Variables affecting the appropriateness of self-disclosure. *Journal of Consulting and Clinical Psychology*, 1974, *42*, 588–593. (b)

Chermayeff, S., and Alexander, C. *Community and privacy*. Garden City, N.Y.: Doubleday, 1963.

Chermayeff, S., and Tzonis, A. *Shape of community: Realization of human potential*. Baltimore: Penguin, 1971.

Cheyne, J. A., and Efran, M. G. The effect of spatial and interpersonal variables on the invasion of group control territories. *Sociometry*, 1972, *35*, 477–489.

Christian, J. J. Hormonal control of population growth. In B. E. Eleftheriou and R. L. Srott (eds.), *Hormonal correlates of behavior* (Vol. 1). New York: Plenum, 1975.

Clark, R. N., Burgess, R. L., and Hendee, J. C. The development of antilitter behavior in a forest campground. *Journal of Applied Behavior Analysis*, 1972, *5*, 1–5.

Cloward, R. A., and Ohlin, L. E. *Delinquency and opportunity: A theory of delinquent gangs*. New York: The Free Press, 1960.

Cofer, C. N., and Appley, M. H. *Motivation: Theory and research*. New York: Wiley, 1964.

Cohen, R., McManus, J., Fox, D., and Kastelnik, C. *Psych City: A simulated community*. New York: Pergamon, 1973.

Cohen, S. Environmental load and the allocation of attention. In A. Baum, J. E. Singer, and S. Valins (eds.), *Advances in environmental psychology*. (Vol. 1). Hillsdale, N.J.: Erlbaum, 1978.

Cohen, S. Aftereffects of stress on human performance and social behavior: A review of research and theory. *Psychological Bulletin*, 1980, *88*, 82–108.

Cohen, S., Evans, G. W., Krantz, D. S., and Stokols, D. Physiological, motivational and cognitive effects of aircraft noise on children: Moving from the laboratory to the field. *American Psychologist*, 1980, *35*, 231–243.

Cohen, S., Evans, G. W., Krantz, D. S., Stokols, D., and Kelly, S. Aircraft noise and children: Longitudinal and cross-sectional evidence on adaptation to noise and the effectiveness of noise abatement. *Journal of Personality and Social Psychology*, 1981, *40*, 331–345.

Cohen, S., Glass, D. C., and Phillips, S. Environment and health. In H. E. Freeman, S. Levine, and L. G. Reeder (eds.), *Handbook of medical sociology*. Englewood Cliffs, N.J.: Prentice-Hall, 1979.

Cohen, S., Glass, D. C., and Singer, J. E. Apartment noise, auditory discrimination, and reading ability in children. *Journal of Experimental Social Psychology*, 1973, *9*, 407–422.

Cohen, S., and Lezak, A. Noise and inattentiveness to social cues. *Environment and Behavior*, 1977, *9*, 559–573.

Cohen, S., and Spacapan, S. The aftereffects of stress: An additional interpretation. *Environmental Psychology and Nonverbal Behavior*, 1978, *3*, 43–57.

Cohen, S., and Weinstein, N. Nonauditory effects of noise on behavior and health. *Journal of Social Issues*, 1981, *37*, 36–70.

Colquhoun, W. P., and Goldman, R. F. Vigilance under induced hyperthermia. *Ergonomics*, 1972, *15*, 621–632.

Comer, R. J., and Piliavin, J. A. The effects of physical deviance upon face-to-face interaction: The other side. *Journal of Personality and Social Psychology*, 1972, *23*, 33–39.

Commoner, B. *The closing circle: Nature, man, and technology*. New York: Knopf, 1971.

Cone, J. D., and Hayes, S. C. Applied behavior analysis and the solution of environmental problems. In I. Altman and J. F. Wohlwill (eds.), *Human behavior and environment: Advances in theory and research* (Vol. II). New York: Plenum, 1977.

Cook, M. Experiments on orientation and proxemics. *Human Relations*, 1970, *23*, 61–76.

Cook, T. D., and Campbell, D. T. *Quasi-experimentation: Design & analysis issues for field settings.* Chicago: Rand McNally, 1979.

Coombs, C. H., Dawes, R. M., and Tversky, A. *Mathematical psychology: An elementary introduction.* Englewood Cliffs, N.J.: Prentice-Hall, 1970.

Cooper, C. St. Francis Square: Attitudes of its residents. *Journal of the American Institute of Architects*, 1971, *58*, 22–27.

Cooper, C. The house as a symbol of the self. In J. Lang, C. Burnette, W. Moleski, and D. Vachon (eds.), *Designing for human behavior: Architecture and the behavioral sciences.* Stroudsburg, Pa.: Dowden, Hutchinson & Ross, 1974.

Corbett, J. A. Are the suites the answer? *Environment and Behavior*, 1973, *5*, 413–420.

Corcoran, D. W. J. Noise and loss of sleep. *Quarterly Journal of Experimental Psychology*, 1962, *14*, 178–182.

Coughlin, R. E. The perception and valuation of water quality: A review of research method and findings. In K. H. Craik and E. H. Zube (eds.), *Perceiving environmental quality: Research and applications*, New York: Plenum, 1976.

Cox, K. R., and Zannaras, G. Designative perceptions of macro-spaces: Concepts, a methodology, and applications. In R. M. Downs and D. Stea (eds.), *Image and environment: Cognitive mapping and spatial behavior.* Chicago: Aldine, 1973.

Cozby, P. C. Self-disclosure: A literature review. *Psychological Bulletin*, 1973, *79*, 73–91.

Craik, K. H. Environmental psychology. In *New directions in psychology*, Vol. 4, New York: Holt, Rinehart & Winston, 1970.

Craik, K. H. Appraising the objectivity of landscape dimensions. In J. V. Krutilla (ed.), *Natural environments: Studies in theoretical and applied analysis.* Baltimore: Johns Hopkins University Press, 1972. (a)

Craik, K. H. Psychological factors in landscape appraisal. *Environment and Behavior*, 1972, *4*, 255–266. (b)

Craik, K. H. The personality paradigm in environmental psychology. In S. Wapner, S. Cohen, and B. Kaplan (eds.), *Experiencing the environment.* New York: Plenum, 1976.

Craik, K. H. Multiple scientific paradigms in environmental psychology. *International Journal of Psychology*, 1977, *12*, 147–157.

Craik, K. H. *Impressions of a place: Effects of media, context, and personality.* Paper presented at annual conference of American Psychological Association, Toronto, 1978.

Craik, K. H., and Appleyard, D. Streets of San Francisco: Brunswick's lens model applied to urban inference and assessment. *Journal of Social Issues*, 1980, *36*, 72–85.

Craik, K. H., and McKechnie, G. E. *Perception of environmental quality: Preferential judgments versus comparative appraisals.* Unpublished manuscript, University of California, Berkeley, 1974.

Craik, K. H., and Zube, E. H. The development of perceived environmental quality indices. In K. H. Craik and E. H. Zube (eds.), *Perceiving environmental quality: Research and applications*, New York: Plenum, 1976. (a)

Craik, K. H., and Zube, E. H. Summary and research strategies. In K. H. Craik and E. H. Zube (eds.), *Perceiving environmental quality: Research and applications*, New York: Plenum, 1976. (b)

Crane, D. A. Lynch: The image of the city—review. *Journal of the American Institute of Planners*, 1961, *27*, 152–155.

Crook, M. A., and Langdon, F. J. The effects of aircraft noise in schools around London Airport. *Journal of Sound and Vibration*, 1974, *34*, 222–232.

Cunningham, M. R. Notes of the psychological basis of environmental design: The right-left dimension in apartment floor plans. *Environment and Behavior*, 1977, *9*, 125–137.

D'Atri, D. A. Psychophysiological responses to crowding. *Environment and Behavior*, 1975, *7*, 237–252.

Dabbs, J. Physical closeness and negative feelings. *Psychonomic Science*, 1971, *23*, 141–143.

Dabbs, J., Fuller, P., and Carr, S. *Personal space when cornered: College students and prison inmates.* Paper presented at annual meeting of American Psychological Association, Montreal, 1973.

Danford, S., and Willems, E. Subjective responses to architectural displays: A question of validity. *Environment and Behavior,* 1975, *7,* 486–516.

Daniel, T. C. Criteria for development and application of perceived environmental quality indices. In K. H. Craik and E. H. Zube (eds.), *Perceiving environmental quality: Research and applications.* New York: Plenum, 1976.

Daniel, T. C., Wheeler, L., Boster, R. S., and Best, P. Quantitative evaluation of landscape: An application of signal detection analysis to forest management alternatives. *Man-Environment Systems,* 1973, *3,* 330–344.

Darley, J. M., and Latané, B. Bystander interventions in emergencies: Diffusion of responsibility. *Journal of Personality and Social Psychology,* 1968, *8,* 377–383.

Davidson, A. R., and Jacard, J. J. Variables that moderate the attitude-behavior relation: Results of a longitudinal survey. *Journal of Personality and Social Psychology,* 1979, *37,* 1364–1376.

Davis, D. E. Physiological effects of continued crowding. In A. H. Esser (ed.), *Behavior and environment: The use of space by animals and men.* New York: Plenum, 1971.

Day, H. Evaluation of subjective complexity, pleasingness and interestingness for a series of random polygons varying in complexity. *Perception and Psychophysics,* 1967, *2,* 281–286.

Dean, A., and Lin, N. The stress-buffering role of social support: Problems and prospects for systematic investigation. *Journal of Nervous and Mental Disease,* 1977, *165,* 403–417.

Dean, L. M., Pugh, W. M., and Gunderson, E. K. Spatial and perceptual components of crowding: Effects on health and satisfaction. *Environment and Behavior,* 1975, *7,* 225–236.

Dean, L. M., Pugh, W. M., and Gunderson, E. K. The behavioral effects of crowding: Definitions and methods. *Environment and Behavior,* 1978, *10,* 419–432.

Deasy, C. M. When architects consult people. *Psychology Today,* 1970, *3,* 54–57, 78–79.

de Jonge, D. Images of urban areas: Their structure and psychological foundations. *Journal of the American Institute of Planners,* 1962, *28,* 266–276.

DeLong, A. J. Dominance-territorial relations in a small group. *Environment and Behavior,* 1970, *2,* 190–191.

DeLong, A. J. Dominance-territorial criteria and small group structure. *Comparative Groups Studies,* 1971, *2,* 235–265.

DeLong, A. J. Kinesic signals at utterance boundaries in preschool children. *Dissertation Abstracts,* 1973, *33.* (a)

DeLong, A. J. Territorial stability and hierarchical formation. *Small Group Behavior,* 1973, *4,* 56–63. (b)

Department of City Planning. *The visual environment of Los Angeles.* Los Angeles: Department of City Planning, 1971.

Derlega, V. J., and Chaiken, A. L. Privacy and self-disclosure in social relationships. *Journal of Social Issues,* 1977, *33,* 102–116.

Derlega, V. J., Wilson, M., and Chaikin, A. L. Friendship and disclosure reciprocity. *Journal of Personality and Social Psychology,* 1976, *34,* 578–582.

Deutsch, M., and Collins, M. E. *Interracial housing: A psychological evaluation of a social experiment.* Minneapolis: University of Minnesota Press, 1951.

Devall, W. B. Conservation: An upper-middle class social movement: A replication. *Journal of Leisure Research,* 1970, *2,* 123–135.

Diener, E., Fraser, S. C., Beaman, A. L., and Kelem, R. T. Effects of deindividuation variables on stealing among Halloween trick–or–treaters. *Journal of Personality and Social Psychology,* 1967, *33,* 178–183.

Dinges, N. G., and Oetting, E. R. Interaction distance anxiety in the counseling dyad. *Journal of Counseling Psychology,* 1972, *19,* 146–149.

Ditton, R. B., and Goodale, T. L. Water quality perception and the recreational users of Green Bay, Lake Michigan. *Water Resources,* 1973, *9,* 569–579.

Dohrenwend, B. S. Social stress and community psychology. *American Journal of Community Psychology,* 1978, *6,* 1–14.

Donnerstein, E., and Wilson, D. W. Effects of noise and perceived control on ongoing and subsequent aggressive behavior. *Journal of Personality and Social Psychology,* 1976, *34,* 774–781.

Dorfman, P. W. Measurement and meaning of recreation satisfaction: A case study of camping. *Environment and Behavior,* 1979, *11,* 483–510.

Dosey, M. A., and Meisels, M. Personal space and self-protection. *Journal of Personality and Social Psychology,* 1969, *11,* 93–97.

Dotson, F. Patterns of voluntary association among urban working-class families. *American Sociological Review,* 1951, *16,* 687–693.

Douglas, T. J. *Environmental problems of the urban elderly: Measuring the quality of life.* Paper presented at annual conference of Environmental Design Research Association, Charleston, S.C., 1980.

Downs, R. M. The cognitive structure of an urban shopping center. *Environment and Behavior,* 1970, *2,* 13–39.

Downs, R. M., and Stea, D. Cognitive maps and spatial behavior: Process and products. In R. M. Downs and D. Stea (eds.), *Image and environment: Cognitive mapping and spatial behavior.* Chicago: Aldine, 1973.

Downs, R. M., and Stea, D. *Maps in minds: Reflections on cognitive mapping.* New York: Harper & Row, 1977.

Draper, P. Crowding among hunter–gatherers: The !Kung Bushmen. *Science,* 1973, *182,* 301–303.

Dubos, R. *Man adapting.* New Haven: Yale University Press, 1965.

Dubos, R. *So human an animal.* New York: Charles Scribner, 1968.

Dubos, R. We can't buy our way out. *Psychology Today,* 1970, *3,* (10), 20–22, 86–87.

DuHamel, T. R., and Jarmon, H. Social schemata of emotionally disturbed boys and their male siblings. *Journal of Consulting and Clinical Psychology,* 1971, *36,* 281–295.

Duhl, L. J. (ed.). *The urban condition.* New York: Basic Books, 1963.

Duke, M. P., and Nowicki, S., Jr. A new measure and social learning model for interpersonal distance. *Journal of Experimental Research in Personality,* 1972, *6,* 1–16.

Duke, M. P., and Wilson, J. The measurement of interpersonal distance in pre–school children. *Journal of Genetic Psychology,* 1973, *123,* 361–362.

Duncan, C. P. Personal communication (1963). Referenced in E. J. Webb, D. T. Campbell, R. D. Schwartz, and L. Sechrest. *Unobtrusive measures: Nonreactive research in the social sciences.* Chicago: Rand McNally, 1966.

Duncan, S. Mental maps of New York. *New York,* December 19, 1977, 51–62.

Dunlap, R. E. The impact of political orientation on environmental attitudes and actions. *Environment and Behavior,* 1975, *7,* 428–455.

Dweck, C. S. The role of expectations and attributions in the alleviation of learned helplessness. *Journal of Personality and Social Psychology,* 1975, *31,* 674–685.

Dyckman, J. W. Environment and behavior: Introduction. *The American Behavioral Scientist,* 1966, *10,* 1–2.

Eastmann, A. A. Colour contrast versus luminance contrast. *Illuminating Engineering,* 1968, *63,* 613–620.

Eaton, W. W. Life events, social supports, and psychiatric symptoms: A reanalysis of the New Haven data. *Journal of Health and Social Behavior,* 1978, *19,* 230–234.

Ebbesen, E. B., Kjos, G. L., and Konecni, V. J. Spatial ecology: Its effects on the choice of friends and enemies. *Journal of Experimental Social Psychology,* 1976, *12,* 505–518.

Eberts, E. H., and Lepper, M. R. Individual consistency in the proxemic behavior of pre-school children. *Journal of Personality and Social Psychology,* 1975, *32,* 481–489.

Edney, J. J. Property, possession and permanence: A field study in human territoriality. *Journal of Applied Social Psychology,* 1972, *3,* 275–282.

Edney, J. J. Human territoriality. *Psychological Bulletin,* 1974, *81,* 959–975.

Edney, J. J. Human territories: Comment on functional properties. *Environment and Behavior*, 1976, *8*, 31–49.

Edney, J. J., and Buda, M. A. Distinguishing territoriality and privacy: Two studies. *Human Ecology*, 1976, *4*, 283–295.

Edney, J. J., and Jordan-Edney, N. L. Territorial spacing on a beach. *Sociometry*, 1974, *37*, 92–103.

Edwards, D. J. A. Approaching the unfamiliar: A study of human interaction distances. *Journal of Behavioral Sciences*, 1972, *1*, 249–250.

EFL. *Profiles of significant schools: Schools without walls.* New York: Educational Facilities Laboratories, 1965.

Efran, M. G., and Cheyne, J. A. Shared space: The cooperative control of spatial areas by two interacting individuals. *Canadian Journal of Behavioral Science*, 1973, *5*, 201–210.

Efran, M. G., and Cheyne, J. A. Affective concomitants of the invasion of shared space: Behavioral, physiological and verbal indicators. *Journal of Personality and Social Psychology*, 1974, *29*, 219–226.

Eibl–Eibesfeldt, I. *Ethology: The biology of behavior.* New York: Holt, Rinehart & Winston, 1970.

Eisemon, T. *Simulations and requirements for citizen participation in public housing: The Truax technique.* Madison, Wisconsin: Institute for Simulation Analysis, 1975.

Ekehammar, B. Intractionism in personality from a historical perspective. *Psychological Bulletin*, 1974, *81*, 1026–1048.

Elms, A. Horoscopes and Ardrey. *Psychology Today*, 1972, *6*, 36.

Endler, N. S., and Magnusson, D. Toward an interactional psychology of personality. *Psychological Bulletin*, 1976, *83*, 956–974.

Epstein, W. *Varieties of perceptual learning.* New York: McGraw-Hill, 1967.

Epstein, Y. M. Crowding stress and human behavior. *Journal of Social Issues*, 1981, *37*, 126–144.

Epstein, Y. M., and Karlin, R. A. Effects of acute experimental crowding. *Journal of Applied Social Psychology*, 1975, *5*, 34–53.

Erikson, E. Ego development and historical change. In *The psychoanalytic study of the child*, Vol. 2. New York: International Universities Press, 1946.

Erikson, E. The problem of ego identity. *Journal of the American Psychoanalytic Association*, 1956, *4*, 56–121.

Eschenbrenner, A. J., Jr. Effects of intermittent noise on the performance of a complex psychomotor task. *Human Factors*, 1971, *13*, 59–63.

Esser, A. H. Dominance hierarchy and clinical course of psychiatrically hospitalized boys. *Child Development*, 1968, *39*, 147–157.

Esser, A. H. A biosocial perspective on crowding. In J. F. Wohlwill and D. H. Carson (eds.), *Environment and the social sciences: Perspectives and applications.* Washington, D.C.: American Psychological Association, 1972.

Esser, A. H. Cottage fourteen: Dominance and territoriality in a group of institutionalized boys. *Small Group Behavior*, 1973, *4*, 131–146.

Esser, A. H. Discussion of papers presented in the symposium *Theoretical and empirical issues with regard to privacy, territoriality, personal space, and crowding. Environment and Behavior*, 1976, *8*, 117–125.

Esser, A. H., Chamberlain, A. S., Chapple, E. D., and Kline, N. S. Territoriality of patients on a research ward. In J. Wortis (ed.), *Recent advances in biological psychiatry.* New York: Plenum, 1965.

Evans, G. W. *The relationship between interpersonal distance and human behavior.* Unpublished manuscript, Cognitive Processes Laboratory, University of Massachusetts, Amherst, 1972.

Evans, G. W. An examination of the information overload mechanism of personal space. *Man-Environment Systems*, 1974, *4*, 61.

Evans, G. W. *Behavioral and physiological consequences of crowding in humans.* Unpublished doctoral dissertation, University of Massachusetts, Amherst, 1975.

Evans, G. W. Crowding and human performance. *Journal of Applied Social Psychology*, 1979, 9, 27–46. (a)

Evans, G. W. Design implications of spatial research. In J. R. Aiello and A. Baum (eds.), *Residential crowding and design.* New York: Plenum, 1979. (b)

Evans, G. W. Environmental cognition. *Psychological Bulletin*, 1980, 88, 259–287.

Evans, G. W., and Eichelman, W. Preliminary models of conceptual linkages among proxemic variables. *Environment and Behavior*, 1976, 8, 87–117.

Evans, G. W., and Howard, R. B. Personal space. *Psychological Bulletin*, 1973, 80, 334–344.

Evans, G. W., and Jacobs, S. V. Air pollution and human behavior. *Journal of Social Issues*, 1981, 37, 95–125.

Evans, G. W., and Lovell, B. Design modification in an open-plan school. *Journal of Educational Psychology*, 1979, 71, 41–49.

Evans, G. W., Marrero, D., and Butler, P. Environmental learning and cognitive mapping. *Environment and Behavior*, 1981, 13, 83–104.

Evans, G. W., and Pezdek, K. Cognitive mapping: Knowledge of real–world distance and location information. *Journal of Experimental Psychology: Human Learning and Memory*, 1980, 6, 13–24.

Evans, G. W., and Wood, K. W. Assessment of environmental aesthetics in scenic highway corridors. *Environment and Behavior*, 1980, 12, 255–273.

Everett, P. B., Hayward, S. C., and Myers, A. W. The effects of a token reinforcement procedure on bus ridership. *Journal of Applied Behavior Analysis*, 1974, 7, 1–9.

Everitt, J., and Cadwallader, M. The home area concept in urban analysis. In W. J. Mitchell (ed.), *Environmental design: Research and practice.* Los Angeles: University of California, 1972.

Falk, S. A., and Woods, N. F. Hospital noise-levels and potential health hazards. *New England Journal of Medicine*, 1973, 289, 774–781.

Fanger, P. O. Improvement of human comfort and resulting effects on working capacity. *Biometeorology*, 1972, 5, 31–41.

Faris, R., and Dunham, H. W. *Mental disorders in urban areas* (2nd ed.). Chicago: Phoenix Books, 1965.

Fearn, R. W. Noise levels in youth clubs (correspondence). *Journal of Sound and Vibration*, 1972, 22, 127–128.

Felipe, N., and Sommer, R. Invasions of personal space. *Social Problems*, 1966, 14, 206–214.

Festinger, L. Architecture and group membership. *Journal of Social Issues*, 1951, 7, 152–163.

Festinger, L. *A theory of cognitive dissonance.* Stanford, Ca.: Stanford University Press, 1957.

Festinger, L., Pepitone, A., and Newcomb, T. M. Some consequences of deindividuation in a group. *Journal of Abnormal and Social Psychology*, 1952, 47, 382–389.

Festinger, L., Schacter, S., and Back, K. *Social pressures in informal groups.* New York: Harper, 1950.

Fiedler, F. E., and Fiedler, J. Port noise complaints: Verbal and behavioral reactions to airport-related noise. *Journal of Applied Psychology*, 1975, 60, 498–506.

Fines, K. D. Landscape evaluation—a research project in East Sussex. *Regional Studies*, 1968, 2, 40–55.

Fines, K. D. Landscape evaluations—a research project in East Sussex: Rejoinder to critique by D. M. Brancher. *Regional Studies*, 1969, 3, 219.

Finkel, J. M., and Glass, D. C. Reappraisal of the relationship between noise and human performance by means of a subsidiary task measure. *Journal of Applied Psychology*, 1970, 54, 211–213.

Finnie, W. C. Field experiments in litter control. *Environment and Behavior*, 1973, 5, 123–145.

Firestone, I. J., Karuza, J., Greenberg, C. I., and Kingma, K. *The perception of crowding: Modality, perspectual, and feedback effects.* Paper presented at annual conference of Environmental Design Research Association, Champaign-Urbana, Ill., 1978.

Fischer, C. S. *The urban experience.* New York: Harcourt Brace Jovanovich, 1976.

Fischer, C. S., Baldassare, M., Gerson, K., Jackson, R. M., Jones, L. M., and Stueve, C. A. *Networks and places: Social relations in the urban setting.* New York: The Free Press, 1977.

Fishbein, M. The prediction of behaviors from attitudinal variables. In C. D. Mortensen and K. K. Sereno (eds.), *Advances in communication research.* New York: Harper & Row, 1973.

Fishbein, M., and Ajzen, I. Attitudes toward objects as predictors of a single and multiple behavioral criteria. *Psychological Review,* 1974, *81,* 59–74.

Fisher, R. L. Social schema of normal and disturbed school children. *Journal of Educational Psychology,* 1967, *58,* 88–92.

Fiske, W., and Maddi, S. R. (eds.). *Functions of varied experience.* Homewood, Ill.: Dorsey Press, 1961.

Fog, H., and Jonsson, E. *Traffic noise in residential areas.* National Swedish Institute for Building Research Report 36E, 1968.

Folkman, S., and Lazarus, R. S. Coping in an adequately functioning middle-aged population. *Journal of Health and Social Behavior,* 1980, *21,* 219–239.

Foote, N., Abu-Lughod, J., Foley, M., and Winnick, L. *Housing choices and housing constraints.* New York: McGraw-Hill, 1970.

Forbes, G., and Gromoll, H. The lost letter technique as a measure of social variables: Some exploratory findings. *Social Forces,* 1971, *50,* 113–115.

Fox, W. F. Human performance in the cold. *Human Factors,* 1967, *9,* 203–220.

Foxx, R. M., and Hake, D. F. Gasoline conservation: A procedure for measuring and reducing the driving of college students. *Journal of Applied Behavior Analysis,* 1977, *10,* 61–74.

Francescato, D., and Mebane, W. How citizens view two great cities: Milan and Rome. In R. M. Downs and D. Stea (eds.), *Image and environment: Cognitive mapping and spatial behavior.* Chicago: Aldine, 1973.

Francescato, G., Weidemann, S., Anderson, J., and Chenoweth, R. Predictors of residents' satisfaction in high-rise and low-rise housing. *Journal of Architectural Research,* 1975, *4,* 4–10.

Frankel, A. S., and Barrett, J. Variations in personal space as a function of authoritarianism, self-esteem, and racial characteristics of a stimulus situation. *Journal of Consulting and Clinical Psychology,* 1971, *37,* 95–98.

Frankenhaeuser, M., and Lundberg, U. The influence of cognitive set on performance and arousal under different noise loads. *Motivation and Emotion,* 1977, *1,* 139–149.

Frede, M. C., Gautney, D. B., and Baxter, J. C. Relationships between body image boundary and interaction patterns on the Maps Test. *Journal of Consulting and Clinical Psychology,* 1968, *32,* 575–578.

Freedman, J. L. A positive view of population density. *Psychology Today,* 1971, *86,* 58–61.

Freedman, J. L. *Crowding and behavior.* San Francisco: W. H. Freeman, 1975.

Freedman, J. L. Current status of work on crowding and suggestions for housing design. In J. R. Aiello and A. Baum (eds.), *Residential crowding and design.* New York: Plenum, 1979. (a)

Freedman, J. L. Theoretical note: Reconciling apparent differences between the responses of humans and other animals to crowding. *Psychological Review,* 1979, *86,* 80–85. (b)

Freedman, J. L., Heshka, S., and Levy, A. Population density and pathology: Is there a relationship? *Journal of Experimental Social Psychology,* 1975, *11,* 539–552.

Freedman, J. L., Klevansky, S., and Ehrlich, P. I. The effect of crowding on human task performance. *Journal of Applied Social Psychology,* 1971, *1,* 7–26.

Freedman, J. L., Levy, A., Buchanan, R. W., and Price, J. Crowding and human aggressiveness. *Journal of Experimental Social Psychology,* 1972, *8,* 528–548.

Fried, M. Grieving for a lost home. In L. J. Duhl (ed.), *The urban condition.* New York: Basic Books, 1963.

Fried, M., and Gleicher, P. Some sources of residential satisfaction in an urban slum. *Journal of the American Institute of Planners,* 1961, *27,* 305–315.

Fried, M., and Levin, J. Some social functions of the urban slum. In B. J. Frieden and R. Morris (eds.), *Urban Planning and Social Policy.* New York: Basic Books, 1968.

Friedmann, A., Zimring, C., and Zube, E. *Environmental design evaluation.* New York: Plenum, 1978.

Fry, A. M., and Willis, F. N. Invasion of personal space as a function of the age of the invader. *Psychological Record*, 1971, *2*, 358–389.

Fry, G. A., Assessment of visual performance, *Illuminating Engineering*, 1962, *57*, 426–437.

Gal, R., and Lazarus, R. S. The role of activity in stress. *Journal of Human Stress*, 1975, *1*, 4–20.

Galle, O. R., Gove, W. R., and McPherson, J. M. Population density and pathology: What are the relationships for man? *Science*, 1972, *176*, 23–30.

Gans, H. J. Planning and social life: Friendship and neighbor relations in suburban communities. *Journal of the American Institute of Planners*, 1961, *27*, 134–140.

Gans, H. J. *The urban villagers.* New York: The Free Press, 1962.

Gans, H. J. *The Levittowners.* New York: Pantheon Books, 1967.

Gardin, H., Kaplan, C. J., Firestone, I. J., and Cowan, G. A. Proxemic effects on cooperation, attitude, and approach-avoidance in a prisoner's dilemma game. *Journal of Personality and Social Psychology*, 1973, *27*, 13–19.

Gardner, G. T. Effects of federal human subjects regulations on data obtained in environmental stressor research. *Journal of Personality and Social Psychology*, 1978, *36*, 628–634.

Garfinkel, H. Studies of the routine grounds of everyday activities. *Social Problems*, 1964, *11*, 225–250.

Gatchel, R. J., and Proctor, J. D. Physiological correlates of learned helplessness in man. *Journal of Abnormal Psychology*, 1976, *85*, 27–34.

Geen, R., and O'Neal, E. Activation of cue elicited aggression by general arousal. *Journal of Personality and Social Psychology*, 1969, *11*, 289–292.

Gelfand, D. M., Hartman, D. P., Walder, P., and Page, B. Who reports shoplifters? A field-experimental study. *Journal of Personality and Social Psychology*, 1973, *25*, 276–285.

Geller, E. S., Witmer, J. F., and Orebaugh, A. L. Instructions as a determinant of paper-disposal behaviors. *Environment and Behavior*, 1976, *8*, 417–441.

Gibson, E. J., and Walk, R. D. The "visual cliff." *Scientific American*, 1960, *202*, 64–71.

Gibson, J. J. Perception as a function of stimulation. In S. Koch (ed.), *Psychology: A study of a science*, Vol. I. New York: McGraw-Hill, 1959.

Gibson, J. J. The concept of the stimulus in psychology. *American Psychologist*, 1960, *15*, 694–703.

Gibson, J. J. The useful dimensions of sensitivity. *American Psychologist*, 1963, *18*, 1–15.

Gibson, J. J. *The senses considered as perceptual systems.* Boston: Houghton Mifflin, 1966.

Gibson, J. J. *An ecological approach to visual perception.* Boston: Houghton Mifflin, 1979.

Ginsburg, H. J., Pollman, V. A., Wauson, M. S., and Hope, M. L. Variation of aggressive interaction among male elementary school children as a function of changes in spatial density. *Environmental Psychology and Nonverbal Behavior*, 1977, *2*, 67–75.

Glass, D. C., Reim, B., and Singer, J. E. Behavioral consequences of adaptation to controllable and uncontrollable noise. *Journal of Experimental Social Psychology*, 1971, *7*, 244–257.

Glass, D. C., and Singer, J. E. Behavioral aftereffects of unpredictable and uncontrollable aversive events. *American Scientist*, 1972, *80*, 457–465. (a)

Glass, D. C., and Singer, J. E. *Urban stress.* New York: Academic Press, 1972. (b)

Glass, D. C., Singer, J. E., and Friedman, L. N. Psychic cost of adaptation to an environmental stressor. *Journal of Personality and Social Psychology*, 1969, *12*, 200–210.

Glass, D. C., Singer, J. E., and Pennebaker, J. W. Behavioral and physiological effects of uncontrollable environmental events. In D. Stokols (ed.), *Perspectives on environment and behavior: Theory, research, and applications.* New York: Plenum, 1977.

Glass, D. C., Snyder, M. L., and Singer, J. E. Periodic and aperiodic noise: The safety-signal hypothesis and noise aftereffects. *Physiological Psychology*, 1973, *1*, 361–363.

Glassman, J. B., Burkhart, B. R., Grant, R. D., and Vallery, G. G. Density, expectation, and extended task performance: An experiment in the natural environment. *Environment and Behavior*, 1978, *10*, 299–316.

Goeckner, D., Greenough, W., and Mead, W. Deficit in learning tasks following chronic overcrowding in rats. *Journal of Personality and Social Psychology*, 1973, *28*, 256–261.

Goffman, E. *The presentation of self in everyday life.* New York: Doubleday, 1959.

Goffman, E. *Asylums: Essays on the social situation of mental patients and other inmates.* Garden City, N.Y.: Anchor, 1961.

Goldberg, G. N., Kiesler, C. A., and Collins, B. E. Visual behavior and face-to-face distance during interaction. *Sociometry,* 1969, *32,* 43–53.

Goldring, P. *Role of distance and posture in the evaluation of interactions.* Paper presented at annual convention of American Psychological Association, 1967.

Goldsmith, J. R. Effects of air pollution on human health. In A. C. Stearn (ed.), *Air pollution* (2nd ed.). New York: Academic Press, 1968.

Golledge, R. G. Methods and methodological issues in environmental cognition research. In G. T. Moore and R. G. Golledge (eds.), *Environmental knowing.* Stroudsburg, Pa.: Dowden, Hutchinson & Ross, 1976.

Golledge, R. G. Multidimensional analysis in the study of environmental behavior and environmental design. In I. Altman and J. Wohlwill (eds.), *Human behavior and environment* (Vol. 2). New York: Plenum, 1977.

Golledge, R. G., and Zannaras, G. *The perception of urban structure: An experimental approach.* Paper presented at annual conference of Environmental Design Research Association, Pittsburgh, 1970.

Golledge, R. G. and Zannaras, G. Cognitive approaches to the analysis of human spatial behavior. In W. Ittelson (ed.), *Environment and cognition,* New York: Seminar Press, 1973.

Good, L. R., Siegal, S. M., and Bay, A. P. *Therapy by design: Implications of architecture for human behavior.* Springfield, Ill.: Charles C. Thomas, 1965.

Goodchild, B. Class differences in environmental perception. *Urban Studies,* 1974, *11,* 59–79.

Goodman, R. F., and Clary, B. B. Community attitudes and action in response to airport noise. *Environment and Behavior,* 1976, *8,* 441–471.

Gore, S. The effect of social support in moderating the health consequences of unemployment. *Journal of Health and Social Behavior,* 1978, *19,* 157–165.

Gottheil, E., Corey, J., and Paredes, A. Psychological and physical dimensions of personal space. *Journal of Psychology,* 1968, *69,* 7–9.

Graham, C. H. *Vision and visual perception.* New York: Wiley, 1965.

Green, D. E., and Rao, V. R. *Applied multidimensional scaling: A comparison of approaches and algorithms.* New York: Holt, Rinehart & Winston, 1972.

Greenbaum, P., and Rosenfeld, H. M. Patterns of avoidance in response to interpersonal staring and proximity: Effects of bystanders on drivers at a traffic intersection. *Journal of Personality and Social Psychology,* 1978, *36,* 575–587.

Greenberg, C. I., and Chambers, D. *Further tests on the validity of the model room simulator in crowding threshold judgments.* Paper presented at annual conference of Environmental Design Research Association, Tucson, Arizona, 1979.

Griffiths, I. D. The thermal environment. In D. Canter and P. Stringer (eds.), *Environmental interaction: Psychological approaches to our physical surroundings.* New York: International Universities Press, 1975.

Griffiths, I. D., and Boyce, P. R. Performance and thermal comfort. *Ergonomics,* 1971, *14,* 457–468.

Griffiths, I. D., and Langdon, F. J. Subjective response to road traffic noise. *Journal of Sound and Vibration,* 1968, *8,* 16–32.

Griffitt, W. Environmental effects on interpersonal affective behavior: Ambient effective temperature and attraction. *Journal of Personality and Social Psychology,* 1970, *15,* 240–244.

Griffitt, W., and Veitch, R. Hot and crowded: Influences of population density and temperature on interpersonal affective behavior. *Journal of Personality and Social Psychology,* 1971, *17,* 92–98.

Gross, H. Privacy and autonomy. In J. R. Pennock and J. W. Chapman (eds.), *Privacy.* New York: Atherton Press, 1971.

Guardo, C. J., and Meisels, M. Child-parent spatial patterns under praise and reproof. *Developmental Psychology,* 1971, *5,* 365. (a)

Guardo, C. J., and Meisels, M. Factor structure of children's personal space schemata. *Child Development,* 1971, *42,* 1307–1312. (b)

Gump, P. V. Ecological psychology and children. In E. M. Heatherington et al. (eds.), *Review of child development research.* Chicago: University of Chicago Press, 1975.

Gump, P. V., and Good, L. R. Environments operating in open space and traditionally designed schools. *Journal of Architectural Research,* 1976, 5, 20–27.

Gustke, L. D., and Hodgson, R. W. Rate of travel along an interpretive trail: The effect of an environmental discontinuity. *Environment and Behavior,* 1980, 12, 53–63.

Gutman, R. Site planning and social behavior. *Journal of Social Issues,* 1966, 22, 103–105.

Gutmann, D. Women and the conception of ego strength. *Merrill-Palmer Quarterly,* 1965, 11, 229–240.

Haase, R. F. The relationship of sex and instructional set to the regulation of interpersonal interaction distance in a counseling analogue. *Journal of Counseling Psychology,* 1970, 17, 233–236.

Haase, R. F., and DiMattia, D. J. Proxemic behavior: Counselor, administrator, and client preference for seating arrangement in dyadic interaction. *Journal of Counseling Psychology,* 1970, 17, 319–325.

Haase, R. F., and Pepper, D. T., Jr. Nonverbal components of empathic communication. *Journal of Counseling Psychology,* 1972, 19, 417–424.

Haber, G. M. Territorial invasion in the classroom: Invadee response. *Environment and Behavior,* 1980, 12, 17–31.

Hall, E. T. The anthropology of manners. *Scientific American,* 1955, 192, 85–89.

Hall, E. T. *The silent language.* New York: Doubleday, 1959.

Hall, E. T. The silent language in overseas business. *Harvard Business Review,* 1960, 38, 87–96.

Hall, E. T. A system for the notation of proxemic behavior. *American Anthropologist,* 1963, 65, 1003–1026. (a)

Hall, E. T. Proxemics, the study of man's spatial relations. In I. Goldston (ed.), *Man's image in medicine and anthropology.* New York: International Universities Press, 1963. (b)

Hall, E. T. *The hidden dimension.* New York: Doubleday, 1966.

Hall, R., Purcell, A. T., Thorne, R., and Metcalfe, J. Multidimensional scaling analysis of interior, designed spaces. *Environment and Behavior,* 1976, 8, 595–611.

Hand, D. J., Tarnopolsky, A., Barker, S. M., and Jenkins, L. M. Relationships between psychiatric hospital admissions and aircraft noise: A new study. In J. V. Tobias (ed.), *The proceedings of the third international congress on noise as a public health problem.* Washington, D.C.: American Speech and Hearing Association, 1980.

Hansen, W. B., and Altman, I. Decorating personal places: A descriptive analysis. *Environment and Behavior,* 1976, 8, 491–505.

Hanson, S., Vitek, J. D., and Hanson, P. O. Natural disaster: Long-range impact on human response to future disaster threats. *Environment and Behavior,* 1979, 11, 268–284.

Hardwick, D. A., McIntyre, C. W., and Pick, H. L. The content and manipulation of cognitive maps in children and adults. *Monographs of the Society for Research in Child Development,* 1976, 41(3, Serial No. 166).

Hare, A. P., and Bales, R. F. Seating position and small group interaction. *Sociometry,* 1963, 26, 480–486.

Harman, E., and Betak, J. F. *Some preliminary findings on the cognitive meaning of external privacy in housing.* Paper presented at annual conference of Environmental Design Research Association, Milwaukee, 1974.

Hart, R. A., and Moore, G. T. The development of spatial cognition: A review. In R. M. Downs and D. Stea (eds.), *Image and environment: Cognitive mapping and spatial behavior.* Chicago: Aldine, 1973.

Hartley, L. R. Effect of prior noise or prior performance on serial reaction. *Journal of Experimental Psychology,* 1973, 101, 255–261.

Hartley, L. R., and Adams, R. G. Effect of noise on the Stroop Test. *Journal of Experimental Psychology,* 1974, 102, 62–66.

Hartman, C. The limitations of public housing: Relocation choices in a working-class community. *Journal of American Institute of Planners,* 1963, 24, 283–296.

Hartman, C. The housing of relocated families. In J. Bellush and M. Hausknecht (eds.), *Urban renewal: People, politics, and planning.* New York: Doubleday, 1967.

Hartman, C. Relocation: Illusory promises and no relief. *Virginia Law Review,* 1971, *57,* 745–817.

Hartman, C. *Housing and social policy.* Englewood Cliffs, N.J.: Prentice-Hall, 1975.

Hartnett, J. J., Bailey, F., and Gibson, W. Personal space as influenced by sex and type of movement. *Journal of Psychology,* 1970, *76,* 139–144.

Hasell, J., Scavo, C., and Moore, R. D. *A gaming/simulation workshop to explore a partnership process for neighborhood revitalization.* Paper presented at annual conference of Environmental Design Research Association, Charleston, S.C., 1980.

Hayduk, L. A. *Proxemics: An investigation into the shape of human personal space.* Unpublished manuscript, University of Western Ontario, London, Ontario, Canada, 1975.

Hayduk, L. A. Personal space: An evaluative and orienting overview. *Psychological Bulletin,* 1978, *85,* 117–134.

Hayes, S. C., and Cone, J. D. Reducing residential electrical energy use: Payments, information, and feedback. *Journal of Applied Behavior Analysis,* 1977, *10,* 425–435.

Hazard, W. R. Predictions of noise disturbance near large airports. *Journal of Sound and Vibration,* 1971, *15,* 425–445.

Hazen, N., Lockman, J., and Pick, H. The development of children's representations of large-scale environment. *Child Development,* 1978, *49,* 623–636.

Hebb, D. O. Drives and the C.N.S. *Psychological Review,* 1955, *62,* 243–254.

Hediger, H. Wild animals in captivity. London: Buttersworth, 1950.

Hediger, H. *Studies of the psychology and behavior of captive animals in zoos and circuses.* London: Buttersworth, 1955.

Hediger, H. The evolution of territorial behavior. In S. L. Washburn (ed.), *Social life of early man.* New York: Wennergren Foundation, 1961.

Held, R., and Rekosh, J. Motor-sensory feedback and the geometry of visual space. *Science,* 1963, *141,* 722–723.

Heller, J. F., Groff, B. D., and Solomon, S. H. Toward an understanding of crowding: The role of physical interaction. *Journal of Personality and Social Psychology,* 1977, *35,* 183–190.

Hempel, D. J., and Tucker, L. R., Jr. Citizen preferences for housing as community social indicators. *Environment and Behavior,* 1979, *11,* 399–428.

Herman, J., and Siegel, A. The development of spatial representations of large-scale environments. *Journal of Experimental Child Psychology,* 1978, *26,* 389–406.

Herzog, T. R., Kaplan, S., and Kaplan, R. The prediction of preference for familiar urban places. *Environment and Behavior,* 1976, *8,* 627–647.

Heslin, R., and Boss, D. Nonverbal intimacy in airport arrival and departure. *Personality and Social Psychology Bulletin,* 1980, *6,* 248–252.

Higgins, J., Peterson, J., and Lise-Lotte, L. Social adjustment and familial schema. *Journal of Abnormal Psychology,* 1969, *74,* 296–299.

Hinshaw, M., and Allott, K. Environmental preferences of future housing consumers. *Journal of the American Institute of Planners,* 1972, *38,* 102–207.

Hiroto, D. S. Locus of control and learned helplessness. *Journal of Experimental Psychology,* 1974, *102,* 187–193.

Hiroto, D. S., and Seligman, M. E. P. Generality of learned helplessness in man. *Journal of Personality and Social Psychology,* 1975, *31,* 311–327.

Hockey, G. J. R. Effect of loud noise on attentional selectivity. *Quarterly Journal of Experimental Psychology,* 1970, *22,* 23–36. (a)

Hockey, G. J. R. Signal probability and spatial location as possible bases for increased selectivity in noise. *Quarterly Journal of Experimental Psychology,* 1970, *22,* 37–42. (b)

Holahan, C. J. Seating patterns and patient behavior in an experimental dayroom. *Journal of Abnormal Psychology,* 1972, *80,* 115–124.

Holahan, C. J. Environmental change in a psychiatric setting: A social systems analysis. *Human Relations,* 1976, *29,* 153–166. (a)

Holahan, C. J. Environmental effects on outdoor social behavior in a low-income urban neighborhood: A naturalistic investigation. *Journal of Applied Social Psychology*, 1976, *6*, 48–63. (b)

Holahan, C. J. Consultation in environmental psychology: A case study of a new counseling role. *Journal of Counseling Psychology*, 1977, *24*, 251–254. (a)

Holahan, C. J. Urban-rural differences in judged appropriateness of altruistic responses: Personal vs. situational effects. *Sociometry*, 1977, *40*, 378–382. (b)

Holahan, C. J. *Environment and behavior: A dynamic perspective.* New York: Plenum, 1978.

Holahan, C. J. Action research in the built environment. In R. H. Price and P. E. Politser, *Evaluation and Action in the Social Environment.* New York: Academic Press, 1980.

Holahan, C. J., Culler, R. E., and Wilcox, B. L. Effects of visual distraction on reaction time in a simulated traffic environment. *Human Factors*, 1978, *20*, 409–413.

Holahan, C. J., and Dobrowolny, M. B. Cognitive and behavioral correlates of the spatial environment: An interactional analysis. *Environment and Behavior*, 1978, *10*, 317–334.

Holahan, C. J., and Holahan, C. K. Sex-related differences in the schematization of the behavioral environment. *Personality and Social Psychology Bulletin*, 1977, *3*, 123–126.

Holahan, C. J., and Kovalic, J. *National planning for water resources management: New prospects for environmental psychology.* Paper presented at annual meeting of Southwestern Psychological Association, El Paso, Texas, 1974.

Holahan, C. J., and Levinger, G. Psychological versus spatial determinants of social schema distance: A methodological note. *Journal of Abnormal Psychology*, 1971, *28*, 232–236.

Holahan, C. J., and Moos, R. H. *Development of qualitative indices of social support.* Unpublished manuscript, Social Ecology Laboratory, Stanford University and VA Medical Centers, Palo Alto, Ca., 1980.

Holahan, C. J., and Moos, R. H. Social support and psychological adjustment: Predictive benefits of social climate indices. *American Journal of Community Psychology*, in press.

Holahan, C. J., and Moos, R. H. Social support and psychological distress: A longitudinal analysis. *Journal of Abnormal Psychology*, 1981, *49*, 365– 370.

Holahan, C. J., and Saegert, S. Behavioral and attitudinal effects of large-scale variation in the physical environment of psychiatric wards. *Journal of Abnormal Psychology*, 1973, *82*, 454–462.

Holahan, C. J., and Slaikeu, K. A. Effects of contrasting degrees of privacy on client self-disclosure in a counseling setting. *Journal of Counseling Psychology*, 1977, *24*, 55–59.

Holahan, C. J., and Wilcox, B. L. Environmental satisfaction in high and low-rise student housing: An interactional analysis. *Journal of Educational Psychology*, 1978, *70*, 237–41.

Holahan, C. J., and Wilcox, B. L. Environmental satisfaction and social coping in university residential settings: A Lewinian perspective. In J. R. Aiello and A. Baum (eds.), *Residential crowding and design.* New York: Plenum, 1979.

Holahan, C. J., Wilcox, B. L., Burnham, M. A., and Culler, R. E. Social adjustment as a function of floor level in high rise student housing. *Journal of Applied Psychology*, 1978, *63*, 527–529.

Holahan, C. K., and Holahan, C. J. Effects of gender and psychological masculinity and femininity on environmental schematization. *Personality and Social Psychology Bulletin*, 1979, *5*, 231–235.

Homans, G. C. *The human group.* New York: Harcourt, Brace, 1950.

Horowitz, M. J. Human spatial behavior. *American Journal of Psychotherapy*, 1965, *19*, 20–28.

Horowitz, M. J. Spatial behavior and psychopathology. *Journal of Nervous and Mental Disease*, 1968, *146*, 24–35.

Horowitz, M. J., Duff, D. F., and Stratton, L. O. Body buffer zone: Exploration of personal space. *Archives of General Psychiatry*, 1964, *11*, 651–656.

House, J. S., and Wolf, S. Effects of urban residence on interpersonal trust and helping behavior. *Journal of Personality and Social Psychology*, 1978, *36*, 1029–1043.

Hovland, C., Janis, I., and Kelley, H. H. *Communication and Persuasion.* New Haven: Yale University Press, 1953.

Howard, E. *Territory and bird life.* London: Collins, 1948.

Howard, J. A. Person–situation interaction models. *Personality and Social Psychology Bulletin,* 1979, 5, 191–195.

Howells, L. T., and Becker, S. W. Seating arrangement and leadership emergence. *Journal of Abnormal and Social Psychology,* 1962, 64, 148–150.

Hummell, C. F., Levitt, L., and Loomis, R. J. Perceptions of the energy crisis: Who is blamed and how do citizens react to environment-lifestyle trade-offs? *Environment and Behavior,* 1978, 10, 37–88.

Hutt, C., and Vaizey, M. J. Differential effects of group density on social behavior. *Nature,* 1966, 209, 1371–1372.

I.G.U. Air Pollution Study Group for the U.K. *Papers on selected social aspects of air pollution in the United Kingdom.* International Geographical Union, Commission on Man and Environment, Calgary, Canada, 1972.

Insel, P. M., and Lindgren, H. C. *Too close for comfort: The psychology of crowding.* Englewood Cliffs, N.J.: Prentice-Hall, 1978.

Insko, C., and Cialdini, R. A test of three interpretations of attitudinal verbal reinforcement. *Journal of Personality and Social Psychology,* 1969, 12, 333–341.

Ittelson, W. H. *Visual space perception.* New York: Springer, 1960.

Ittelson, W. H. Perception of the large-scale environment. *Transactions of the New York Academy of Sciences,* 1970, 32, 807–815.

Ittelson, W. H. (ed.), *Environment and cognition.* New York: Seminar Press, 1973.

Ittelson, W. H. Some issues facing a theory of environment and behavior. In H. M. Proshansky, W. H. Ittelson, and L. G. Rivlin (eds.), *Environmental psychology: People and their physical settings.* New York: Holt, Rinehart & Winston, 1976.

Ittelson, W. H., Franck, K. A., and O'Hanlon, T. J. The nature of environmental experience. In S. Wapner, S. B. Cohen, and B. Kaplan, (eds.), *Experiencing the environment.* New York: Plenum, 1976.

Ittelson, W. H., and Kilpatrick, F. P. Experiments in perception. *Scientific American,* 1952, 185, 50–55.

Ittelson, W. H., and Krawetz, N. *Further research in the novel environment laboratory.* Unpublished manuscript, Environmental Psychology Program, City University of New York, 1975.

Ittelson, W. H., Proshansky, H. M., and Rivlin, L. G. Bedroom size and social interaction of the psychiatric ward. *Environment and Behavior,* 1970, 2, 255–270.

Ittelson, W. H., Proshansky, H. M., and Rivlin, L. G. The environmental psychology of the psychiatric ward. In H. M. Proshansky, W. H. Ittelson, and L. G. Rivlin (eds.), *Environmental psychology: People and their physical settings.* New York: Holt, Rinehart & Winston, 1976.

Ittelson, W. H., Rivlin, L. G., and Proshansky, H. M. The use of behavioral maps in environmental psychology. In H. M. Proshansky, W. H. Ittelson, and L. G. Rivlin (eds.), *Environmental psychology: People and their physical settings.* New York: Holt, Rinehart & Winston, 1976.

Izumi, K. Psychosocial phenomena and building design. *Building Research,* 1965, 2, 9–11.

Jacobs, J. *The death and life of great American cities.* New York: Random House, 1961.

James, L. R. and Singh, B. K. An introduction to the logic, assumptions, and basic analytic procedures of two-stage least squares. *Psychological Bulletin,* 1978, 85, 1104–1122.

Jansen, G. Effects of noise on physiology. *American Speech and Hearing Association Review,* 1969, 89–98.

Jeffery, C. *Crime prevention through environmental design.* Beverly Hills, Ca.: Sage, 1971.

Jerison, H. J. Effects of noise on human performance. *Journal of Applied Psychology,* 1959, 43, 96–101.

Johnston, A. W. and Cole, B. L. Investigations of distraction by irrelevant information. *Australian Road Research,* 1976, 6, 3–23.

Jones, D. M., Smith, A. P., and Broadbent, D. E. Effects of moderate intensity noise on the Bakan vigilance task. *Journal of Applied Psychology,* 1979, 64, 627–634.

Jones, H. H., and Cohen, A. Noise as a health hazard at work, in the community, and in the home. *USPHS, Public Health Reports*, July, 1968, *83*, 533–536.

Jones, S. E. A comparative proxemics analysis of dyadic interaction in selected subcultures of New York City. *Journal of Social Psychology*, 1971, *84*, 35–44.

Jones, S. E., and Aiello, J. R. Proxemic behavior of black and white first, third, and fifth grade children. *Journal of Personality and Social Psychology*, 1973, *25*, 21–27.

Jonsson, A., and Hansson, L. Prolonged exposure to a stressful stimulus (noise) as a cause of raised blood pressure in man. *Lancet*, 1977, *1*, 86–87.

Jöreskog, K. G. A general method for analysis of covariance structures. *Biometrika*, 1970, *57*, 239–257.

Jorgenson, D. O. Field study of the relationship between status discrepancy and proxemic behavior. *Journal of Social Psychology*, 1975, *97*, 173–179.

Jorgenson, D. O., and Dukes, F. O. Deindividuation as a function of density and group membership. *Journal of Personality and Social Psychology*, 1976, *34*, 24–39.

Jourard, S. M. An exploratory study of body accessibility. *British Journal of Social and Clinical Psychology*, 1966, *5*, 221–231. (a)

Jourard, S. M. Some psychological aspects of privacy. *Law and Contemporary Problems*, 1966, *31*, 307–318. (b)

Jourard, S. M. Experimenter-subject "distance" and self-disclosure. *Journal of Personality and Social Psychology*, 1970, *15*, 278–282.

Jourard, S. M. *Self-disclosure: An experimental analysis of the transparent self.* New York: Wiley, 1971.

Jung, C. *Memories, dreams, reflections.* London: Collins, 1969.

Kahle, L. R., and Berman, J. J. Attitudes cause behaviors: A cross-lagged panel analysis. *Journal of Personality and Social Psychology*, 1971, *37*, 315–321.

Kahneman, D. *Attention and effort.* Englewood Cliffs, N.J.: Prentice-Hall, 1973.

Kalven, H., Jr. Privacy in tort law: Were Warren and Brandeis wrong? *Law and Contemporary Problems*, 1966, *31*, 326–341.

Kaplan, R. Patterns of environmental preference. *Environment and Behavior*, 1977, *9*, 195–217. (a)

Kaplan, R. Preference and everyday nature: Method and application. In D. Stokols (ed.), *Perspectives on environment and behavior: Theory, research, and applications.* New York: Plenum, 1977. (b)

Kaplan, S. Review of defensible space. *Architectural Forum*, 1973, *138* (4), 8.

Kaplan, S. Cognitive maps in perception and thought. In R. M. Downs and D. Stea (eds.), *Image and environment: Cognitive mapping and spatial behavior.* Chicago: Aldine, 1973. (a)

Kaplan, S. An information model for the prediction of preference. In E. H. Zube, R. O. Brush, and J. G. Fabos (eds.), *Landscape assessment.* Stroudsburg, Pa.: Dowden, Hutchinson & Ross, 1975.

Kaplan, S. Participation in the design process: A cognitive approach. In D. Stokols (ed.), *Perspectives on environment and behavior: Theory, research, and applications.* New York: Plenum, 1977.

Kaplan, S., Kaplan, R., and Wendt, J. S. Rated preference and complexity for natural and urban visual material. *Perception and Psychophysics*, 1972, *12*, 354–356.

Karabenick, S., and Meisels, M. Effects of performance evaluation on interpersonal distance. *Journal of Personality*, 1972, *40*, 257–286.

Karan, P. P., Bladen, W. A., and Singh, G. Slum dwellers' and squatters' images of the city. *Environment and Behavior*, 1980, *12*, 81–100.

Karlin, R. A., McFarland, D., Aiello, J. R., and Epstein, Y. M. Normative mediation of reactions to crowding. *Environmental Psychology and Nonverbal Behavior*, 1976, *1*, 30–40.

Karlin, R. A., Rosen, L. S., and Epstein, Y. M. Three into two doesn't go: A follow-up on the effects of overcrowded dormitory rooms. *Personality and Social Psychology Bulletin*, 1979, *5*, 391–395.

Kasl, S. Physical and mental health effects of involuntary relocation and institutionalization—a review. *American Journal of Public Health*, 1972, *62*, 379–384.

Kasl, S., and Harburg, E. Perceptions of the neighborhood and the desire to move out. *Journal of the American Institute of Planners*, 1972, *38*, 318–324.

Kasmar, J. V. The development of a usable lexicon of environmental descriptors. *Environment and Behavior*, 1970, *2*, 153–170.

Kates, R. W. Experiencing the environment as hazard. In H. M. Proshansky, W. H. Ittelson, and L. G. Rivlin (eds.), *Environmental psychology: People and their physical settings.* New York: Holt, Rinehart & Winston, 1976.

Kaufmann, J. H. Is territoriality definable? In A. H. Esser (ed.), *Behavior and environment: The use of space by animals and men.* New York: Plenum, 1971.

Keller, S. *The urban neighborhood: A sociological perspective.* New York: Random House, 1968.

Kelly, F. D. Communicational significance of therapist proxemic cues. *Journal of Consulting and Clinical Psychology*, 1972, *39*, 345.

Kelly, J. G. The mental health agent in the urban community. In *Urban America and the planning of mental health services.* New York: Group for the Advancement of Psychiatry, 1964.

Kenrick, D. T., and Johnson, G. A. Interpersonal attraction in aversive environments: A problem for the classical conditioning paradigm? *Journal of Personality and Social Psychology*, 1979, *37*, 572–579.

Keyfitz, N. Population density and the style of social life. In R. H. Moos and P. M. Insel (eds.), *Issues in social ecology.* Palo Alto, Ca.: National Press Books, 1974.

Kiesler, C. A., Collins, B. E., and Miller, N. *Attitude change: A critical analysis of theoretical approaches,* New York: Wiley, 1969.

Kilpatrick, F. P. (ed.), *Human behavior from the transactional point of view.* Hanover, N.H.: Institute for Associated Research, 1952.

Kilpatrick, F. P. Two processes in perceptual learning. *Journal of Experimental Psychology*, 1954, *47*, 362–370.

King, M. G. Interpersonal relations in preschool children and average approach distance. *The Journal of Genetic Psychology*, 1966, *109*, 109–116.

Kinzel, A. F. Body buffer zone in violent prisoners. *American Journal of Psychiatry*, 1970, *127*, 59–64.

Kira, A. *The bathroom: Criteria for design.* Ithaca, N.Y.: Center for Housing and Environmental Studies, Cornell University, 1966.

Kirmeyer, S. L. Urban density and pathology. *Environment and Behavior*, 1978, *10*, 247–270.

Kleck, R. E., Buck, P. L., Goller, W. C., London, R. S., Pfeiffer, J. R., and Vukcevic, D. P. Effect of stigmatizing conditions on the use of personal space. *Psychological Reports*, 1968, *23*, 111–118.

Kleinman, A., and Maxim, P. (eds.), *Theoretical bases for psychopathology.* New York: Spectrum, 1980.

Klopfer, P. H. From Ardrey to altruism: A discourse on the biological basis of human behavior. *Behavioral Science*, 1968, *13*, 399–401.

Knipmeyer, J., and Prestholdt, P. *The influence of environmental noise upon group aggression.* Paper presented at annual meeting of Southeastern Psychological Association, New Orleans, April 1973.

Knowles, E. S. Boundaries around group interaction: The effect of group size and member status on boundary permeability. *Journal of Personality and Social Psychology*, 1973, *26*, 327–332.

Knowles, E. S., Kreuser, B., Haas, S., Hyde, M., and Schuchart, G. Group size and the extension of social space boundaries. *Journal of Personality and Social Psychology*, 1976, *33*, 647–654.

Koenig, D. J. Additional research on environmental activism. *Environment and Behavior*, 1975, *7*, 472–485.

Koffka, K. *Principles of Gestalt psychology.* New York: Harcourt, Brace & World, 1935.

Kohlenberg, R., Phillips, T., and Proctor, W. A behavioral analysis of peaking in residential electrical energy consumers. *Journal of Applied Behavior Analysis*, 1976, *9*, 13–18.

Kohler, I. Experiments with goggles. *Scientific American*, 1962, *206*, 62–72.

Kohler, W. *Gestalt psychology.* New York: Liveright, 1929.

Konecni, V. The mediation of aggressive behavior: Arousal level vs. anger and cognitive labeling. *Journal of Personality and Social Psychology*, 1975, *32*, 706–712.

Konecni, V. J., Libuser, L., Morton, H., and Ebbeson, E. B. Effects of a violation of personal space on escape and helping responses. *Journal of Experimental Social Psychology*, 1975, *11*, 288–299.

Koneya, M. Location and interaction in row-and-column seating arrangements. *Environment and Behavior*, 1976, *8*, 265–283.

Korte, C., and Kerr, N. Responses to altruistic opportunities under urban and rural conditions. *Journal of Social Psychology*, 1975, *95*, 183–184.

Korte, C., Ypma, I., and Toppen, A. Helpfulness in Dutch society as a function of urbanization and environmental input level. *Journal of Personality and Social Psychology*, 1975, *32*, 996–1003.

Kosslyn, K., and Pomerantz, J. P. Imagery, propositions, and the form of internal representations. *Cognitive Psychology*, 1977, *9*, 52–76.

Kosslyn, S. M., Heldmeyer, K. H., and Locklear, E. P. Children's drawings as data about internal representations. *Journal of Experimental Child Psychology*, 1977, *23*, 191–211.

Kozlowski, L. T., and Bryant, K. J. Sense of direction, spatial orientation and cognitive maps. *Journal of Experimental Psychology: Human Perception and Performance*, 1977, *3*, 590–598.

Krantz, D. S., Glass, D. C., and Snyder, M. L. Helplessness, stress level and the coronary prone behavior pattern. *Journal of Experimental Social Psychology*, 1974, *10*, 284–300.

Krauss, R. M., Freedman, J. L., and Whitcup, M. Field and laboratory studies of littering. *Journal of Experimental Social Psychology*, 1978, *14*, 109–122.

Krech, D., Crutchfield, R. S., and Ballachey, E. L. *Individual in society*. New York: McGraw-Hill, 1962.

Kritzer, H. M. Political protest and political violence: A nonrecursive causal model. *Social Forces*, 1977, *55*, 630–640.

Krovetz, M. L. Who needs what when: Design of pluralistic learning environments. In D. Stokols (ed.), *Perspectives on environment and behavior: Theory, research, and applications*. New York: Plenum, 1977.

Kryter, K. D. The effects of noise on man. *Journal of Speech and Hearing Disorders*, 1950, *15* (Monograph Supplement 1).

Kryter, K. D. *The effects of noise on man*. New York: Academic Press, 1970.

Kubzansky, P. E., Salter, L. R., and Porter, G. *"Ideology" of open-office design: Issues and remedies*. Paper presented at annual conference of Environmental Design Research Association, Charleston, S.C., 1980.

Kuethe, J. L. Social schemas. *Journal of Abnormal and Social Psychology*, 1962, *64*, 31–38. (a)

Kuethe, J. L. Social schemas and the reconstruction of social object displays from memory. *Journal of Abnormal and Social Psychology*, 1962, *65*, 71–74. (b)

Kuethe, J. L. Prejudice and aggression: A study of specific social schemata. *Perceptual and Motor Skills*, 1964, *18*, 107–115.

Kuper, L. *Living in towns*. London: Cresset Press, 1953.

Kushler, M. G. *Energy conservation attitudes and behaviors of high school youth: Results of a two year study*. Paper presented at annual conference of Environmental Design Research Association, Charlestown, S.C., 1980.

Ladd, F. Black youths view their environments. *Environment and Behavior*, 1970, *2*, 74–99.

Lamanna, R. A. Value consensus among urban residents. *Journal of the American Institute of Planners*, 1964, *30*, 317–323.

Lang, J. Theories of perception and "formal" design. In J. Lang, C. Burnette, W. Moleski, and D. Vaschon (eds.), *Designing for human behavior: Architecture and the behavioral sciences*. Stroudsburg, Pa.: Dowden, Hutchinson & Ross, 1974.

Langer, E. J., and Rodin, J. The effects of choice and enhanced personal responsibility for the aged: A field experiment in an institutional setting. *Journal of Personality and Social Psychology*, 1976, *34*, 191–198.

Langer, E. J., and Saegert, S. Crowding and cognitive control. *Journal of Personality and Social Psychology*, 1977, *35*, 175–182.

Langner, S., and Michael, S. T. *Life stress and mental health*. New York: The Free Press, 1963.

Lassen, C. L. Effects of proximity on anxiety and communication in the initial psychiatric interview. *Journal of Abnormal Psychology*, 1973, *81*, 226–232.

Latané, B., Eckman, J., and Joy, V. Shared stress and interpersonal attraction. *Journal of Experimental Social Psychology*, 1966, *2*, 80–94.

Latta, R. M. Relation of status incongruence to personal space. *Personality and Social Psychology Bulletin*, 1978, *4*, 143–146.

Laufer, R. S., Proshansky, H. M., and Wolfe, M. Some analytic dimensions of privacy. In H. M. Proshansky, W. H. Ittelson, and L. G. Rivlin (eds.), *Environmental psychology: People and their physical settings*. New York: Holt, Rinehart & Winston, 1976.

Laufer, R. S., and Wolfe, M. Privacy as a concept and a social issue: A multi-dimensional development theory. *Journal of Social Issues*, 1977, *33*, 22–42.

Lawson, B. R., and Walters, D. The effects of a new motorway on an established residential area. In D. Canter and T. Lee (eds.), *Psychology and the built environment*. New York: Wiley, 1974.

Lawton, M. P. The human being and the institutional building. In C. Burnette, J. Lang, and D. Vaschon (eds.), *Architecture for human behavior: Collected papers from a mini-conference*. Philadelphia: Philadelphia Chapter/The American Institute of Architects, 1971.

Lawton, M. P. Competence, environmental press, and the adaptation of older people. In P. G. Windley and G. Ernst (eds.), *Theory development in environment and aging*. Washington, D.C.: Gerontological Society, 1975.

Lawton, M. P. The impact of the environment on aging and behavior. In J. E. Birren and K. W. Schaie (eds.), *Handbook of the psychology of aging*. New York: Van Nostrand Reinhold, 1977.

Lawton, M. P. Therapeutic environments for the aged. In D. Canter and S. Canter (eds.), *Designing for therapeutic environments: A review of research*. Chichester, England: Wiley, 1979.

Lawton, M. P., and Cohen, J. Environment and the well-being of elderly inner-city residents. *Environment and Behavior*, 1974, *6*, 194–211.

Lawton, M. P., and Nahemow, L. Ecology and the aging process. In C. Eisdorfer and M. P. Lawton (eds.), *The psychology of adult development and aging*. Washington, D.C.: APA, 1973.

Lazarsfeld, P., and Merton, R. Friendship as a social process: A substantive and methodological analysis. In M. Berger, T. Abel, and C. Page (eds.), *Freedom and control in modern society*. New York: Van Nostrand, 1954.

Lazarus, R. S. *Psychological stress and the coping process*. New York: McGraw-Hill, 1966.

Lazarus, R. S. Emotions and adaptation: Conceptual and empirical relations. In W. J. Arnold (ed.), *Nebraska Symposium on Motivation*. Lincoln: University of Nebraska Press, 1968, 175–265.

Lazarus, R. S. The concepts of stress and disease. In L. Levi (ed.), *Society, stress and disease*. Vol. 1. London: Oxford University Press, 1971.

Lazarus, R. S., and Cohen, J. Environmental stress. In I. Altman and J. Wohlwill (eds.), *Human behavior and environment: Advances in theory and research*. New York: Plenum, 1977.

Lazarus, R. S., Cohen, J., Folkman, S., Kanner, A., and Schaefer, C. Psychological stress and adaptation: Some unresolved issues. In H. Selye (ed.), *Guide to stress research*. New York: Van Nostrand, 1979.

Lazarus, R. S., and Launier, R. Stress-related transactions between person and environment. In L. A. Pervin and M. Lewis (eds.), *Interaction between internal and external determinants of behavior*. New York: Plenum, 1978.

Lebowitz, M. D., Cassell, E. J., and McCarroll, J. R. Health and the urban environment: XV. Acute respiratory episodes as reactions by sensitive individuals to air pollution and weather. *Environmental Research*, 1972, *5*, 135–141.

Leckart, B. T., and Bakan, P. Complexity judgments of photographs and looking time. *Perceptual and Motor Skills*, 1965, *21*, 16–18.

Lee, T. R. "Brennan's Law" of shopping behavior. *Psychological Reports*, 1962, *11*, 662.

Lee, T. R. The optimum provision and siting of social clubs. *Durham Research Review*, 1963, 4, 53–61.

Lee, T. R. Urban neighborhood as a socio–spatial schema. *Human Relations*, 1968, *21*, 241–267.

Lee, T. R. Perceived distance as a function of direction in the city. *Environment and Behavior*, 1970, *2*, 39–51.

Lee, T. R. Psychology and living space. In R. M. Downs and D. Stea (eds.), *Image and environment: Cognitive mapping and spatial behavior*. Chicago: Aldine, 1973.

Leff, H. L. *Experience, environment, and human potentials*. New York: Oxford University Press, 1978.

Leithead, C. S., and Lind, A. R. *Heat stress and heat disorders*. London: Cassell, 1964.

Leuba, C. Toward some integration of learning theories: The concept of optimal stimulation. *Psychological Reports*, 1955, *1*, 27–33.

Levy, L., and Herzog, A. Effects of population density and crowding on health and social adaptation in the Netherlands. *Journal of Health and Social Behavior*, 1974, *15*, 228–240.

Lewin, K. *Principles of topological psychology*. New York: McGraw-Hill, 1936.

Lewin, K. Group decision and social change. In T. M. Newcomb and E. L. Hartley (eds.), *Readings in social psychology*. New York: Holt, Rinehart & Winston, 1947.

Lewis, O. *The children of Sanchez*. New York: Random House, 1961.

Lewis, O. *La Vida*. New York: Random House, 1966.

Ley, D., and Cybriwsky, R. The spatial ecology of stripped cars. *Environment and Behavior*, 1974, *6*, 53–68.

Liebow, E. *Tally's corner*. Boston: Little, Brown, 1967.

Lin, L., Simeone, R. S., Ensel, W. M., and Kuo, W. Social support, stressful life events, and illness: A model and an empirical test. *Journal of Health and Social Behavior*, 1979, *20*, 108–119.

Lingwood, D. A. Environmental education through information-seeking: The case of an environmental teach-in. *Environment and Behavior*, 171, *3*, 220–262.

Lipman, A., and Slater, R. Homes for old people: Towards a positive environment. In D. Canter and S. Canter (eds.), *Designing for therapeutic environments: A review of research*. Chichester, England: Wiley, 1979.

Little, K. B. Personal space. *Journal of Experimental Social Psychology*, 1965, *1*, 237–247.

Little, K. B. Cultural variations in social schemata. *Journal of Personality and Social Psychology*, 1968, *10*, 1–7.

Little, K. B., Ulehla, F. J., and Henderson, C. Value congruence and interaction distance. *Journal of Social Psychology*, 1968, *75*, 249–253.

Lockhart, J. M., and Kiess, H. O. Auxiliary heating of the hands during cold exposure and manual performance. *Human Factors*, 1971, *13*, 457–465.

Lofstedt, B., Ryd, H., and Wyon, D. How classroom temperatures affect performance on school work. *BUILD International*, 1969, *2*, 23–34.

Logan, H. L., and Berger, E. Measurement of visual information cues. *Illuminating Engineering*, 1961, *56*, 393–403.

Long, B. H., Ziller, R. C., and Bankes, J. Self-other orientations of institutionalized behavior-problem adolescents. *Journal of Consulting and Clinical Psychology*, 1970, *34*, 43–47.

Loo, C. M. The effect of spatial density on the social behavior of children. *Journal of Applied Social Psychology*, 1972, *2*, 372–381.

Loo, C. M. Important issues in researching the effects of crowding on humans. *Representative Research in Social Psychology*, 1973, *4*, 219–227. (a)

Loo, C. M. The effect of spatial density on the social behavior of children. *Journal of Applied Social Psychology*, 1973, *2*, 372–381. (b)

Loo, C. M. Beyond the effects of crowding: Situational and individual differences. In D. Stokols (ed.), *Perspectives on environment and behavior*. New York: Plenum, 1977.

Loo, C. M. Behavior problem indices: The differential effects of spatial density on low and high scorers. *Environment and Behavior*, 1978, *10*, 489–510.

Loo, C. M. The effects of spatial density on children: "Fishing with a net rather than a pole." In A. Baum and Y. M. Epstein (eds.), *Human response to crowding.* Hillsdale, N.J.: Lawrence Erlbaum, 1978.

Lorenz, K. *On aggression.* New York: Harcourt Brace Jovanovich, 1966.

Lorenz, K. Analogy as a source of knowledge. *Science,* 1974, *185,* 229–234.

Los Angeles Department of City Planning. *The visual environment of Los Angeles.* Los Angeles: Department of City Planning, 1971.

Lott, A. J. Bright, M. A., Weinstein, P., and Lott, B. E. Liking for persons as a function of incentive and drive during acquisition. *Journal of Personality and Social Psychology,* 1970, *14,* 66–75.

Lott, A. J., and Lott, B. E. A learning theory approach to interpersonal attitudes. In A. G. Greenwald, T. C. Brock, and T. M. Ostrom (eds.), *Psychological foundations of attitudes.* New York: Academic Press, 1968.

Lott, B. S., and Sommer, R. Seating arrangements and status. *Journal of Personality and Social Psychology,* 1967, *7,* 90–95.

Love, K. D., and Aiello, J. R. Using projective techniques to measure interaction distance: A methodological note. *Personality and Social Psychology Bulletin,* 1980, *6,* 102–104.

Lowenthal, D. Research in environmental perception and behavior: Perspectives on current problems. *Environment and Behavior,* 1972, *4,* 333–342.

Lowenthal, D., and Riel, M. The nature of perceived and imagined environments. *Environment and Behavior,* 1972, *4,* 189–202.

Lucas, R. C. The contribution of environmental research to wilderness policy decisions. *Journal of Social Issues,* 1966, *22,* 116–126.

Lundberg, U. Emotional and geographical phenomena in psychophysical research. In R. M. Downs and D. Stea (eds.), *Image and environment: Cognitive mapping and spatial behavior.* Chicago: Aldine, 1973.

Lundberg, U. Urban commuting: Crowdedness and catecholamine excretion. *Journal of Human Stress,* 1976, *2,* 26–32.

Lyman, S. M., and Scott, M. B. Territoriality: A neglected sociological dimension. *Social Problems,* 1967, *15,* 235–249.

Lynch, K. *The image of the city.* Cambridge, Mass.: M.I.T. Press, 1960.

Lynch, K. The city as environment. *Scientific American,* 1965, *213,* 209–219.

Lynch, K., and Rivkin, M. A walk around the block. *Landscape,* 1959, *8,* 24–34.

Lynch, K., and Rodwin, L. A theory of urban form. *Journal of the American Institute of Planners,* 1958, *24,* 201–214.

Mackworth, N. H. Researches on the measurement of human performance. In H. W. Sinaiko (ed.), *Selected papers on human factors in the design and use of control systems.* New York: Dover Publications, 1961.

Maderthaner, R., Guttmann, G., Swaton, E., and Otway, H. Effect of distance upon risk perception. *Journal of Applied Psychology,* 1978, *63,* 380–382.

Maier, S. F., and Seligman, M. E. P. Learned helplessness: Theory and evidence. *Journal of Experimental Psychology: General,* 1976, *105,* 3–46.

Maloney, M. P., and Ward, M. O. Ecology: Let's hear from the people. *American Psychologist,* 1973, *28,* 583–586.

Maloney, M. P., Ward, M. O., and Braucht, C. N. A revised scale for the measurement of ecological attitudes and knowledge. *American Psychologist,* 1975, *30,* 787–790.

Mann, L. Learning to live with lines. In J. Helmer and N. A. Eddington (eds.), *Urban man: The psychology of urban survival.* New York: The Free Press, 1973.

Manning, P. *Office design: A study of environment.* Liverpool, England: The Pilkington Research Unit, 1965.

Marans, R. W. Perceived quality of residential environments: Some methodological issues. In K. H. Craik and E. H. Zube (eds.), *Perceiving environmental quality: Research and applications.* New York: Plenum, 1976.

Marans, R. W., and Rodgers, W. Toward an understanding of community satisfaction. In A. Hawley and V. Rock (eds.), *Metropolitan America in contemporary perspective.* New York: Halstead Press, 1975.

Margulis, S. T. Conceptions of privacy: Current status and next steps. *Journal of Social Issues,* 1977, *33,* 5–21.

Marris, P. A report on urban renewal in the U.S. In L. J. Duhl (ed.), *The urban condition.* New York: Simon & Schuster, 1963.

Marsella, A. J., Escudero, M., and Gordon, P. The effects of dwelling density on mental disorders in Filipino men. *Journal of Health and Social Behavior,* 1970, *11,* 288–294.

Marshall, N. *Environmental components of orientations toward privacy.* Paper presented at annual conference of Environmental Design Research Association, Pittsburgh, Pa., 1970.

Marshall, N. Privacy and environment. *Human Ecology,* 1972, *1,* 93–110.

Maruyama, M. The second cybernetics: Deviation-amplifying mutual causal process. *American Scientist,* 1963, *51,* 164–179.

Mathews, K. E., and Canon, L. K. Environmental noise level as a determinant of helping behavior. *Journal of Personality and Social Psychology,* 1975, *32,* 571–577.

Matthews, K. A., Scheier, M. F., Brunson, B. I., and Carducci, B. Attention, unpredictability, and reports of physical symptoms: Eliminating the benefits of predictability. *Journal of Personality and Social Psychology,* 1980, *38,* 525–537.

Maurer, R., and Baxter, J. D. Images of the neighborhood and city among Black-, Anglo-, and Mexican-American children. *Environment and Behavior,* 1972, *4,* 351–388.

McBride, G., King, M. G., and James, J. W. Social proximity effects on galvanic skin responses in adult humans. *Journal of Psychology,* 1965, *61,* 153–157.

McCain, G., Cox, V. C., and Paulus, P. B. The relationship between illness complaints and degree of crowding in a prison environment. *Environment and behavior,* 1976, *8,* 283–290.

McCallum, R., Rusbult, C.E., Hong, G. K., Walden, T.A., and Schopler, J. Effects of resource availability and importance of behavior on the experience of crowding. *Journal of Personality and Social Psychology,* 1979, *37,* 1304–1313.

McCauley, C., Coleman, G., and DeFusco, P. Commuters' eye contact with strangers in city and suburban train stations: Evidence of short-term adaptation to interpersonal overload in the city. *Environmental Psychology and Nonverbal Behavior,* 1978, *2,* 215–225.

McClelland, L. *Saving energy in "utilities included" rental housing.* Paper presented at annual conference of Environmental Design Research Association, Charleston, S.C., 1980.

McClelland, L., and Auslander, N. Perceptions of crowding and pleasantness in public settings. *Environment and Behavior,* 1978, *10,* 535–554.

McClelland, L., and Cook, S. W. Promoting energy conservation in master-metered apartments through group financial incentives. *Journal of Applied Social Psychology,* 1980, *10,* 20–31.

McCormick, E. J. *Human factors in engineering and design.* New York: McGraw-Hill, 1976.

McEvoy, J., III. The American concern with environmental activism. In W. R. Burch, Jr. et al. (eds.), *Social behavior, natural resources and the environment.* New York: Harper & Row, 1972.

McGrath, J. E. (ed.), *Social and psychological factors in stress.* New York: Holt, Rinehart & Winston, 1970.

McGrew, P. L. Social and spatial density effects on spacing behavior in preschool children. *Journal of Child Psychology and Psychiatry,* 1970, *11,* 197–205.

McGuire, W. J. The nature of attitudes and attitude change. In G. Lindzey and A. Aronson (eds.), *Handbook of social psychology,* Vol. 3, 2nd ed. Reading, Mass.: Addison-Wesley, 1969.

McKechnie, G. E. *Manual for the Environmental Response Inventory.* Palo Alto, Ca.: Consulting Psychologists Press, 1974.

McKechnie, G. E. Simulation techniques in environmental psychology. In D. Stokols (ed.), *Perspectives on environment and behavior: Theory, research, and application.* New York: Plenum, 1977. (a)

McKechnie, G. E. The environmental response inventory in application. *Environment and Behavior,* 1977, *9,* 255–276. (b)

McKennell, A. C., and Hunt, E. A. *Noise annoyance in central London.* London: Building Research Station, 1966.

McLean, E. K., and Tarnopolsky, A. Noise, discomfort and mental health: A review of the socio-medical implications of disturbance by noise. *Psychological Medicine*, 1977, *7*, 19–62.

McNall, P. E., Jaax, J., Rohles, F. H., Nevins, R. G., and Springer, W. Thermal comfort (thermally neutral) conditions for three levels of activity. *ASHRAE Transactions*, 1967, No. 73.

McNees, M. P., Schnelle, J. F., Gendrich, J., Thomas, M. M., and Beagle, G. McDonalds litter hunt: A community litter control system for youth. *Environment and Behavior*, 1979, *11*, 131–138.

Medalie, J., and Goldbourt, U. Angina pectoris among 10,000 men: II. Psychosocial and other risk factors as evidenced by a multivariate analysis of a five year incidence study. *American Journal of Medicine*, 1976, *60*, 910–921.

Meecham, W. C., and Smith, H. G. Effects of jet aircraft noise on mental hospital admissions. *British Journal of Audiology*, 1977, *11*, 81–85.

Mehrabian, A. Inference of attitudes from the posture, orientation and distance of a communicator. *Journal of Consulting and Clinical Psychology*, 1968, *32*, 296–308. (a)

Mehrabian, A. Relationships of attitude to seated posture, orientation and distance. *Journal of Personality and Social Psychology*, 1968, *10*, 26–30. (b)

Mehrabian, A., and Diamond, S. G. Effects of furniture arrangement, props, and personality on social interaction. *Journal of Personality and Social Psychology*, 1971, *20*, 18–30.

Mehrabian, A., and Russell, J. A. *An approach to environmental psychology.* Cambridge, Mass.: M.I.T. Press, 1974.

Mehrabian, A., and Williams, M. Nonverbal concomitants of perceived and intended persuasiveness. *Journal of Personality and Social Psychology*, 1969, *13*, 37–58.

Meisels, M., and Guardo, C. J. Development of personal space schemata. *Child Development*, 1969, *49*, 1167–1178.

Mercer, G. W., and Benjamin, M. L. Spatial behavior of university undergraduates in double-occupancy residence rooms: An inventory of effects. *Journal of Applied Social Psychology*, 1980, *10*, 32–44.

Messer, M. The possibility of an age-concentrated environment becoming a normative system. *Gerontologist*, 1967, *7*, 247–251.

Michelson, W. Urban sociology as an aid to urban physical development: some research strategies. *Journal of the American Institute of Planners*, 1968, *34*, 105–108.

Michelson, W. *Environmental choice, human behavior, and residential satisfaction.* New York: Oxford University Press, 1976. (a)

Michelson, W. *Man and his urban environment: A sociological approach.* Reading, Mass.: Addison-Wesley, 1976. (b)

Michelson, W. From congruence to antecedent conditions: A search for the basis of environmental improvement. In D. Stokols (ed.), *Perspectives on environment and behavior: Theory, research, and applications.* New York: Plenum, 1977.

Micklin, M. (ed.), *Population, environment, and social organization: Current issues in human ecology.* Hinsdale, Ill.: Dryden Press, 1973.

Milgram, S. The experience of living in cities. *Science*, 1970, *167*, 1461–1468.

Milgram, S. Chapter II, Introduction. In W. H. Ittelson (ed.), *Environment and cognition.* New York: Seminar Press, 1973.

Milgram, S. Psychological maps of Paris. In H. M. Proshansky, W. H. Ittelson, and L. G. Rivlin (eds.), *Environmental psychology: People and their physical settings.* New York: Holt, Rinehart & Winston, 1976.

Milgram, S. *The individual in a social world: Essays and experiments.* Reading, Mass.: Addison-Wesley, 1977.

Milgram, S., Greenwald, J., Kessler, S., McKenna, W., and Waters, J. A psychological map of New York City. *American Scientist*, 1972, *60*, 194–200.

Miller, A. R. *The assault on privacy.* New York: New American Library, 1972.

Miller, D. C. The allocation of priorities to urban and environmental problems by powerful leaders and organizations. In W. R. Burch, Jr., N. H. Cheek, Jr., and L. Taylor (eds.), *Social behavior, natural resources, and the environment.* New York: Harper & Row, 1972.

Miller, G. A., Galanter, E., and Pribram, K. H. *Plans and the structure of behavior*. New York: Holt, Rinehart & Winston, 1960.

Miller, I. W., III, and Norman, W. H. Learned helplessness in humans: A review and attribution-theory model. *Psychological Bulletin*, 1979, *86*, 93–118.

Miller, S., and Nardini, K. M. Individual differences in the perception of crowding. *Environmental Psychology and Nonverbal Behavior*, 1977, *2*, 3–12.

Miller, W. R., and Seligman, M. E. P. Depression and learned helplessness in man. *Journal of Abnormal Psychology*, 1975, *84*, 228–238.

Mischel, W. Toward a cognitive social learning reconceptualization of personality. *Psychological Review*, 1973, *80*, 252–283.

Mitchell, R. Some social implications of higher density housing. *American Sociological Review*, 1971, *36*, 18–29.

Moore, G. T. Developmental differences in environmental cognition. In W. F. E. Preiser (ed.), *Environmental design research*. Stroudsburg, Pa.: Dowden, Hutchinson & Ross, 1973.

Moore, G. T. Developmental variations between and within individuals in the cognitive representation of large-scale spatial environments. *Man-Environment Systems*, 1974, *4*, 55–57.

Moore, G. T. Knowing about environmental knowing: The current state of theory and research on environmental cognition. *Environment and Behavior*, 1979, *11*, 33–70.

Moos, R. H. Conceptualizations of human environments. *American Psychologist*, 1973, *28*, 652–655.

Moos, R. H. *The human context: Environmental determinants of behavior*. New York: Wiley, 1976.

Moos, R. H. *Evaluating educational environments*. San Francisco: Jossey-Bass, 1979.

Moos, R. H. *The environmental quality of residential care settings*. Paper presented at annual conference of Environmental Design Research Association, Charleston, S.C., 1980.

Moos, R. H. A social-ecological perspective on health. In G. Stone, F. Cohen, and N. Adler (eds.), *Health psychology*. San Francisco: Jossey-Bass, 1981.

Moos, R. H., and Lemke, S. *The multiphasic environmental assessment procedure (MEAP): Preliminary manual*. Social Ecology Laboratory, VA Medical Center and Stanford University, Palo Alto, Ca., 1979.

Morgan, C. J. Bystander intervention: Experimental test of a formal model. *Journal of Personality and Social Psychology*, 1978, *36*, 43–55.

Morrison, D. E., Hornback, K. E., and Warner, W. K. The environmental movement: Some preliminary observations and predictions. In W. R. Burch, Jr., et al. (eds.), *Social behavior, natural resources and the environment*. New York: Harper & Row, 1972.

Mullins, P., and Robb, J. H. Residents' assessment of a New Zealand public-housing scheme. *Environment and Behavior*, 1977, *9*, 573–625.

Nahemow, L. Research in a novel environment. *Environment and Behavior*, 1971, *3*, 81–103.

Nahemow, L., and Lawton, M. P. Toward an ecological theory of adaptation and aging. In W. F. E. Preiser (ed.), *Environmental Design Research*. Stroudsburg, Pa.: Dowden, Hutchinson & Ross, 1973.

Nahemow, L., and Lawton, M. P. Toward an ecological theory of adaptation and aging. In H. M. Proshansky, W. H. Ittelson, and L. G. Rivlin (eds.), *Environmental psychology: People and their physical settings*. New York: Holt, Rinehart & Winston, 1976.

Nasar, J. L. *On determining dimensions of environmental perception*. Paper presented at annual conference of Environmental Design Research Association, Charleston, S.C., 1980.

Nemecek, J., and Grandjean, E. Results of an ergonomic investigation of large-space offices. *Human Factors*, 1973, *15*, 111–124.

Newman, J., and McCauley, C. Eye contact with strangers in city, suburb and small town. *Environment and Behavior*, 1977, *9*, 547–559.

Newman, O. *Defensible space: Crime prevention through urban design*. New York: Macmillan, 1972.

Newman, R. C., and Pollack, D. Proxemics in deviant adolescents. *Journal of Consulting and Clinical Psychology*, 1973, *40*, 6–8.

Novaco, R. W., Stokols, D., Campbell, J., and Stokols, J. Transportation, stress, and community psychology. *American Journal of Community Psychology*, 1979, 7, 361–380.

Oechsli, F., and Buechley, R. Excess mortality associated with three Los Angeles September hot spells. *Environmental Research*, 1970, 3, 277–284.

Ohta, R., Walsh, D., and Krauss, I. *Spatial perspective taking ability in young and elderly adults.* Paper presented at annual conference of American Psychological Association, San Francisco, 1977.

Olszewski, D. A., Rotton, J., and Solor, E. A. *Conversation, conglomerate noise, and behavioral aftereffects.* Paper presented at annual meeting of Midwestern Psychological Association, Chicago, May 1976.

O'Neal, E., Caldwell, C., and Gallup, G. G., Jr. Territorial invasion and aggression in young children. *Environmental Psychology and Nonverbal Behavior*, 1977, 2, 14–22.

Onibokun, A. G. Social system correlates of residential satisfaction. *Environment and Behavior*, 1976, 8, 323–345.

OPCS. *Second survey of aircraft noise annoyance around London (Heathrow) Airport.* HMSO: London, 1971.

O'Riordan, T. Attitudes, behavior, and environmental policy issues. In I. Altman and J. F. Wohlwill (eds.), *Human behavior and environment: Advances in theory and research* (Vol. I). New York: Plenum, 1976.

Orleans, P. Differential cognition of urban residents: Effects of social scale on mapping. In R. M. Downs and D. Stea (eds.), *Image and environment: Cognitive mapping and spatial behavior.* Chicago: Aldine, 1973.

Orleans, P., and Schmidt, S. Mapping the city: Environmental cognition of urban residents. In W. J. Mitchell (ed.), *Environmental design: Research and practice.* Los Angeles: University of California, 1972.

Osgood, C. E., Suci, G. J., and Tannenbaum, P. H. *The measurement of meaning.* Urbana, Ill.: University of Illinois Press, 1957.

Osmond, H. Function as the basis of psychiatric ward design. *Mental Hospitals* (Architectural Supplement), 1957, 8, 23–29.

Osmond, H. The relationship between architect and psychiatrist. In C. Goshen (ed.), *Psychiatric Architecture,* Washington, D.C.: American Psychiatric Association, 1959.

Osmond, H. Design must meet patients' human needs. *The Modern Hospital*, 1966, 106, 98–170.

Otway, H. J., and Pahner, P.D. Risk assessment. *Futures*, 1976, 8, 122–134.

Page, R. A. Noise and helping behavior. *Environment and Behavior*, 1977, 9, 311–335.

Parke, R. D., and Sawin, D. B. Children's privacy in the home: Developmental, ecological, and childrearing determinants. *Environment and Behavior*, 1979, 11, 87–104.

Parsons, H. M. What happened at Hawthorne? *Science*, 1974, 183, 922–932.

Parsons, H. M. Work environments. In I. Altman and J. F. Wohlwill (eds.), *Human behavior and environment* (Vol. I). New York: Plenum, 1976.

Pastalan, L. A. Privacy as an expression of human territoriality. In L. A. Pastalan and D. H. Carson (eds.), *Spatial behavior of older people.* Ann Arbor: University of Michigan Press, 1970.

Patterson, A. H. Methodological developments in environment–behavioral research. In D. Stokols (ed.), *Perspectives on environment and behavior.* New York: Plenum, 1977.

Patterson, A. H. Territorial behavior and fear of crime in the elderly. *Environmental Psychology and Nonverbal Behavior*, 1978, 2, 131–144.

Patterson, M. L. Compensation and nonverbal immediacy behaviors: A review. *Sociometry*, 1973, 36, 237–253. (a)

Patterson, M. L. Stability of nonverbal immediacy behaviors. *Journal of Experimental Social Psychology*, 1973, 9, 97–109. (b)

Patterson, M. L. *Factors affecting interpersonal spatial proximity.* Paper presented at annual meeting of American Psychological Association, New Orleans, 1974.

Patterson, M. L. An arousal model of interpersonal intimacy. *Psychological Review*, 1976, *83*, 235–245.

Patterson, M. L. Arousal change and the cognitive labeling: Pursuing the mediators of intimacy exchange. *Environmental Psychology and Nonverbal Behavior*, 1978, *3*, 17–22.

Patterson, M. L., and Holmes, D. S. Social interaction correlates of MMPI extroversion-introversion scale. *American Psychologist*, 1966, *21*, 724–725.

Patterson, M. L., Mullens, S., and Romano, J. Compensatory reactions to spatial intrusion. *Sociometry*, 1971, *34*, 114–121.

Patterson, M. L., and Sechrest, L. B. Interpersonal distance and impression formation. *Journal of Personality*, 1970, *38*, 161–166.

Paulus, P. B., Annis, A. B., Seta, J. J., Schkade, J. K., and Matthews, R. W. Density does affect task performance. *Journal of Personality and Social Psychology*, 1976, *34*, 248–253.

Paulus, P. B., and Matthews, R. W. When density affects task performance. *Personality and Social Psychology Bulletin*, 1980, *6*, 119–124.

Paulus, P. B., McCain, G., and Cox, V. C. Death rates, psychiatric commitments, blood pressure, and perceived crowding as a function of institutional crowding. *Environmental Psychology and Nonverbal Behavior*, 1978, *3*, 107–116.

Pedersen, D. M. Developmental trends in personal space. *Journal of Psychology*, 1973, *83*, 3–9.

Pedersen, D. M., and Shears, L. M. A review of personal space research in the framework of general systems theory. *Psychological Bulletin*, 1973, *80*, 367–388.

Pellegrini, R. J., and Empey, J. Interpersonal spatial orientation in dyads. *Journal of Psychology*, 1970, *76*, 67–70.

Pempus, E., Sawaya, C., and Cooper, R. E. "*Don't fence me in*": Personal space depends on architectural enclosure. Paper presented at annual meeting of American Psychological Association, Chicago, 1975.

Pepler, R. D. Performance and well-being in heat. *Temperature: Its measurement and control in science and industry*, 3, Part 3. New York: Reinhold, 1963.

Pervin, L. A. Performance and satisfaction as a function of individual-environment fit. *Psychological Bulletin*, 1968, *69*, 56–68.

Petrinovich, L. Probabilistic functionalism: A conception of research method. *American Psychologist*, 1979, *5*, 373–390.

Pezdek, K., and Evans, G. W. Visual and verbal memory for objects and their spatial location. *Journal of Experimental Psychology: Human Learning and Memory*, 1979, *5*, 360–373.

Piaget, J. *The child's construction of reality*. New York: Basic Books, 1954. (a)

Piaget, J. Perceptual and cognitive (or operational) structures in the development of the concept of space in the child (Summary). *Proceedings of the 14th International Congress of Psychology*. Amsterdam: North Holland, 1954. (b)

Piaget, J. *The psychology of intelligence*. Totowa, N.J.: Littlefield, Adams, 1963.

Piaget, J. and Inhelder, B. *The child's conception of space*. New York: Norton, 1967.

Piaget, J., Inhelder, B., and Szeminska, A. *The child's conception of geometry*. New York: Basic Books, 1960.

Pick, H. L., Acredolo, L. P., and Gronseth, M. *Children's knowledge of the spatial layout of their homes*. Paper presented at the Society for Research in Child Development meetings, Philadelphia, 1973.

Pierce, J. C. Water resource preservation: Personal values and public support. *Environment and Behavior*, 1979, *11*, 147–161.

Pogell, S., Balling, J., Passoneau, J., and Valadez, J. *Does public participation optimize the environment?* Paper presented at annual conference of Environmental Design Research Association, Charleston, S.C., 1980.

Porteous, J. D. *Environment and behavior: Planning and everyday urban life*. Reading, Mass.: Addison-Wesley, 1977.

Poulton, E. C. *Environment and human efficiency*. Springfield, Ill.: Charles C. Thomas, 1970.

Poulton, E. C. Arousing environmental stresses can improve performance whatever people say. *Aviation, Space, and Environmental Medicine*, 1976, *47*, 1193–1204.

Poulton, E. C. Continuous intense noise masks auditory feedback and inner speech. *Psychological Bulletin*, 1977, *84*, 977–1001.

Poulton, E. C. A note on the masking of acoustic clicks. *Applied Ergonomics*, 1978, 9, 103.

Poulton, E. C. Composite model for human performance in continuous noise. *Psychological Review*, 1979, 86, 361–375.

Poulton, E. C., and Edwards, R. S. Interactions and range effects in experiments on pairs of stresses: Mild heat and low-frequency noise. *Journal of Experimental Psychology*, 1974, 102, 621–628.

Poulton, E. C., and Kerslake, D. Effect of warmth on perceptual efficiency. *Aerospace Medicine*, 1965, 36, 29–34.

President's Commission on Mental Health. *Report to the President from the President's Commission on Mental Health*. Washington, D.C.: U.S. Government Printing Office, 1978.

Price, J. L. *The effects of crowding on the social behavior of children.* Unpublished doctoral dissertation, Columbia University, 1971.

Priest, R. F., and Sawyer, J. Proximity and peership: Base of balance in interpersonal attraction. *American Journal of Sociology*, 1967, 72, 633–649.

Proshansky, H. M. For what are we training our graduate students? *American Psychologist*, 1972, 27, 205–212. (a)

Proshansky, H. M. Methodology in environmental psychology: Problems and issues. *Human Factors*, 1972, 14, 451–460. (b)

Proshansky, H. M. Theoretical issues in environmental psychology. *Representative Research in Social Psychology*, 1973, 4, 93–109.

Proshansky, H. M. Environmental psychology and the real world. *American Psychologist* 1976, 31, 303–310.

Proshansky, H. M. The city and self-identity. *Environment and Behavior*, 1978, 10, 147–170.

Proshansky, H. M., and Altman, I. Overview of the field. In W. P. White (ed.), *Resources in environment and behavior*. Washington, D.C.: American Psychological Association, 1979.

Proshansky, H. M., Ittelson, W. H., and Rivlin, L. G. The influence of the physical environment on behavior: Some basic assumptions. In H. M. Proshansky, W. H. Ittelson, and L. G. Rivlin (eds.), *Environmental psychology: Man and his physical settings*. New York: Holt, Rinehart & Winston, 1970.

Proshansky, H. M., Ittelson, W. H., and Rivlin, L. G. Freedom of choice and behavior in a physical setting. In H. M. Proshansky, W. H. Ittelson, and L. G. Rivlin, (eds.), *Environmental psychology: People and their physical settings*. New York: Holt, Rinehart & Winston, 1976.

Provins, K. A. Environmental heat, body temperature, and behavior. *Australian Journal of Psychology*, 1966, 18, 118–129.

Pylyshyn, Z. W. What the mind's eye tells the mind's brain: A critique of mental imagery. *Cognitive Psychology*, 1973, 80, 1–24.

Pyron, B. Form and diversity in human habitats: Judgmental and attitude responses. *Environment and Behavior*, 1972, 4, 87–121.

Rabkin, J. G., and Struening, E. L. Life events, stress, and illness. *Science*, 1976, 194, 1013–1020.

Rainwater, L. Fear and the house-as-haven in the lower class. *Journal of the American Institute of Planners*, 1966, 32, 23–31.

Rankin, R. E. Air pollution control and public apathy. *Journal of Air Pollution Control Association*, 1969, 19, 565–569.

Rapoport, A. Toward a redefinition of density. *Environment and Behavior*, 1975, 7, 133–158.

Rapoport, A., and Hawkes, R. The perception of urban complexity. *Journal of the American Institute of Planners*, 1970, 36, 106–111.

Rapoport, A., and Kantor, R. E. Complexity and ambiguity in environmental design. *Journal of the American Institute of Planners*, 1976, 33, 210–221.

Rawls, J. R., Trego, R. E., McGaffey, C. N., and Rawls, D. J. Personal space as a predictor of performance under close working conditions. *Journal of Social Psychology*, 1972, 86, 261–267.

Regnier, V. A., and Rausch, K. J. *Spatial and temporal neighborhood use patterns of older low-in-*

come central city dwellers. Paper presented at annual conference of Environmental Design Research Association, Charleston, S.C., 1980.

Reiter, S. M., and Samuel, W. Littering as a function of prior litter and the presence or absence of prohibitive signs. *Journal of Applied Social Psychology,* 1980, *10,* 45–55.

Rent, G. S., and Rent, C. S. Low-income housing: Factors related to residential satisfaction. *Environment and Behavior,* 1978, *10,* 459–488.

Report of the National Advisory Commission on Civil Disorders. New York: Bantam, 1968.

Rivlin, L. G., and Rothenberg, M. The use of space in open classrooms. In H. M. Proshansky, W. H. Ittelson, and L. G. Rivlin (eds.), *Environmental psychology: People and their physical settings.* New York: Holt, Rinehart & Winston, 1976.

Rivlin, L. G., Rothenberg, M., Justa, F., Wallis, A., and Wheeler, F. *Children's conceptions of open classrooms through use of scaled models.* Paper presented at annual conference of Environmental Design Research Association, Washington, D.C., 1974.

Robins, J. Highway massacre: Nowadays, carnage isn't all accidental. *Wall Street Journal,* October 20, 1978, *1,* 20.

Rock, I. *The nature of perceptual adaptation.* New York: Basic Books, 1966.

Rodin, J. Crowding, perceived choice and response to controllable and uncontrollable outcomes. *Journal of Experimental Social Psychology,* 1976, *12,* 564–578.

Rodin, J., and Langer, E. J. Long-term effects of a control-relevant intervention with the institutionalized aged. *Journal of Personality and Social Psychology,* 1977, *35,* 897–902.

Rodin, J., Solomon, S. K., and Metcalf, J. Role of control in mediating perceptions of density. *Journal of Personality and Social Psychology,* 1978, *36,* 988–999.

Roethlisberger, F., and Dickson, W. *Management and the worker.* Cambridge, Mass.: Harvard University Press, 1939.

Rohe, W., and Patterson, A. H. *The effects of varied levels of resources and density on behavior in a day care center.* Paper presented at annual meeting of Environmental Design Research Association, Milwaukee, 1974.

Rohles, F. H. Thermal sensations of sedentary man in moderate temperatures. *Human Factors,* 1971, *13,* 553–560.

Roos, P. D. Jurisdiction: An ecological concept. *Human Relations,* 1968, *21,* 75–84.

Rosenbaum, W. A. *The politics of environmental concern.* New York: Praeger, 1973.

Rosenberg, M. J., and Hovland, C. I. Cognitive, affective and behavioral components of attitudes. In M. J. Rosenberg, C. I. Hovland, W. J. McGuire, R. P. Abelson, and J. W. Brehm (eds.), *Attitude organization and change: An analysis of consistency among attitude components.* New Haven: Yale University Press, 1960.

Rosenfeld, H. M. Effect of an approval seeking induction on interpersonal proximity. *Psychological Reports,* 1965, *17,* 120–122.

Rosenthal, A. M. *Thirty-eight Witnesses.* New York: McGraw-Hill, 1964.

Rosow, I. The social effects of the physical environment. *Journal of the American Institute of Planners,* 1961, *27,* 127–133.

Rosow, I. *Social integration of the aged.* New York: The Free Press, 1967.

Ross, H. E. *Behavior and perception in strange environments.* New York: Basic Books, 1975.

Ross, M., Layton, B., Erickson, B., and Schopler, J. Affect facial regard, and reactions to crowding. *Journal of Personality and Social Psychology,* 1973, *28,* 69–76.

Rothenberg, M., and Rivlin, L. G. *An ecological approach to the study of open classrooms.* Paper presented at a conference on ecological factors in human development, University of Surrey, England, 1975.

Rotter, J. B. Generalized expectancies for internal versus external control of reinforcement. *Psychological Monographs,* 1966, *80,* 1–28.

Rotton, J., Barry, T., Frey, J., and Soler, E. Air pollution and interpersonal attraction. *Journal of Applied Social Psychology,* 1978, *8,* 57–71.

Rotton, J., Olewski, D., Charleton, M., and Soler, E. Loud speech, conglomerate noise, and behavioral aftereffects. *Journal of Applied Psychology,* 1978, *63,* 360–365.

Rozelle, R. M., and Bazer, J. C. Meaning and value in conceptualizing the city. *Journal of the American Institute of Planners,* 1972, *38,* 116–122.

Rule, J. B. *Private lives and public surveillance.* New York: Schocken, 1974.

Russell, J. A., and Mehrabian, A. Approach-avoidance and affiliation as functions of the emotion-eliciting quality of an environment. *Environment and Behavior*, 1978, *10*, 355–388.

Russell, J. A., and Pratt, G. A description of the affective quality attributed to environments. *Journal of Personality and Social Psychology*, 1980, *38*, 311–322.

Ryan, E. Personal identity in an urban slum. In L. J. Duhl (ed.), *The urban condition*. New York: Basic Books, 1963.

Ryan, W. *Blaming the victim*. New York: Vintage Books, 1971.

Rylander, R., Sorensen, S., and Kajland, A. Annoyance reactions from aircraft noise exposure. *Journal of Sound and Vibration*, 1972, *24*, 419–444.

Saarinen, T. F. *Image of the Chicago Loop*. Unpublished manuscript, University of Chicago, 1964.

Saarinen, T. F. *Image of the University of Arizona campus*. Unpublished manuscript, University of Arizona, Tucson, 1967.

Saarinen, T. F. *Perception of environment* (Resource Paper No. 5). Washington, D.C.: Association of American Geographers, Commission on College Geography, 1969.

Saarinen, T. F. Student views of the world. In R. M. Downs and D. Stea (eds.), *Image and environment: Cognitive mapping and spatial behavior*. Chicago: Aldine, 1973.

Saegert, S. Crowding: Cognitive overload and behavioral constraint. In W. F. E. Preiser (ed.), *Environmental design research*. Stroudsburg, Pa.: Dowden, Hutchinson & Ross, 1973.

Saegert, S. *Effects of spatial and social density on arousal, mood, and social orientation*. Unpublished doctoral dissertation, University of Michigan, Ann Arbor, 1974.

Saegert, S. Stress-inducing and reducing qualities of environments. In H. M. Proshansky, W. H. Ittelson, and L. G. Rivlin (eds.), *Environmental psychology: People and their physical settings*. New York: Holt, Rinehart & Winston, 1976.

Saegert, S. High density environments: Their personal and social consequences. In A. Baum and Y. M. Epstein (eds.), *Human response to crowding*. Hillsdale, N.J.: Erlbaum, 1978.

Saegert, S., Mackintosh, E., and West, S. Two studies of crowding in urban public spaces. *Environment and Behavior*, 1975, *7*, 159–184.

Sadalla, E. K. Population size, structural differentiation, and human behavior. *Environment and Behavior*, 1978, *10*, 271–292.

Sadalla, E. K., Burroughs, W. J., and Staplin, L. J. The experience of crowding. *Personality and Social Psychology Bulletin*, 1978, *4*, 304–308.

Sadalla, E. K., Burroughs, W. J., and Quaid, M. *House form and social identity: A validity study*. Paper presented at annual conference of Environmental Design Research Association, Charleston, S.C., 1980.

Sadalla, E. K., and Magel, S. G. The perception of traversed distance. *Environment and Behavior*, 1980, *12*, 65–79.

Sadalla, E. K., and Staplin, L. J. An information storage model for distance cognition. *Environment and Behavior*, 1980, *12*, 183–193. (a)

Sadalla, E. K., and Staplin, L. J. The perception of traversed distance: Interactions. *Environment and Behavior*, 1980, *12*, 167–182. (b)

Safdie, M. Habitat '67. *Habitat*, 1966, *8*, 2–6.

Sainsbury, P. *Suicide in London*. New York: Basic Books, 1956.

Sanders, A. F. The influence of noise on two discrimination tasks. *Ergonomics*, 1961, *4*, 253–258.

Sanoff, H., & Cohn, S. Preface. In H. Sanoff and S. Cohn (eds.), *Proceedings of the first annual environmental design research association conference*. Raleigh: North Carolina State University, 1970.

Savinar, J. The effect of ceiling height on personal space. *Man-Environment Systems*, 1975, *5*, 321–324.

Schaps, E. Cost, dependency, and helping. *Journal of Personality and Social Psychology*, 1972, *21*, 74–78.

Scherer, S. E. Proxemic behavior of primary school children as a function of their socioeconomic class and subculture. *Journal of Personality and Social Psychology*, 1974, *29*, 800–805.

Schiffenbauer, A. Designing for high-density living. In J. R. Aiello and A. Baum (eds.), *Residential crowding and design*. New York: Plenum, 1979.

Schiffenbauer, A., Brown, J., Perry, P., Shulack, L., and Zanzala, A. The relationship between destiny and crowding: Some architectural modifiers. *Environment and Behavior*, 1977, *9*, 3–14.

Schmid, C. Completed and attempted suicides. *American Sociological Review*, 1955, *20*, 273.

Schmid, C. Urban crime areas: Part I. *American Sociological Review*, 1969, *25*, 527–542.

Schmid, C. Urban crime areas: Part II. *American Sociological Review*, 1970, *25*, 655–678.

Schmidt, D. E., Goldman, R. D., and Feimer, N. R. Perceptions of crowding: Predicting at the residence, neighborhood and city levels. *Environment and Behavior*, 1979, *11*, 105–130.

Schmidt, D. E., and Keating, J. P. Human crowding and personal control: An integration of the research. *Psychological Bulletin*, 1979, *86*, 680–700.

Schmitt, R. C. Density, delinquency and crime in Honolulu. *Sociology and Social Research*, 1957, *41*, 274–276.

Schmitt, R. C. Implications of density in Hong Kong. *Journal of the American Institute of Planners*, 1963, *29*, 210–214.

Schmitt, R. C. Density, health and social disorganization. *Journal of the American Institute of Planners*, 1966, *32*, 38–40.

Schooler, K. K. Response of the elderly to environment: A stress-theoretic perspective. In P. G. Windley and G. Ernst (eds.), *Theory development in environment and aging*. Washington, D.C.: Gerontological Society, 1975.

Schopler, J., and Stockdale, J. E. An interference analysis of crowding. *Journal of Environmental Psychology and Nonverbal Behavior*, 1977, *1*, 81–88.

Schulz, R. Effects of control and predictability on the physical and psychological well-being of the institutionalized aged. *Journal of Personality and Social Psychology*, 1976, *33*, 563–573.

Schulz, R., and Hanusa, B. H. Long-term effects of control and predictability-enhancing interventions: Findings and ethical issues. *Journal of Personality and Social Psychology*, 1978, *38*, 1194–1201.

Schuman, S. Patterns of urban heat-wave deaths and implications for prevention: Data from New York and St. Louis during July, 1966. *Environmental Research*, 1972, *5*, 59–75.

Schwartz, B. The social psychology of privacy. *American Journal of Sociology*, 1968, *73*, 741–752.

Schwebel, A. I., and Cherlin, D. L. Physical and social distancing in teacher-pupil relationships. *Journal of Educational Psychology*, 1972, *63*, 543–550.

Seashore, S. M., and Bowers, D. G. *Changing the structure and functioning of an organization: Report of a field experiment*. Ann Arbor: University of Michigan Press, 1963.

Seaver, W. B., and Patterson, A. H. Decreasing fuel-oil consumption through feedback and social commendation. *Journal of Applied Behavior Analysis*, 1976, *9*, 147–152.

Segall, M. H., Campbell, D. T., and Herskevits, M. J. *The influence of culture on visual perception*. New York: Bobbs-Merrill, 1966.

Seligman, C., and Darley, J. M. Feedback as a means of decreasing residential energy consumption. *Journal of Applied Psychology*, 1977, *62*, 363–368.

Seligman, M. E. P. Fall into helplessness. *Psychology Today*, June, 1973, 43–48.

Seligman, M. E. P. Depression and learned helplessness. In R. J. Friedman and M. M. Katz (eds.), *The psychology of depression: Contemporary theory and research*. New York: Wiley, 1974.

Seligman, M. E. P. *Helplessness*. San Francisco: Freeman, 1975.

Selye, H. *The stress of life*. New York: McGraw-Hill, 1956.

Selye, H. The evolution of the stress concept. *American Scientist*, 1973, *61*, 692–699.

Selye, H. *Stress in health and disease*. Woburn, Mass.: Butterworth, 1976.

Shapiro, R. A., and Berland, T. Noise in the operating room. *New England Journal of Medicine*, 1972, *287*, 1236–1238.

Shaw, C., and McKay, H. D. *Juvenile delinquency and urban areas.* Chicago: University of Chicago Press, 1942.

Shemyakin, F. N. Orientation in space. In B. G. Ananyev, et al., (eds.), *Psychological Science in the U.S.S.R.,* Vol. I, Washington: Office of Technical Services, Report 62-11083, 1962.

Sherrod, D. R., and Cohen, S. Density, personal control, and design. In J. R. Aiello and A. Baum (eds.), *Residential crowding and design.* New York: Plenum, 1979.

Sherrod, D. R., and Downs, R. Environmental determinants of altruism: The effects of stimulus overload and perceived control on helping. *Journal of Experimental Social Psychology,* 1974, *10,* 468–479.

Sherrod, D. R., Hage, J. N., Halpern, P. L., and Moore, B. S. Effects of personal causation and perceived control on responses to an aversive environment: The more control, the better. *Journal of Experimental Social Psychology,* 1977, *13,* 14–27.

Siegel, A. W., and Schadler, M. Young children's cognitive maps of their classroom. *Child Development,* 1977, *48,* 388–394.

Siegel, A. W., and White, S. H. The development of spatial representations of large-scale environments. In H. W. Reese (ed.), *Advances in child development and behavior,* (Vol. 10). New York: Academic Press, 1975.

Siegel, J. M., and Steele, C. M. Noise level and social discrimination. *Personality and Social Psychology Bulletin,* 1979, *5,* 95–99.

Silverman, I. Nonreactive methods and the law. *American Psychologist,* 1975, *30,* 764–769.

Simmel, G. Secrecy and group communication. In K. H. Wolff (ed. and trans.), *The sociology of Georg Simmel.* New York: The Free Press, 1950.

Simonson, E., and Brozek, J. Effect of illumination level on visual performance and fatigue. *Journal of the Optical Society of America,* 1948, *38,* 384–397.

Sivadon, P. Space as experienced: Therapeutic implications. In H. M. Proshansky, W. H. Ittelson, and L. G. Rivlin (eds.), *Environmental psychology: Man and his physical setting.* New York: Holt, Rinehart & Winston, 1970.

Skaburskis, J. V. Territoriality and neighborhood design. *Journal of Architectural Research,* 1974, *3,* 39–43.

Smetana, J., Bridgeman, D. L., and Bridgeman, B. A field study of interpersonal distance in early childhood. *Personality and Social Psychology Bulletin,* 1978, *4,* 309–313.

Smith, A. J., and Broadbent, D. E. Effects of noise on performance on embedded figures tasks. *Journal of Applied Psychology,* 1980, *65,* 246–248.

Smith, F. J., and Lawrence, E. S. Alone and crowded: The effects of spatial restriction on measures of affect and simulation response. *Personality and Social Psychology Bulletin,* 1978, *4,* 139–142.

Smith, G. C., and Alderdice, D. Public responses to national park environmental policy. *Environment and Behavior,* 1979, *11,* 329–350.

Smith, M. B., and Hobbs, N. The community and the community mental health center. *American Psychologist,* 1966, *21,* 499–509.

Smith, R. J., and Knowles, E. S. Attributional consequences of personal space invasions. *Personality and Social Psychology Bulletin,* 1978, *4,* 429–433.

Smith, R. J., and Knowles, E. S. Affective and cognitive mediators of reactions to spatial invasions. *Journal of Experimental Social Psychology,* 1979, *15,* 437–452.

Smith, S., and Haythorn, W. W. The effects of compatibility, crowding, group size, and leadership seniority on stress, anxiety, hostility, and annoyance in isolated groups. *Journal of Personality and Social Psychology,* 1972, *22,* 67–69.

Snyder, F. W., and Pronko, N. H. *Vision with spatial inversion.* Wichita, Kan.: University of Wichita Press, 1952.

Sommer, R. Studies in personal space. *Sociometry,* 1959, *22,* 247–260.

Sommer, R. Leadership and group geography. *Sociometry,* 1961, *24,* 99–110.

Sommer, R. The distance for comfortable conversation: A further study. *Sociometry,* 1962, *25,* 111–116.

Sommer, R. Further studies of small group ecology. *Sociometry,* 1965, *28,* 337–348.

Sommer, R. Man's proximate environment. *Journal of Social Issues,* 1966, *22,* 59–70.

Sommer, R. Small group ecology. *Psychological Bulletin,* 1967, *67,* 145–152.

Sommer, R. Intimacy ratings in five countries. *International Journal of Psychology*, 1968, *3*, 109–114.

Sommer, R. *Personal space*. Englewood Cliffs, N.J.: Prentice-Hall, 1969.

Sommer, R. *Design awareness*. San Francisco: Rinehart Press, 1972.

Sommer, R. *Tight spaces: Hard architecture and how to humanize it*. Englewood Cliffs, N.J.: Prentice-Hall, 1974.

Sommer, R. Action research. In D. Stokols (ed.), *Perspectives on environment and behavior: Theory, research, and application*. New York: Plenum, 1977.

Sommer, R., and Becker, F. D. Territorial defense and the good neighbor. *Journal of Personality and Social Psychology*, 1969, *11*, 85–92.

Sommer, R., and Ross, H. Social interaction on a geriatrics ward. *International Journal of Social Psychiatry*, 1958, *4*, 128–133.

Southwick, C. H. An experimental study of intragroup agonistic behavior in rhesus monkeys (*Macaca mulatta*). *Behavior*, 1967, *28*, 182–209.

Southworth, M. The sonic environment of cities. *Environment and Behavior*, 1969, *1*, 49–70.

Special Task Force. *Standards for planning water and land resources*. Washington, D.C.: U.S. Water Resources Council, 1970.

Spivack, M. Sensory distortions in tunnels and corridors. *Hospital and Community Psychiatry*, January 1967, 24–30.

Staats, A. W., Minke, K. A., Martin, C. H., and Higa, W. R. Deprivation-satiation and strength of attitude conditioning: A test of attitude-reinforcer-discriminative theory. *Journal of Personality and Social Psychology*, 1972, *24*, 178–185.

Staats, C. K., Staats, A. W., and Heard, W. G. Attitude development and ratio of reinforcement. *Sociometry*, 1960, *23*, 338–350.

Stea, D. The measurement of mental maps: An experimental model for studying conceptual spaces. In Cox, K. R., and Golledge, R. G. (eds.), *Behavioral Problems in Geography: A Symposium*. Evanston, Ill.: Northwestern University Press, 1969.

Stea, D. Architecture in the head: Cognitive mapping. In J. Lang, C. Burnette, W. Moleski, and D. Vachon (eds.), *Designing for human behavior: Architecture and the behavioral sciences*. Stroudsburg, Pa.: Dowden, Hutchinson & Ross, 1974.

Stea, D., and Blaut, J. M. Some preliminary observations on spatial learning in school children. In R. M. Downs and D. Stea (eds.), *Image and environment*. Chicago: Aldine, 1973. (a)

Stea, D., and Blaut, J. M. Notes toward a developmental theory of spatial learning. In R. M. Downs and D. Stea (eds.), *Image and environment*. Chicago: Aldine, 1973. (b)

Stea, D. and Downs, R. M. From the outside looking in at the inside looking out. *Environment and Behavior*, 1970, *2*, 3–13.

Stea, D., and Taphanel, S. Theory and experiment on the relation between environmental modeling (toy-play) and environmental cognition. In D. Canter and T. Lee (eds.), *Psychology and the built environment*. London: Architectural Press, 1974.

Stea, D., and Wood, D. *A cognitive atlas: The psychological geography of four Mexican cities*. In press.

Steele, F. I. *Physical settings and organization development*. Reading, Mass.: Addison-Wesley, 1973.

Steinzor, B. The spatial factor in face-to-face discussion groups. *Journal of Abnormal and Social Psychology*, 1950, *45*, 552–555.

Sterling, T. D., Phair, J. J., Pollack, S. V., Schumsky, D. A., and DeGroot, I. Urban morbidity and air pollution. *Archives of Environmental Health*, 1966, *13*, 158–170.

Sterrett, J. H. The job interview: Body language and perceptions of potential effectiveness. *Journal of Applied Psychology*, 1978, *63*, 388–390.

Stevens, W. R., and Foxell, C. A. P. Visual acuity. *Light and Lighting*, 1955, *48*, 419–424.

Stires, L. Classroom seating location, student grades, and attitudes: Environment or selection? *Environment and Behavior*, 1980, *12*, 241–254.

Stobaugh, R., and Yergin, D. *Energy future: Report of the energy project at the Harvard Business School*. New York: Random House, 1979.

Stockbridge, H. C. W., and Lee, M. The psycho-social consequences of aircraft noise. *Applied Ergonomics*, 1973, *4*, 44–45.

Stokols, D. On the distinction between density and crowding: Some implications for future research. *Psychological Review*, 1972, *79*, 275–278. (a)

Stokols, D. A social psychological model of human crowding phenomena. *American Institute of Planners Journal*, 1972, *38*, 72–83. (b)

Stokols, D. The experience of crowding in primary and secondary environments. *Environment and Behavior*, 1976, *8*, 49–86.

Stokols, D. Environmental psychology. *Annual Review of Psychology*, 1978, *29*, 253–295.

Stokols, D. Group x place transactions: Some neglected issues in psychological research on settings. In D. Magnusson (ed.), *The situation: An interactional perspective*. Hillsdale, N.J.: Erlbaum, 1981.

Stokols, D. A congruence analysis of human stress. In I. G. Sarason and C. D. Spielberger (eds.), *Stress and anxiety*, Vol. 6. Washington, D.C.: Hemisphere, 1979.

Stokols, D. A typology of crowding experiences. In A. Baum and Y. M. Epstein (eds.), *Human response to crowding*. Hillsdale, N.J.: Erlbaum, 1978.

Stokols, D., and Novaco, R. W. Transportation and well-being: An ecological perspective. In I. Altman, J. Wohlwill, and P. Everett (eds.), *Human behavior and environment: Advances in theory and research*, Vol. 5. New York: Plenum, in press.

Stokols, D., Novaco, R. W., Stokols, J., and Campbell, J. Traffic congestion, Type A behavior, and stress. *Journal of Applied Psychology*, 1968, *63*, 467–480.

Stokols, D., Ohlig, W., and Resnick, S. M. Perception of residential crowding, classroom experiences, and student health. *Human Ecology*, 1978, *6*, 233–252.

Stokols, D., Rall, M., Pinner, B., and Schopler, J. Physical, social, and personal determinants of the perception of crowding. *Environment and Behavior*, 1973, *5*, 87–117.

Stokols, D., and Shumaker, S. A. People and places: A transactional view of settings. In J. Harvey (ed.), *Cognition, social behavior, and the environment*. Hillsdale, N.J.: Erlbaum, 1981.

Stratton, L. O., Tekippe, D. J., and Flick, G. L. Personal space and self-concept. *Sociometry*, 1973, *36*, 424–429.

Strauss, A. *Images of the American city*. New York: Free Press, 1961.

Studer, R. G. Man-environment relations: Discovery or design. In W. F. E. Preiser (ed.), *Environmental design research* (Vol. 2). Stroudsburg, Pa.: Dowden, Hutchinson & Ross, 1973.

Studer, R. G., and Stea, D. Architectural programming, environmental design, and human behavior. *Journal of Social Issues*, 1966, *22*, 127–136.

Sundstrom, E. An experimental study of crowding: Effects of room size, intrusion, and goal-blocking on nonverbal behaviors, self-disclosure, and self-reported stress. *Journal of Personality and Social Psychology*, 1975, *32*, 645–654.

Sundstrom, E. Interpersonal behavior and the physical environment. In L. Wrightsman (ed.), *Social psychology* (2nd ed.). Monterey, Ca.: Brooks/Cole, 1976.

Sundstrom, E. A test of equilibrium theory: Effects of topic intimacy and proximity on verbal and nonverbal behavior in pairs of friends and strangers. *Environmental Psychology and Nonverbal Behavior*, 1978, *3*, 3–16. (a)

Sundstrom, E. Crowding as a sequential process: Review of research on the effects of population density on humans. In A. Baum and Y. M. Epstein (eds.), *Human response to crowding*. Hillsdale, N.J.: Erlbaum, 1978. (b)

Sundstrom, E., and Altman, I. Field study of dominance and territorial behavior. *Journal of Personality and Social Psychology*, 1974, *30*, 115–125.

Sundstrom, E., Burt, R. E., and Kamp, D. Privacy at work: Architectural correlates of job satisfaction and job performance. *Academy of Management Journal*, 1980, *23*, 101–117.

Sundstrom, E., and Sundstrom, M. G. Personal space invasions: What happens when the invader asks permission? *Environmental Psychology and Nonverbal Behavior*, 1977, *2*, 76–82.

Suttles, G. D. *The social order of the slum*. Chicago: University of Chicago Press, 1968.

Swanson, D., Swanson, L.A., and Dukes, M. J. *The evaluation and design of neighborhood preservation strategies to minimize the negative aspects of displacement*. Paper presented at annual conference of Environmental Design Research Association, Charleston, S.C., 1980.

Taylor, R. B., and Stough, R. R. Territorial cognition: Assessing Altman's typology. *Journal of Personality and Social Psychology*, 1978, *36*, 418–423.

Teaff, J. D., Lawton, M. P., Nahemow, L., and Carlson, D. Impact of age integration on the well-being of elderly tenants in public housing. *Journal of Gerontology*, 1978, *33*, 126–133.

Teichner, W. H., and Wehrkamp, P. F. Visual motor performance as a function of short-duration ambient temperature. *Journal of Experimental Psychology*, 1954, *47*, 447–450.

Tennen, H., and Eller, S. J. Attributional components of learned helplessness and facilitation. *Journal of Personality and Social Psychology*, 1977, *35*, 265–271.

Tennis, G. H., and Dabbs, J. M. Sex, setting and personal space: First grade through college. *Sociometry*, 1975, *38*, 385–394.

Terry, R. L., and Lower, M. Perceptual withdrawal from an invasion of personal space. *Personality and Social Psychology Bulletin*, 1979, *5*, 396–397.

Tesch, F. E., Huston, L., and Indebaum, E. A. Attitude similarity, attraction, and physical proximity in a dynamic state. *Journal of Applied Social Psychology*, 1973, *3*, 63–72.

Thalhofer, N. N. Violation of a spacing norm in high social density. *Journal of Applied Social Psychology*, 1980, *10*, 175–183.

Theologus, G. C., Wheaton, G. R., and Fleishman, E. A. Effects of intermittent, moderate intensity noise on human performance. *Journal of Applied Psychology*, 1974, *59*, 539–547.

Thiel, P. A sequence-experience notation for architectural and urban space. *Town Planning Review*, 1961, *32*, 33–52.

Thompson, B. J., and Baxter, J. C. Interpersonal spacing in two-person cross-cultural interaction. *Man-Environment Systems*, 1973, *3*, 115–117.

Tichauer, E. R. The effects of climate on working efficiency. *Impetus*, 1962, *1*(5), 24–31.

Tiger, L. *Men in groups*. New York: Random House, 1969.

Tognacci, L. N., Weigel, R. H., Wideen, M. F., and Vernon, D. T. Environmental quality: How universal is public concern? *Environment and Behavior*, 1972, *4*, 73–87.

Tolman, E. C. Cognitive maps in rats and men. *Psychological Review*, 1948, *55*, 189–208.

Tolor, A., and Salafia, W. R. The social schemata technique as a projective device. *Psychological Reports*, 1971, *28*, 423–429.

Valins, S., and Baum, A. Residential group size, social interaction and crowding. *Environment and Behavior*, 1973, *5*, 421–440.

Ventre, F. T. Toward a science of environment. *The American Behavioral Scientist*, 1966, *10*, 28–31.

Venturi, R. *Complexity and contradiction in architecture*. New York: Museum of Modern Art, 1966.

Venturi, R., Brown, D. S., and Izenour, S. Learning from Las Vegas. In *Environment and cognition*. W. H. Ittelson (ed.). New York: Seminar Press, 1973.

Von Bertalanffy, L. *General Systems Theory*. New York: Braziller, 1968.

Von Eckardt, W. Double Manhattan. *The New Republic*, January 21, 1978, 26–28.

Waldbott, G. L. *Health effects of environmental pollutants*. St. Louis: Mosby, 1973.

Walk, R. D., and Gibson, E. J. A comparative and analytical study of visual depth perception. *Psychological Monographs*, 1961, *75* No. 15.

Walker, E. L. Psychological complexity as a basis for a theory of motivation and choice. *Nebraska Symposium on Motivation*, 1964, *12*, 47–95.

Walker, J. M. Energy demand behavior in a master-metered apartment complex: An experimental analysis. *Journal of Applied Psychology*, 1979, *64*, 190–196.

Wall, G. Public response to air pollution in South Yorkshire, England. *Environment and Behavior*, 1973, *5*, 219–248.

Walsh, D. P. Noise levels in open plan classrooms. *Journal of Architectural Research*, 1975, *4*, 5–15.

Wandersman, A. A conceptual framework of user participation in planning environments. *Environment and Behavior*, 1979, *11*, 185–208.

Ward, B., and Dubos, R. *Only one earth*. New York: W. W. Norton, 1972.

Ward, L. M., and Suedfeld, P. Human responses to highway noise. *Environmental Research,* 1973, *6,* 306–326.

Warner, H. D. Effects of intermittent noise on human target detection. *Human Factors,* 1969, *11,* 245–250.

Warren, C., and Laslett, B. Privacy and secrecy: A conceptual comparison. *Journal of Social Issues,* 1977, *33,* 43–52.

Watson, O. M., and Graves, T. D. Quantitative research in proxemic behavior. *American Anthropologist,* 1966, *68,* 971–985.

Webb, E. J., Campbell, D. T., Schwartz, R. D., and Sechrest, L. *Unobtrusive measures: Non-reactive research in the social sciences.* Chicago: Rand McNally, 1966.

Webber, M. M. Order in diversity: Community without propinquity. In L. Wingo, Jr. (ed.), *Cities and space.* Baltimore: The Johns Hopkins Press, 1963.

Weigel, R. H., and Newman, L. S. Increasing attitude-behavior correspondence by broadening the scope of the behavioral measure. *Journal of Personality and Social Psychology,* 1976, *33,* 793–802.

Weigel, R. H., and Weigel, J. Environmental concern: The development of a measure. *Environment and Behavior,* 1978, *10,* 3–16.

Weiner, B., Frieze, I., Kukla, A., Reed, L., Rest, S., and Rosenbaum, R. M. *Perceiving the causes of success and failure.* Morristown, N.J.: General Learning Press, 1971.

Weiner, F. H. Altruism, ambiance, and action: The effects of rural and urban rearing on helping behavior. *Journal of Personality and Social Psychology,* 1976, *34,* 112–124.

Weinstein, C. S. Modifying student behavior in an open classroom through changes in the physical design. *American Educational Research Journal,* 1977, *14,* 249–262.

Weinstein, L. Social schemata of emotionally disturbed boys. *Journal of Abnormal Psychology,* 1965, *70,* 457–461.

Weinstein, L. The mother-child schema, anxiety, and academic achievement in elementary school boys. *Child Development,* 1968, *39,* 257–264.

Weinstein, N. D. Human evaluations of environmental noise. In K. H. Craik and E. H. Zube (eds.), *Perceiving environmental quality: Research and applications.* New York: Plenum, 1976.

Weinstein, N. D. Individual differences in reactions to noise: A longitudinal study in a college dormitory. *Journal of Applied Psychology,* 1978, *63,* 458–466.

Weitzer, W. Environmental images and behavior. Paper presented at annual conference of Environmental Design Research Association, Charleston, S.C., 1980.

Wellman, B., Craven, P., Whitaker, M., Stevens, H., Shorter, A., Dutoit, S., and Baker, H. Community ties and support systems: From intimacy to support. In L. S. Bourne, R. D. Mackinnon, and J. S. Simmons (eds.), *The form of cities in central Canada: Selected papers.* Toronto: University of Toronto Press 1973.

Wells, B. W. P. Subjective responses to the lighting installation in a modern office building and their design implications. *Building Science,* 1965, *1,* 153–165.

Wertheimer, M. *Productive thinking.* New York: Harper & Row, 1945.

Westin, A. *Privacy and freedom.* New York: Atheneum, 1967.

Weston, H. C. The relation between illumination and visual performance. *Industrial Health Research Board Report No. 87,* 1947, HMSO, London.

Wexner, L. B. The degree to which colors (hues) are associated with mood-tones. *Journal of Applied Psychology,* 1954, *38,* 432–435.

Whitaker, J. D. Fear stalks Tyler House hallways. *Washington Post,* February 16, 1976, 1, 3.

White, M. Interpersonal distance as affected by room size, status, and sex. *Journal of Social Psychology,* 1975, *95,* 241–249.

White, R. W. Strategies of adaptation: An attempt at systematic description. In G. V. Coelho, D. A. Hamburg, and J. E. Adams (eds.), *Coping and adaptation.* New York: Basic Books, 1974.

Whyte, W. F. *Street corner society.* Chicago: University of Chicago Press, 1943.

Whyte, W. H. How the new suburbia socializes. *Fortune,* August, 1953, 120–122, 186, 190.

Whyte, W. H. *The organization man.* New York: Doubleday, 1956.

Whyte, W. H. Please, just a nice place to sit. *New York Times Magazine,* December 3, 1972, 20, 22, 30.

Wicker, A. W. Undermanning, performance, and students' subjective experiences in behavior settings of large and small high schools. *Journal of Personality and Social Psychology,* 1968, *10,* 255–261.

Wicker, A. W. Attitudes versus actions: The relationship of verbal and overt behavioral responses to attitude objects. *Journal of Social Issues,* 1969, *24,* 41–78. (a)

Wicker, A. W. Cognitive complexity, school size and participation in school behavior settings: A test of the frequency of interaction hypothesis. *Journal of Educational Psychology,* 1969, *60,* 200–203. (b).

Wicker, A. W. Size of church membership and members' support of church behavior settings. *Journal of Personality and Social Psychology,* 1969, *13,* 278–285. (c)

Wicker, A. W. Processes which mediate behavior-environment congruence. *Behavioral Science,* 1972, *17,* 265–277.

Wicker, A. W. Undermanning theory and research: Implications for the study of psychological and behavioral effects of excess human populations. *Representative Research in Social Psychology,* 1973, *4,* 185–206.

Wicker, A. W. *An introduction to ecological psychology.* Monterey, Ca.: Brooks/Cole, 1979.

Wicker, A. W., and Kirmeyer, S. L. What the rangers think. *Parks and Recreation,* October, 1976, 28–30, 42–43.

Wicker, A. W., McGrath, J. E., and Armstrong, G. E. Organization size and behavior setting capacity as determinants of member participation. *Behavioral Science,* 1972, *17,* 499–513.

Wicker, A. W., and Mehler, A. Assimilation of new members in a large or small church group. *Journal of Applied Psychology,* 1971, *55,* 151–156.

Widgery, R., and Stackpole, C. Desk position, interviewee anxiety, and interviewer credibility: An example of cognitive balance in a dyad. *Journal of Counseling Psychology,* 1972, *19,* 173–177.

Wilcox, B. L., and Holahan, C. J. The social ecology of the megadorm in university student housing. *Journal of Educational Psychology,* 1976, *68,* 453–458.

Wilkinson, R. T. Interaction of noise with knowledge of results and sleep deprivation. *Journal of Experimental Psychology,* 1963, *66,* 332–337.

Wilkinson, R. T., Fox, R. H., Goldsmith, R., Hampton, I. F., and Lewis, H. E. Psychological and physical responses to raised body temperature. *Journal of Applied Physiology,* 1964, *29,* 287–292.

Willis, F. N. Initial speaking distance as a function of the speakers' relationship. *Psychonomic Science,* 1966, *5,* 221–222.

Willmott, P. *The evolution of a community.* London: Routledge & Kegan Paul, 1962.

Willmott, P., and Young, M. *Family and class in a London suburb.* London: Routledge & Kegan Paul, 1960.

Wilner, D. M., Walkley, R. P., Pinkerton, T. C., and Tayback, M. *The housing environment and family life.* Baltimore: Johns Hopkins Press, 1962.

Wilson, G. D. Arousal properties of red versus green. *Perceptual and Motor Skills,* 1966, *23,* 947–949.

Wilson, J. Q. Planning and politics: Citizen participation in urban renewal. *Journal of the American Institute of Planners,* 1963, *29,* 242–249.

Winkel, G. The nervous affair between behavior scientists and designers. *Psychology Today,* 1970, *3,* 31–35.

Winkel, G., Malek, R., and Theil, P. The role of personality differences in judgments of roadside quality. *Environment and Behavior,* 1969, *1,* 199–223.

Winkel, G., and Sasanoff, R. An approach to an objective analysis of behavior in architectural space. In H. M. Proshansky, W. H. Ittelson, and L. G. Rivlin (eds.), *Environmental psychology: People and their physical settings.* New York: Holt, Rinehart & Winston, 1976.

Winsborough, H. H. The social consequences of high population density. *Law and Contemporary Problems,* 1965, *30,* 120–126.

Wirth, L. Urbanism as a way of life. *American Journal of Sociology,* 1938, *44,* 1–24.

Wittig, M. A., and Skolnick, P. Status versus warmth as determinants of sex differences in personal space. *Sex Roles,* 1978, *4,* 493–503.

Wohlwill, J. F. The physical environment: A problem for a psychology of stimulation. *Journal of Social Issues*, 1966, *22*, 29–38.

Wohlwill, J. F. Amount of stimulus exploration and preference as differential functions of stimulus complexity. *Perception and Psychophysics*, 1968, *4*, 307–312. (a)

Wohlwill, J. F. *Man as a seeker and neutralizer of stimulation.* Paper presented at the Sanitarian's Institute of Environmental Quality Management, University of Connecticut, 1968. (b)

Wohlwill, J. F. The emerging discipline of environmental psychology. *American Psychologist*, 1970, *25*, 303–312.

Wohlwill, J. F. Environmental aesthetics: The environment as a source of affect. In I. Altman and J. F. Wohlwill (eds.), *Human behavior and environment: Advances in theory and research* (Vol. 1). New York: Plenum, 1976.

Wohlwill, J. F. The social and political matrix of environmental attitudes: An analysis of the vote on the California coastal zone regulation act. *Environment and Behavior*, 1979, *11*, 71–86.

Wolfe, M. Room size, group size and density: Behavior patterns in a children's psychiatric facility. *Environment and Behavior*, 1975, *7*, 199–225.

Wolfe, M., and Golan, M. B. *Privacy and institutionalization.* Paper presented at annual conference of the Environmental Design Research Association, Vancouver, 1976.

Wolfe, M., and Laufer, R. S. The concept of privacy in childhood and adolescence. In D. H. Carson (ed.), *Man-environment interactions: Evaluations and applications.* Stroudsburg, Pa.: Dowden, Hutchinson & Ross, 1975.

Wolfe, M., and Proshansky, H. M. The physical setting as a factor in group function and process. In J. Lang, C. Burnette, W. Moleski, and D. Vachon (eds.), *Designing for human behavior: Architecture and the behavioral sciences.* Stroudsburg, Pa.: Dowden, Hutchinson & Ross, 1974.

Wolfgang, J., and Wolfgang, A. Explanation of attitudes via physical interpersonal distance toward the obese, drug users, homosexuals, police, and other marginal figures. *Journal of Clinical Psychology*, 1971, *27*, 510–512.

Wood, E. Housing design: A social theory. In G. Bell and J. Tyrwhitt, (eds.), *Human identity in the urban environment.* Harmondsworth, England: Penguin, 1972.

Woodhead, M. M. Visual searching in intermittent noise. *Journal of Sound Vibration*, 1964, *1*, 157–161.

Woodhead, M. M. Effect of noise on the distribution of attention. *Journal of Applied Psychology*, 1966, *50*, 296–299.

Worchel, S., and Teddlie, C. The experience of crowding: A two-factor theory. *Journal of Personality and Social Psychology*, 1976, *34*, 30–40.

Worchel, S., and Yohai, S. M. L. The role of attribution in experience of crowding. *Journal of Experimental Social Psychology*, 1979, *15*, 91–104.

Worthington, M. Personal space as a function of the stigma effect. *Environment and Behavior*, 1974, *6*, 289–295.

Wright, H. F. Psychological development in the midwest. *Child Development*, 1956, *27*, 265–286.

Wright, H. F. Observational child study. In P. H. Mussen (ed.), *Handbook of research methods in child development.* New York: Wiley, 1960.

Wyndham, C. Adaptation to heat and cold. *Environmental Research*, 1969, *2*, 442–469.

Wynne-Edwards, V. C. *Animal dispersion in relation to social behavior.* New York: Hafner, 1962.

Wynne-Edwards, V. C. Self-regulating systems in populations of animals. *Science*, 1965, *147*, 1543–1548.

Yancy, W. L. Architecture, interaction and social control. *Environment and Behavior*, 1971, *3*, 3–21.

Yerkes, R. M., and Dodson, J. D. The relation of strength of stimulus to rapidity of habit-formation. *Journal of Comparative and Neurological Psychology*, 1908, *18*, 459–482.

Young, M., and Willmott, P. *Family and kinship in East London.* New York: The Free Press, 1957.

Zanardelli, H. A. Life in a landscape office. In N. Polites (ed.), *Improving the office environment.* Elmhurst, Ill.: The Business Press, 1969.

Zanna, M. P., Kiesler, C. A., and Pilkonis, P. A. Positive and negative attitudinal affect established by classical conditioning. *Journal of Personality and Social Psychology,* 1970, *14,* 321–328.

Zanna, M. P., Olson, J. M., and Fazio, R. H. Attitude-behavior consistency: An individual difference perspective. *Journal of Personality and Social Psychology,* 1980, *38,* 432–440.

Zeidberg, L. D., Prindle, R. A., and Landau, E. The Nashville Air Pollution Study: I. Sulfur dioxide and bronchial asthma. A preliminary report. *American Review of Respiratory Diseases,* 1961, *84,* 489–503.

Zeidberg, L. D., Prindle, R. A., and Landau, E. The Nashville Air Pollution Study: III. Morbidity in relation to air pollution. *American Journal of Public Health,* 1964, *54,* 85–97.

Zeisel, J. Fundamental values in planning with the non-paying client. In C. Burnette, J. Lang, and D. Vaschon (eds.), *Architecture for human behavior: Collected papers from a miniconference.* Philadelphia: Philadelphia Chapter/The American Institute of Architects, 1971.

Zeisel, J. Symbolic meaning of space and the physical dimension of social relations: A case study of sociological research as the basis of architectural planning. In J. Walton and D. Carns (eds.), *Cities in change: Studies on the urban condition.* Boston: Allyn & Bacon, 1963.

Ziller, R. C., and Grossman, S. A. A developmental study of the self-social constructs of normals and the neurotic personality. *Journal of Clinical Psychology,* 1967, *23,* 15–21.

Ziller, R. C., Megas, J., and DiCencio, D. Self-social constructs of normals and acute neuropsychiatric patients. *Journal of Consulting Psychology,* 1964, *28,* 59–63.

Zimbardo, P. G. The human choice: Individuation, reason and order vs. deindividuation, impulse and chaos. In W. J. Arnold and D. Levine (eds.), *Nebraska symposium on motivation.* Lincoln: University of Nebraska Press, 1969.

Zlutnick, S., and Altman, I. Crowding and human behavior. In J. F. Wohlwill and D. H. Carson (eds.), *Environment and the social sciences: Perspectives and applications.* Washington, D.C.: American Psychological Association, 1972.

Zube, E. H. Cross-disciplinary and inter-mode agreement on the description and evaluation of landscape resources. *Environment and Behavior,* 1974, *6,* 69–89.

Zube, E. H. Perception of landscape and land use. In I. Altman and J. Wohlwill (eds.), *Human behavior and the environment: Advances in theory and research.* New York: Plenum, 1976.

Zube, E. H., Brush, R., and Fabos, J. *Landscape assessment: Value, perceptions and resources.* Stroudsburg, Pa.: Dowden, Hutchinson & Ross, 1975.

Zube, E. H., Pitt, D., and Anderson, T. Perception and prediction of scenic resource values of the Northeast. In E. H. Zube, R. Brush, and J. Fabos (eds.), *Landscape assessment: Values, perceptions, and resources.* Stroudsburg, Pa.: Dowden, Hutchinson & Ross, 1975.

Index

About the Author

Charles J. Holahan is associate professor of psychology at the University of Texas at Austin, and has been visiting associate professor at the Social Ecology Laboratory in the Department of Psychiatry at Stanford University. He held a postdoctoral fellowship in the Environmental Psychology Program at the City University of New York after receiving his Ph.D. from the University of Massachusetts at Amherst. Dr. Holahan has published more than thirty-five professional articles and a book, *Environment and Behavior: A Dynamic Perspective,* presenting his research in a wide variety of urban, residential, and organizational settings. He has served as a consultant to many governmental, educational, and medical institutions.